# Lucky's Collectors Guide

To 20th Century

# Yo·Yos

History and Values

By Lucky J. Meisenheimer, M.D.

This book is dedicated to my family:

My Grandparents:
Kermit & Alice Meisenheimer
Cameron & Genevieve Robinson
Who I honor with pride...

My Parents:
John & Alice Meisenheimer
Who guide and encourage with love...

My Wife:
Jacquie
Who owns my heart...

My Sons:
John & Jake
Who makes it all worthwhile...

# Lucky's Collectors Guide
To 20th Century
# Yo·Yos
History and Values

All Rights Reserved © 1999

AUTHOR
Lucky J. Meisenheimer, M.D.

HISTORICAL CONSULTANT
Dan Volk

COVER DESIGN © 1999
& DESKTOP PUBLISHING
T. Brown & Associates

TYPESETTER & ASSISTANT
Sandee Crowther-Combs

EDITING & PROOFREADING
T. Brown & Associates
Jacquie Meisenheimer
Pete Richmond
Laura Adams

ISBN  0-9667612-0-0 PAPERBACK
ISBN  0-9667612-1-9 HARDCOVER

© 1999 Lucky J.'s Swim & Surf, Inc.

No part of this book may be used for reproduction, transmission, storage in archival systems, electronic generation, or translated into any language, recorded in any form, by any other means without written authorization from this publisher, Lucky J.'s Swim & Surf, Inc. All rights reserved © 1999.

This edition may reference various companies or products. These references are not intend to convey any endorsement of or to imply affiliation with this publication, publisher, or author.

For additional information or distribution, contact:

PUBLISHER
Lucky J.'s Swim & Surf, Inc.

ADDRESS
7300 Sandlake Commons Blvd.• Suite #103
Orlando • Fl  32819

PHONE NUMBER
1.877.969.6728

BY

Lucky J. Meisenheimer, M.D.

# TABLE OF CONTENTS

## INTRODUCTION

Preface .................................................. 13
Acknowledgments And Photo Credits ................ 15
How To Use This Book .............................. 16

## COLLECTING AND HISTORY

Grading Yo-Yos ...................................... 20
Grading System ...................................... 20
Yo-Yo Values (See The Appendix For Estimated $ Values) 23
Dating Yo-Yos ....................................... 23
Counterfeits, Fakes And Reproductions ............. 24
Repair And Restoration ............................. 24
Glossary Of Collecting Terms ...................... 25
Yo-Yo Shapes ........................................ 28
Yo-Yo Sizes ......................................... 31
Frequently Asked Questions ........................ 32
History Of The Yo-Yo ............................... 34
History Of Yo-Yo Contests ......................... 36

## YO-YO LISTINGS

Adult ................................................ 40
Advertising Metal ................................... 40
Advertising Plastic ................................. 41
Advertising Wood .................................... 43
Alox ................................................. 45
American Spinners ................................... 46
BC ................................................... 46
Bird In Hand ........................................ 47
Bob Allen ........................................... 47
Character ........................................... 48
Cheerio ............................................. 50
Chemtoy ............................................. 53
Chico ............................................... 54
Clown ............................................... 54
Coca-Cola ........................................... 55
Dakin ............................................... 57
Damert .............................................. 57
Dell ................................................ 58
Disney .............................................. 59
Duncan: ............................................. 61
  • General Information ........................... 61
  • Tips On Dating ................................ 65
  • Packaging ..................................... 65
  • Yo-Yo Man Logo (Mr. Yo-Yo) ................... 66
Duncan Advertising .................................. 67
Duncan Beginner ..................................... 69
Duncan Butterfly .................................... 70
Duncan Cattle Brand ................................. 72
Duncan Gold Award ................................... 72
Duncan Imperial ..................................... 73
Duncan Imperial Jr. ................................. 74
Duncan Jewels ....................................... 75

*Continued ...* ▶

## TABLE OF CONTENTS

**YO-YO LISTINGS** *Continued ...*

| | |
|---|---|
| Duncan Jumbo | 76 |
| Duncan Junior | 77 |
| Duncan Light-Up And Glow | 77 |
| Duncan Miscellaneous Items | 79 |
| Duncan (Miscellaneous Plastic) | 80 |
| Duncan (Miscellaneous Wood) | 82 |
| Duncan O-Boy | 83 |
| Duncan "Olympics" Yo-Yolympics | 84 |
| Duncan Professional | 85 |
| Duncan Rainbow | 85 |
| Duncan Satellite | 86 |
| Duncan Special | 87 |
| Duncan Sportsline | 87 |
| Duncan Super Hero | 88 |
| Duncan Tins | 89 |
| Duncan Tournament | 91 |
| Duncan Wheels | 94 |
| Festival | 95 |
| Fli-Back | 98 |
| Flores | 99 |
| Foreign | 101 |
| Forester | 103 |
| Franklin Mint | 103 |
| Goody | 104 |
| Hanna Barbera Characters | 105 |
| Hasbro | 106 |
| Hi-Ker | 107 |
| Holiday: | 108 |
| • Christmas | 108 |
| • Easter | 109 |
| • Halloween | 109 |
| • New Year | 110 |
| • St. Patrick's Day | 110 |
| • Thanksgiving | 111 |
| • Valentine's Day | 111 |
| Hummingbird | 111 |
| Humphrey | 113 |
| Imperial Toys | 118 |
| JA-RU | 119 |
| Kaysons | 120 |
| Ka-Yo (Cayo) | 120 |
| Kusan | 121 |
| Light-Up | 122 |
| McDonald's | 123 |
| Medalist | 124 |
| Metal (Miscellaneous Manufacturers) | 124 |
| National Championship | 125 |
| Novelty | 126 |
| Oliver Toys | 129 |
| Oriental Trading Company | 129 |
| Parker-National-Canada Games | 130 |
| Party Favor | 131 |
| Peanuts | 132 |
| Plastic Miscellaneous | 133 |
| Plastic (Miscellaneous - Unnamed) | 135 |

*Continued ...*

# TABLE OF CONTENTS

| | |
|---|---|
| Playmaxx | 136 |
| Royal: | 137 |
| • Crownless Royal | 139 |
| • Five Point Crown With Chevron Royal | 139 |
| • Five Point Crown Without Chevron Royal | 140 |
| • One Point Crown Royal | 140 |
| • Crown of England Royal | 141 |
| • Jewel Royal | 141 |
| Russ Berrie And Company Inc. | 142 |
| Russell | 142 |
| Smothers Brothers | 143 |
| Soft Drink (Non-Coke) | 144 |
| Souvenir | 145 |
| Spectra Star | 147 |
| Sport Related | 151 |
| Sports Balls | 152 |
| Star Return Top | 153 |
| Style 55 Champion | 153 |
| Tin Miscellaneous | 153 |
| Tom Kuhn | 155 |
| Toy Tinkers (Toy-O-Balls) | 157 |
| U.S. Toy Company | 158 |
| Wallace Toy Company | 158 |
| Warner Brothers | 158 |
| Whirl-King | 159 |
| Wood (Carved) | 159 |
| Wood Miscellaneous | 160 |
| Wood (Miscellaneous - Unnamed) | 161 |
| World's Fair | 162 |
| Yomega | 163 |

## PLATES

| | |
|---|---|
| Photographs And Yo-Yo Item Numbers | 165 • P |

## MEMORABILIA CATEGORIES

| | |
|---|---|
| Awards | 226 |
| Counter Display Boxes | 226 |
| Miscellaneous Items | 228 |
| Patches | 230 |
| Pins | 231 |
| Posters And Sheet Music | 232 |
| String Packs | 233 |
| Trick Books | 235 |

## APPENDIX

| | |
|---|---|
| Resources | 240 • A |
| Values | 241 • A |
| About The Author | 255 • A |
| Autographs (For Collecting) | 256 • A |

# Introduction

# INTRODUCTION

## PREFACE

Welcome to the world of yo-yo collecting. I am continually fascinated by the variety of people who collect yo-yos: doctors, lawyers, ministers, rock and roll stars, children, retirees, demonstrators, and former champions.

This book arises, in part, from the frustrating lack of available knowledge regarding the hobby. It is my attempt to share what I know with other yo-yo enthusiasts. Early yo-yo manufacturers never considered that hobbyists would continue to have an interest in their product many years later. Few company records exist, and accurately dating and identifying yo-yos can be a difficult process for the novice collector.

This book is the first major attempt to catalog collectible yo-yos and provide guidance to collectors. There are undoubtedly errors, omissions, and debatable "facts" within this book, and I am always happy to receive comments and information that will improve future editions.

It is always a thrill to hear from collectors around the world about their latest yo-yo finds. Individuals involved in the sport of yo-yoing and collecting are the greatest, and I look forward to hearing from you.

*Happy Hunting!*

*- Lucky J. Meisenheimer, M.D.*

# INTRODUCTION

## ACKNOWLEDGMENTS

*For their assistance and information, special thanks to:*

| | | |
|---|---|---|
| American Yo-Yo Association | Duncan/Flambeau Co. | Alan Nagao |
| Frank Alonso | Rich Falls | Richy Nye |
| Bill Alton | Fiend Magazine | Dale Oliver |
| Ted Anderson | John Frier Jr. | Rick Osborne |
| Andrew Arvesen | Vve and Jim Galloway | Tom Parks |
| Bob Baab | Alex Garcia | Tom Radovan |
| Jerry Bakke | Dave Hall | Bob Rule |
| Jennifer Baybrook | Bill Halon | Larry Sayegh |
| Bird in Hand | Eileen Hartle | Judith Schultz |
| Ken Blumenthal | Roger Hernandez | John Seivers |
| Gary Bosch | Infinite Illusions | Linda Sengpiel |
| Kevin Bosko | Jim Johnson | Chuck Short |
| Bob Bowmen | Tim Kelly | Spectra Star Corporation |
| JoAnne Brown-Radovan | Knave Productions | Steve Speegle |
| Steve Brown | Bob Koschoreck | Spin Offs |
| Dave Cabin | Tom Kuhn | Charles Stallions |
| Mike Caffrey | Dennis Lesley | John Stangle |
| Bill Caswell | Gerald Linden | Dick Stohr |
| Robert Cayo | Fred Lindsey | Craig Strange |
| Ray Cebula | Harvey Lowe | Johnny Tillotson |
| Greg Cohen | Bob Lundy | Jason Tracy |
| Cliff Colman | Bud Lutz | Hans Van Dan Elzen |
| Chris Cook | George Malko | Jennifer Van Der Molen |
| Brad Countryman | Bob Malowney | Dan Volk |
| Bill Cress | Jim Marvy | Don Watson |
| Sandee Crowther-Combs | Dennis McBride | David Welch |
| Stuart Crump | Chester Miltenberger | Yomega Corporation |
| Bill De Boisblanc | Monogram Products Inc. | Yo-Yo Times |
| Don and Donna Duncan | Clyde Mortensen | Bob Zeuschel |
| | Dale Myrberg | |

*Thanks to the following collectors who allowed me to photograph items from their personal collections for this book.*

| | | |
|---|---|---|
| Ted Anderson | Bob Malowney | JoAnne Brown-Radovan |
| Don Duncan | Richy Nye | Tom Radovan |
| Jim Johnson | Dale Oliver | Bud Lutz |
| Bob Koschoreck | Rick Osborne | Bob Baab |

## PHOTO CREDITS

Special thanks to all who provided yo-yos for use in this publication. All photography for the "Plate" pages and for sections' titles, except for the following, were taken by the author. Many thanks to these noted collectors who provided photos from their personal collections:

Cabin, Dave .................................................................................... 94, 478, 478.1, 607
Lindsey, Fred ............................................................... 1054, 476.1, 507.1, 436.2, 476
Volk, Dan .......................................................................................... 1025, 674, 619
Hall, Dave .................................................................................................. 627.1, 1907
Blumenthal, Ken ..... 195, 202, 204, 204.1, 205.1, 205.2, 205.3, 207.1, 208, 208.1, 208.2
............ 209, 209.1, 209.2, 666, 1101, 1610, 1611, 1623, 1628.1, 1640.1, 1752, 1770.1
.................................................................... 1771, 1790.1, 1790.2, 1790.3, 1790.4, 1793

## INTRODUCTION

● **HOW TO USE THIS BOOK**

**Y**o-Yo descriptions are listed under the company of production. Duncan, because of the large numbers, has been broken into subcategories. If a company only produced a few yo-yos or the maker's origin is unknown, look under miscellaneous sections. Some special interest categories have been given precedence over the manufacturer categories. For example, Disney and Coca-Cola will be listed under their own categories, regardless of the manufacturer. If you cannot find a yo-yo in the manufacturer listing, check the special interest categories.

Two numbers precede each description. The first is the item number; the second is a two-digit number in italics which references the "Plate Number." If there is no two-digit number, the item is not pictured in the "Plates" section. Descriptions of each yo-yo are listed in the following order: the name, visual description, type of seal, material the yo-yo is made of, manufacturer or distributor's name, yo-yo shape and size (other than standard). Other information that may also be listed includes three-piece versus one-piece wood yo-yos, type of axle, whether the string is pegged, and approximate dates of production.

Photos of yo-yos and memorabilia are numbered and located in the Plates section of this book. The Plate Number is located at the top of each plate page. Item description numbers are listed beneath each object and appear in numerical order throughout the book. These item numbers can be used to look up the yo-yo descriptions and values. A guide for estimated yo-yo values is listed in the Appendix section of this book. Values are listed in three grades: packaged, near mint, and good. For yo-yos not listed, a value can be estimated by comparing the values of similar yo-yos from the same category. If the yo-yo was on a card or in a box, values are based on both the yo-yo and packaging being in near mint condition. Packaging unknown to this author are not listed. Recently produced yo-yos are not considered collectible unless they are in near mint condition. There will be no price listed in the "Good" category for recently produced yo-yos.

Because tens of thousands of different yo-yos have been produced over the decades, no yo-yo guide can be all-inclusive. This book concentrates on yo-yos marketed in the United States, though there are a limited number of foreign yo-yos listed.

# INTRODUCTION

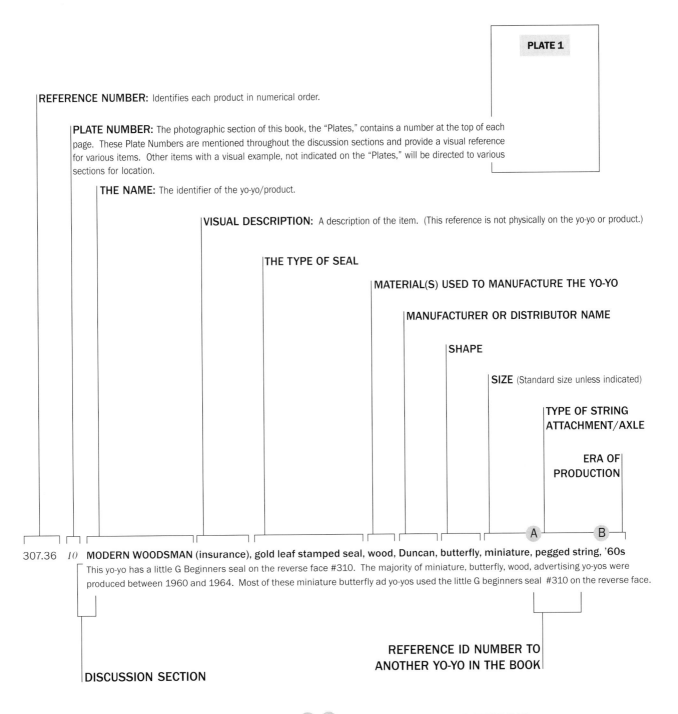

REFERENCE NUMBER: Identifies each product in numerical order.

PLATE NUMBER: The photographic section of this book, the "Plates," contains a number at the top of each page. These Plate Numbers are mentioned throughout the discussion sections and provide a visual reference for various items. Other items with a visual example, not indicated on the "Plates," will be directed to various sections for location.

THE NAME: The identifier of the yo-yo/product.

VISUAL DESCRIPTION: A description of the item. (This reference is not physically on the yo-yo or product.)

THE TYPE OF SEAL

MATERIAL(S) USED TO MANUFACTURE THE YO-YO

MANUFACTURER OR DISTRIBUTOR NAME

SHAPE

SIZE (Standard size unless indicated)

TYPE OF STRING ATTACHMENT/AXLE

ERA OF PRODUCTION

Ⓐ   Ⓑ

307.36   10   MODERN WOODSMAN (insurance), gold leaf stamped seal, wood, Duncan, butterfly, miniature, pegged string, '60s

This yo-yo has a little G Beginners seal on the reverse face #310. The majority of miniature, butterfly, wood, advertising yo-yos were produced between 1960 and 1964. Most of these miniature butterfly ad yo-yos used the little G beginners seal #310 on the reverse face.

REFERENCE ID NUMBER TO ANOTHER YO-YO IN THE BOOK

DISCUSSION SECTION

Ⓐ  Ⓑ  MISCELLANEOUS YO-YO FEATURES:
Various attributes to aid in referencing a particular yo-yo
(Not shown on this example)

*Yo-Yo Values are listed in the Appendix*

*The History and Values of Yo-Yos*

# Collecting
& History

## COLLECTING AND HISTORY

### ● GRADING YO-YOS

*Refer to Plate, Grading Yo-Yos, for illustrated examples.*
Grading is done to determine value. Yo-Yos in a high state of preservation are rare, thereby invoking the laws of supply and demand which absolutely links condition with value. Condition is very subjective. What one collector considers acceptable damage is unacceptable to another. A dealer that says a few paint chips or little rim wear won't affect the value obviously has some yo-yos with paint chips and rim wear for sale.

If this hobby is compared to other mature collecting fields, we can envision what will happen to yo-yo collecting. In the world of comic book collecting, when the hobby was new, comics were either considered collectible or not collectible. Yo-Yo collecting is at this stage right now. Initially with comics, a grading system was developed which consisted of a five-point scale: mint, fine, good, fair, and poor. After a few years, the scale was expanded with in-between grades such as very fine, and very good. Later, "near mint" was added to the evaluation system. Currently, comic collectors have a one hundred point grading system used in evaluating their collectibles. Entire books have been published solely on how to grade comic books. Without a doubt, yo-yo grading will have more importance to the collector in the future as the hobby advances. The current grading system, regarding condition, is provided to give standardization to yo-yo collecting.

### ● GRADING SYSTEM

*Refer to Plate, Grading Yo-Yos, for illustrated examples.*
The prime rule of yo-yo collecting is, "A yo-yo that has lost its seal has lost its collectible value."

#### MINT
Perfect in every way. Looks new. Seal must be centered and have all letters sharply visible with complete paint in each letter. Face has no paint chips, age cracks, fading, or wear. Rim is totally unmarked and paint should have its original shiny luster. Original string untied and never played. Should be in original packaging. It is unusually rare for pre-1965 yo-yos to be considered in mint condition.

* *Example: Yo-Yo is perfect in every way. The seal strike is full and complete. All letters are complete without any parts missing. The jewels are in a straight line and cross only the edge of one line. This yo-yo has not been played with and the coat is shiny and new.*

#### NEAR MINT
Like mint but may have manufacturing flaws such as slightly off centered seal. Letters may not have perfectly sharp impressions. The seal must be completely legible. Should look like new store stock. Although the string may be tied, the yo-yo appears to have not been played. Tin yo-yos must not have any rust.

* *Example A: Only minor manufacturing flaws. The letter "N" has only a partial strike in the word "Duncan." There is some break in the line in the circle, otherwise the yo-yo appears almost mint.*
* *Example B: There has been a partial strike in the letter "A" in the word "Duncan." Part of the bottom half of the word "top" is a partial strike as well.*

#### EXCELLENT
Outstanding eye appeal and looks almost new at first glance. May have been played, but shows no signs of wear. Seal is complete and fully readable without chips. No paint chips on face, but may have minute age cracks. May have fine scratches on rim, but no paint loss. Pegged string yo-yo should have complete unsoiled string, but can have age yellowing. Tin yo-yos can have minor flecks of rust in string slot, but none on rim or face.

* *Example A: There is fading of the letter "T" in the word "tournament," although upon close inspection it is present. The jewels cross both upper and lower lines.*
* *Example B: There is loss of the bottom and tops halves of the letters and partial loss of the circle, otherwise the paint is still shiny and new with no chips.*

*Continued ...* ▶

**COLLECTING AND HISTORY**

Mint

A ——— Near Mint ——— B

A ——— Excellent ——— B

Fine

A ——————— Good ——————— B ——— C

A ——— Fair ——— B

Poor

*Continued ...* ▶

GRADING YO-YOS EXAMPLES *(See Plates Section For The Full Color Example of "Grading Yo-Yos")*

The History and Values of Yo-Yos

## COLLECTING AND HISTORY

● GRADING SYSTEM *Continued...*

### FINE
Better than average yo-yo. Some small paint chips in seal are allowed, but no loss of any complete letters. Some letters may be faded, but are visible. Less than five age paint chips on the face. Minor rim wear with scratches and slight paint loss. Tin yo-yos may have some minor rust on rims and in string slot, but not on face.

> *Example: Notice some mild scuffing on the face of the yo-yo, but only one or two small paint chips. The seal is intact, but there is smudging on the letter "N" in the word "Duncan."*

### GOOD
Up to 10% of the seal missing or unreadable. Several paint chips may be present on the face with some fading of paint color. Moderate rim wear with loss of some, but not all paint. Name written or scratched across non-seal face. Pegged string yo-yos may have lost their string. Tin yo-yos may have some rust on face, but no more than 10% of surface area.

> *Example A: There is loss of almost 10% of the seal. There is some complete letter loss in some areas.*
> *Example B: There is general fading of the overall paint gloss. The seal is not intact, with up to 10% loss. There are several paint chips.*
> *Example C: Although the seal is intact, the stamp is very faded and has to be examined closely. There is rim wear as well as several paint chips.*

### FAIR
Up to 50% of seal missing or unreadable. Multiple paint chips on face. Heavy rim wear with almost complete loss of paint. Some minor wood chips may be present. Name written or scratched across seal face. Jewels may be missing. Tin yo-yos may have rust over 10% of yo-yo face, but not more than 30%.

> *Example A: Some letters are lost in the word "tournament." There are paint chips impeding on several of the letters in the word "Duncan." There are moderately severe paint chips over the face of the yo-yo.*
> *Example B: There are numerous paint chips over the yo-yo. Although the seal is fairly intact, it is not as sharp as we would like to see.*

### POOR
Yo-Yo can be identified, but has almost no collectable value. Greater than 50% of seal is missing. Heavy loss of paint and/or chips in the wood. Tin yo-yos with over 30% of the face rusted.

> *Example: Severe paint chipping. There is some loss of several of the letters, even though much of the seal is readable. The severe paint chipping of almost 50% of the yo-yo would make this a poor yo-yo.*

---

### VALUE FORMULA
Value breakdown can be calculated from the near mint price by using the following formulas:

| | |
|---|---|
| Mint | 120% or more of near mint price |
| Near Mint | 100% |
| Excellent | 80-100% of near mint price |
| Fine | 50-80% of near mint price |
| Good | 20-50% of near mint price |
| Fair | 10-20% of near mint price |
| Poor | 5-10% of near mint price |

\* *As the condition of a yo-yo decreases, plastic models drop in value faster than wood or tin.*

## COLLECTING AND HISTORY

*In recent years, the values of collectible yo-yos have dramatically increased. Some yo-yos sell for hundreds and sometimes even thousands of dollars.*

### YO-YO VALUES

*Values are listed in the Appendix.*
Values listed are intended only to be guidelines of what a buyer might expect to pay or a dealer might expect to receive for an item at a typical flea market or antique shop. They do not reflect the highest or lowest price ever paid for an item. The values listed are not offers to buy or sell yo-yos by the author nor does the author or publisher assume any responsibility for losses that may be incurred as a result of consulting this guide.

Valuation is a difficult task. Dealers like prices on the high side and collectors like them low, that is until a collector possesses a yo-yo. Then they like prices on the high side. If you ask ten different dealers and collectors the value of a specific yo-yo, you'll end up with ten different answers. The values listed in this book reflect the opinion of the author, based upon personal experience, discussions with other experienced collectors, sales lists, and auction prices. They are also based on several other factors including age, rarity, historical importance, how cross collectible the yo-yo is, and the all important "what's hot" influence which reflects how much current collectors desire the yo-yo. Condition also dramatically affects the value of any particular yo-yo. (See the Grading System section.)

Be aware that dealers typically offer only 40% to 50% of book value, or less, when buying toys. A dealer's goal is to get the best price to make a profit. Remember! Nobody is limited to the values quoted in this book. Yo-Yos will continue to be bought and sold at prices different than those listed in this book. Prices will differ from one area of the country to another and personal desires and needs often supersede suggested value. Prices at large toy auctions, internet auctions such as "eBay," and mailing lists directed towards yo-yo collectors, can exceed values in this guide. Use your own judgement to determine what the value of a particular piece is to you. How much you are willing to pay or sell a yo-yo for is the ultimate determining factor in its value.

*Some yo-yos can be difficult to date based on appearance alone. This Beatles yo-yo has graphics suggesting it originally dates from the '60s, however, this yo-yo was manufactured and retailed in 1998.*

### DATING YO-YOS

One of the difficulties in dating yo-yos is that some of the popular ones were produced for decades without a change in the molds or seals. Due to surpluses in warehouses, some yo-yos were sold retail years after production stopped. Therefore, some copyright dates do not match the dates of distribution or retail of the yo-yo. Additionally, dates may correspond to the copyright of characters or graphics instead of the yo-yos.

## COLLECTING AND HISTORY

### COUNTERFEITS, FAKES, AND REPRODUCTIONS

The sale of counterfeit collectible yo-yos is of great concern to collectors. As yo-yo prices continue to rise, the importance of spotting counterfeits, fakes, and reproductions becomes essential. Collectors should make themselves familiar with the differences between counterfeits, reproductions, and fakes. When collecting yo-yos, information and expertise are the greatest safeguards against fraud.

*Both of the above yo-yos are fakes. The first yo-yo has a paper sticker applied from a Power Rangers sticker set. The second is a Hallmark heart shaped yo-yo with a Disney decal applied.*

### COUNTERFEITS

These are fraudulent imitations designed to deceive the buyer. Counterfeits may be made to capitalize on the popularity of a currently produced name brand yo-yo. This violates trademark rights. Of more importance to collectors are counterfeits made to look like antique collectible yo-yos no longer in production. Although this illegal practice is limited, there are some known cases and these are identified in this book.

### FAKES

Yo-Yos with seals added that were never produced by the manufacturer or licensed for use on yo-yos. Beware of unconfirmed character yo-yos, especially Disney. Stickers from other sources are easily applied to blank yo-yos and can reap extra profits for the unscrupulous dealer.

### REPRODUCTIONS

These are legitimate duplications meant to copy the excellence of an item from the past without deceiving the consumer. Legitimate reproductions are designed to enable the collector to easily distinguish them from originals. The Franklin Mint series is a good example of reproduction yo-yos that can be easily identified from the originals. Limited series reproductions frequently develop their own collectible value.

### REPAIR AND RESTORATION

Toy collectors in all fields debate about how much repair and restoration should be allowed. The current axiom is: a restored yo-yo is never worth as much as a yo-yo of the same grade that has not been restored. There are many collectors who feel any restoration attempts, beyond a general cleaning of the yo-yo, decreases its value.

There are some repairs that may not affect the price of the yo-yo. For example, the replacement of a broken axle with another original axle of the same age, or replacing a damaged foil sticker seal with a seal from an identical severely damaged yo-yo, may be considered acceptable restoration. Certainly replacement of a worn string would be acceptable, but removal of a pegged string with replacement by a slip string would be unacceptable.

Adding paint to stamped seals which have lost their paint or did not receive a firm stamp, is not recommended and decreases the value of the yo-yo. Adding coatings of lacquers to protect chipping seals is also not recommended as this may result in more damage to the seal. Collectors and dealers should give a full disclosure of any restoration done when the yo-yo is traded or sold.

*Yo-Yos with broken axles, such as this Duncan Tournament, can be repaired.*

# COLLECTING AND HISTORY

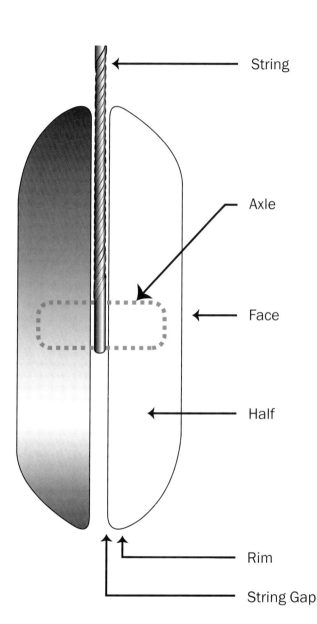

### ● GLOSSARY OF COLLECTING TERMS

Axle, Halves, Novelty Attachments, Packaging and Sales, Seals, Specialty Paints and Specialty Plastic are explained as follows:

## AXLE

The rod connecting the two yo-yo halves together, and is the location for attaching the string. The best playing fixed axle yo-yos typically have wooden axles.

- **Anchored String:** Also referred to as a fixed string. The string is permanently attached to the axle either by being tied to a hole through the axle or by being pegged. (See also Pegged String.)
- **Clutch:** A mechanism that alters the yo-yo's performance by tightening down on the string as rotation slows resulting in an easy return of the yo-yo to the hand. Example: Yomega's Phantom (Brain) #1668.
- **Drilled Axle:** An axle with a hole drilled through the center. The string is passed through the hole and tied to the axle. Sometimes this refers to a fixed or anchored string yo-yo.
- **Metal String Slot Axle:** Both axle and string slot are metal. The halves, usually of plastic, are snapped onto the metal string slot disk to make the yo-yo.
- **One-piece Yo-Yos:** A wooden yo-yo where the axle and halves were turned from one-piece of wood. Example: Duncan Gold Seal #491.
- **Pegged String:** A non-slip string yo-yo. String has been permanently fixed to the axle during manufacturing by lodging the string in the axle hole when the axle is inserted into the yo-yo halves. Sometimes this is referred to as a fixed string or anchored string yo-yo. Pegged string yo-yos should be collected with the original string intact.
- **Rabbit Ear String:** A string with two finger loops rather than one. In the late '20s, Flores sometimes recommended this style of string.
- **Riveted Disk:** A yo-yo without a true axle. A rivet holds the two yo-yo halves together. Usually these are tin yo-yos. Example: PEZ #7.
- **Slip String:** A string which is looped around the axle, but not attached. This allows for the yo-yo to be free spinning.
- **Take Apart Yo-Yos:** These are yo-yos with a replaceable axle. Yo-Yo halves can be unscrewed from the axle to aid in the release of tangled string. Example: Kuhn's No Jive 3 in 1 #1568.
- **Three-piece Yo-Yos:** Yo-Yos that have axles glued into the halves. Example: Duncan Beginners #308.
- **Transaxle:** A mechanism to reduce string friction on the axle. String is in contact with the transaxle that wraps around the axle, allowing for long sleep times. Some transaxles have ball bearings and usually have the longest sleep time. Example: Kuhn's Silver Bullet II #1587.

## HALVES

The body of a yo-yo is made of two pieces that are mirror images of each other.

- **Face:** The outermost side of the yo-yo half. The face is where the manufacturer's logo or ad is placed. From ivory to candy, a multitude of materials have been used over the centuries to make yo-yo halves. Wood was the most common

*Continued ...* ▶

# COLLECTING AND HISTORY

**GLOSSARY OF COLLECTING TERMS** *Continued...*

material used to manufacture yo-yos, until the 1960s, when plastic became predominate. Metal yo-yos have been made, in limited numbers, for decades. Collectors refer to most early metal yo-yos as "tins" even though they are made of other metals or alloys.

- **View Lens:** A plastic yo-yo that has a transparent face, with a logo or artwork underneath.
- **Internal Fins:** The "internal face," or string slot side, may not be solid, but may have plastic dividers called fins. Plastic yo-yos with solid halves were difficult to produce because the drying time was too long. In order to add mass to the yo-yo, but speed drying time, fins were added. Duncan experimented with fins in the '50s, but never produced any yo-yos other than prototypes because air friction from the fins slowed spin time. Examples of early Duncan fin prototypes can be seen in the National Yo-Yo Museum in Chico, CA. Other companies, such as Royal, did produce yo-yos with internal fins.

## NOVELTY ATTACHMENTS

Yo-Yos frequently have parts or mechanisms added to increase the yo-yo's uniqueness.

- **Embedments:** Items embedded into transparent plastic. Example: Stinger Yo-Yo #1132.
- **External Bells:** Bells attached to the outside of a yo-yo that make noise as the yo-yo is played. Example: #1124.
- **Finger Ring:** A ring on the finger end of string usually made of plastic or metal. Example: #1072.
- **Internal Bells:** Small balls or BBs inside a yo-yo to make a sound when played.
- **Jewel:** Many collectors' favorite form of ornamentation. Adding rhinestones to the yo-yo face produces what collectors call a "Jewel Yo-Yo." The standard size for rhinestones was 4mm. Example: Duncan Jeweled #360.
- **Light-Up:** Battery operated yo-yos that light-up while spinning. Example: Duncan Satellite #382.
- **Musical:** Battery operated yo-yos that play music while spinning. Example: Duncan Mel•Yo•Dee #422.
- **Pinball:** Small holes are cut in the face and pinballs are added to make a pinball puzzle game. Found with some plastic models that have a view lens. Example: Festival Toe•Tac•Tic #544.
- **Whistler Holes:** Holes drilled in the rims of the yo-yo. Air currents cause a whistling sound when spun. Example: Duncan Whistling #477.

## PACKAGING AND SALES

The following are terms used in the display and retail of yo-yos.

- **Blister Card:** Any card that holds the yo-yo in place with a clear plastic material.
- **Bubble Blister Card:** Type of blister card that has a hard plastic "bubble" which holds the yo-yo in place.
- **Counter Display Box:** A box used to display yo-yos, which typically sits on a counter or shelf. Early boxes, that held loose yo-yos, frequently had eye-catching graphics. More recent boxes hold carded yo-yos and often have less impressive art.
- **Demonstrator:** Any individual who performs with a yo-yo before an audience in a scheduled public appearance. This does not include contests. Demonstrators typically receive a fee, but this is not a requirement to be considered a demonstrator.
- **Display Card:** Any card used to display a yo-yo for retail sale. This includes notched cards, saddle header cards, and blister cards.
- **Dump Bin:** Large floor or counter display box that typically comes with a variety of yo-yo styles and string packs. These bins often hold more than one gross of yo-yos.
- **Jobber:** A wholesaler who purchases product to supply retail stores and small chains. Large chains typically had their own warehouse buyers and did not use jobbers.
- **Manufacturer:** A company that produces yo-yos, typically for sale to wholesalers and retailers. Some manufacturers may participate in retail sales.
- **Notched Card:** A display card with a cut or "notch" so it can be slid down through the yo-yo's string slot and attached around the axle. The notch usually has a widened hole at the end which fits around the axle to hold the card in place.
- **Polybag:** A polyvinyl plastic bag used to hold and protect yo-yos in storage and transportation before retail sale. Two types

*Continued...*

## COLLECTING AND HISTORY

**GLOSSARY OF COLLECTING TERMS** *Continued ...*

of polybags are common. One has a saddle header card with the yo-yo loose in the bag. The other is heat-sealed and holds both the yo-yo and a notched display card.
- **Rack Display:** A display on which yo-yos are hung or displayed with metal wires or rods. Display cards come with holes pre-punched in the center top for rack display.
- **Retailer:** A business or individual that sells yo-yos to the consumer. Toy stores, drug stores, chain stores, etc. are all retailers. Yo-Yo professionals are considered retailers if they sell yo-yos directly to the consumer.
- **Saddle Header Card:** A cardboard or paper display card which fits over the opening of a polybag. The card is typically folded over the opening of the polybag and stapled closed. The card usually has a hole in the center for rack hanging. Some cards may have graphics and pre-printed prices.
- **Skin Pack Blister Card:** A type of blister card that has a pliable polyvinyl plastic form fitted over the yo-yo.
- **Stock Number:** A number typically found on the display card that identifies ordering information for the yo-yo. Yo-Yos are ordered by retailers and jobbers according to the stock number assigned to the item. Most products have a stock number, but it may or may not be printed on the packaging or display. Some collectors may try to use stock numbers to name a yo-yo, but this does not work well, as many different yo-yos may have had the same stock numbers over time.
- **Wholesaler:** Company which buys yo-yos directly from the factory and supplies retailers. Wholesalers typically sell only to businesses and do not participate in retail sales.

## SEALS

These are company logos, designs or ads placed on the face of the yo-yo. The prime rule of yo-yo collecting is, "A yo-yo that has lost its seal has lost its collectible value."

- **Decal Seal:** Sometimes referred to as decal transfers. Decals come on strips, (See #1725) and when moistened they can be transferred to the face of the yo-yo. Unlike sticker seals, they cannot be removed from the yo-yo without destroying the seal. Example: Duncan 77 #488.
- **Debossed Seal:** Image is depressed into material such as leather, paper or suede, so that the image is below the surface. This seal is then glued to the yo-yo face. Example: Doonesbury #166.
- **Embossed Seal:** Logo or design raised in relief from the surface. Example: Cracker Jack #162.
- **Flasher Seal:** Moving the yo-yo results in a diffraction of light so the images on the seal appear to move. This is also referred to as a Flicker seal. Example: Duncan Spider-Man #468.
- **Gold Leaf Stamped:** A heated metal die is used to stamp the seal on the yo-yo face. Gold leaf on wax paper is inserted between the die and the yo-yo face. Common in wooden yo-yos. Example: Duncan O-Boy #446.1.
- **Hot Stamped:** A heated die that leaves indentations in plastic yo-yos. Example: Duncan Imperial Tenite #347.
- **Imprint Seal:** Seals imprinted on plastic yo-yos. Unlike hot stamped seals, imprint seals don't leave indentations in the plastic. Example: Duncan Flambeau Imperial #352.
- **Ink Stamp Seal:** Logo or lettering inked (usually black) on the face of a yo-yo using a rubber stamp. Sometimes this is called a pad stamped seal. Example: Buster Brown Shoes #75.
- **Large Disk Insert Seal:** Some ad and souvenir yo-yos have imprinted seals on a circular plastic piece that is inserted into a slot on the yo-yo's face. This saves on production costs. Disk size is typically 1-13/16th inches. Example: 1986 World Expo. #1648.
- **Laser Carved Seal:** Distinctive cutting process for carving designs in wood. Example: Flying Camel Yo-Yo #1556.
- **Lithograph Seal:** A type of printing process used on metal yo-yos to create multi-colored designs. Example: Ka-Yo #1024.
- **Medallion Seal:** A non-removable plastic piece which is glued to, and elevated from the face of the yo-yo.
- **Molded Seal:** This seal is created on the face of the yo-yo during the manufacturing process. The logo or lettering is molded into the yo-yo. Example: Ronald McDonald Championship #1060.
- **Paint Seal:** Color painted on wooden yo-yos, usually applied by a silk-screening process. Example: Goody Winner #634.
- **Paper Insert Seal:** A paper seal inserted beneath a clear lens.
- **Removable Disk Seal:** A thin plastic disk, imprinted with an image or lettering, that can be inserted or removed from the face of the Yo-Yo. Example: Playmaxx Pro Yo #1251.
- **Sculpted:** The face of the yo-yo is sculpted in relief. This also refers to a yo-yo that is formed in the shape of the item that it is advertising. Example: Spectra Star Ghostbusters #1434.
- **Small Disk Insert Seal:** Same principle as the large disk insert seal, but a smaller size. The inserted disk is typically 1-3/16th inches in size. Example: Hawaii The 50th State #1371.
- **Stamped Seals or Die Stamped Seals:** Dies are used to make an impression of a logo on the face of a wooden yo-yo. Example Cheerio Big Chief #203.
- **Sticker Seals:** Adhesive labels attached to the face of the yo-yo. Older yo-yos with sticker seals are harder to find due to the glue's tendency to dry and crack causing loss of the seal.

    The two main classifications for sticker seals are:
    a) Foil Sticker Seals - metallic adhesive labels. Example: Cheerio Kitchener Buttons #206.
    b) Paper Seals. Example: Hallmark Disney Yo-Yo #270.

*Continued ...*

## COLLECTING AND HISTORY

● GLOSSARY OF COLLECTING TERMS *Continued...*

### SPECIALTY PAINTS

Wooden yo-yos are usually painted a single lacquer color enamel. Many also have a decorative stripe of a contrasting color airbrushed across the face of the yo-yo. Listed below are some terms used to describe other "specialty" finishes used on yo-yos.

- **Candy Swirls:** Multiple paint colors in a swirl pattern. Example: Fli-Back "Candy Swirl" #573.
- **Crackle Enamel:** Effect produced by two layers of paint. The outer layer is a fast drying paint which results in cracking, letting the base paint show through. Example: Duncan Litening #444.
- **Day-Glo:** Fluorescent Paint. Example: Duncan's 77 Day-Glo #490.
- **Fish Scale:** Unusual paint style where thin filaments of nylon are embedded in the paint. Briefly used on some Duncan models in the late '50s. Example: #360.1
- **Flocked Covering:** Paint which creates a textured surface. Example: Duncan Suede #503.
- **Glitter Enamel:** Flecks of aluminum added to the enamel to produce a glitter-like appearance. Example: Duncan Butterfly #331.
- **Lithograph:** A printing process which captures a greater range of tonality for multi-colored designs. Mostly used on tin yo-yos with multi-colored designs. Example: Festival Screamer #552.
- **Natural Wood:** A wood yo-yo stained, or not, with a clear lacquer coat. Example: BC Natural #140.
- **Pearlescence:** A high gloss paint that has a metallic or pearl-like luster. Example: Duncan Pearlescence #506.
- **Photo Lithographed:** Tin yo-yos with photographs lithographed on the face. Example: Yellow Stone Park #1416.
- **Rainbow Rings:** Concentric circles of multi-color paint on the face. Seen on some styles of wooden souvenir yo-yos. Example: Canada Souvenir #1355.
- **Splatter Paint:** Contrasting paint sprayed onto the base enamel leaving a splattered appearance. Example: Duncan Splatter #507.1.

### SPECIALTY PLASTICS

The following are some of the different plastics used in the production of yo-yos.

- **A.B.S. Plastic (Acrylonitrile-Butadiene-Styrene):** A common opaque plastic. Example: Duncan Special #453.
- **Acetate or Tenite:** Translucent material. Example: Duncan Fleur-de-lis Imperial #349.
- **Glow:** A phosphorescent material. Example: Duncan Glow #383.
- **Marbleized or Variegated Plastic:** Multi-colored plastic that has a marble-like appearance. Both a hard brittle plastic and a softer plastic have been used. Example: #350.
- **Metal Flake** (also called glitter embedments): Metal flakes embedded in plastic. Example: Duncan Gold Award #343.
- **Styrene:** Another translucent plastic material used in the '50s. Example: Duncan Pony Boy #428.

● YO-YO SHAPES

*Refer to Plate, Yo-Yo Shapes, for illustrated examples.*

The side views of the yo-yos shown represent the shapes of over 90% of the yo-yos produced. There are also many different yo-yos with sculpted designs. These are referred to as sculpted yo-yos. This guide will help in understanding the variations in yo-yo shapes as listed in this book. The descriptive term, which collectors most commonly refer to the shape, is listed below each yo-yo. Yo-Yos are grouped according to rim design. (See the rim styles as shown on the Yo-Yo Shapes Plate.)

First Row .................................................................................................................. *Half Round Rims*
Second Row ....................................................................................................................... *Round Rims*
Third Row .............................................................................................................................. *Flat Rims*
Fourth Row ................................................................................................................. *Beveled or No Rim*

### TOURNAMENT

This is the "classic" yo-yo design. There are many variations to the tournament shape. Tournament shape indicates design only and does not reflect quality or playability of the yo-yo. There are three variations of the tournament shape shown. Although tournament would be the generic term to describe all three variations, some collectors tend to break these three styles down into subcategories (Flores, Standard and Imperial tournament shapes). In this book, all these variations are referred to as "tournament shape" and more detailed distinctions are not made. (See the Yo-Yo Shapes Plate.)

Tournament: { Example A ................................................................. *Flores Tournament Shape*
Example B ............................................................. *Standard Tournament Shape*
Example C ............................................................. *Imperial Tournament Shape*

*Continued ...* ▶

## COLLECTING AND HISTORY

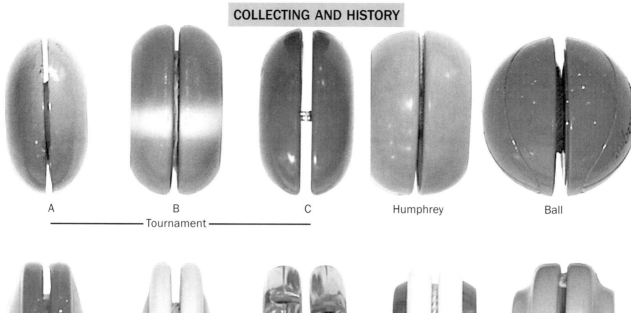

A ——— Slimline ——— B         Flywheel         A ——— Bulge Face ——— B

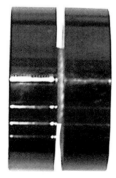  

Puck         Satellite         Coaster

Riveted Disk         Butterfly         Concave Riveted Disk

*Continued...* ▶

The History and Values of Yo-Yos

## COLLECTING AND HISTORY

● YO-YO SHAPES *Continued...*

### HUMPHREY SHAPE
This is named for the company that produces this shape yo-yo. Although Humphrey makes yo-yos of other designs, this design (The All Pro) is their flagship yo-yo. The Humphrey style is the most common style of plastic advertising yo-yo produced. Although Humphrey is best known for this shape, a few other companies use a similar shaped mold.

### BALL
This is a globe shaped yo-yo. Used frequently in making sports ball yo-yos. Any round or globe shaped yo-yo is referred to as "ball" shaped.

*This is an example of a sculpted shape. This yo-yo's form has been molded to simulate a Hostess Cupcake.*

### SLIMLINE
This style frequently has a view lens and is sometimes referred to as a professional. Under the slimline style, example "A" has opaque sides and example "B" is the view lens, slimline style.

### SCULPTED
This is any yo-yo that is molded in relief to represent a character or product.

### FLYWHEEL
Collectors sometimes refer to the flywheel shape as the "modern" style. It has a hollowed out face, usually with extra material on the rounded rim. The extra mass on the rim increases spin time. Some may have axles that protrude from the face. Many flywheel styles have thin plastic removable disk inserts that hide the hollowed out face.

### BULGE FACE
The center part of the face on the yo-yo is flat and protrudes outwardly, creating a bulging shape. Some collectors refer to the "bulge face" as the "Russell" style of yo-yo, but other companies besides Russell do make this style. (See Yo-Yo Shapes Plate, Example A.) This book uses the term "bulge face" to describe this shape. "Bulge face F-210" refers to a specific style of bulge face that is used in advertising. (See Yo-Yo Shapes Plate, Example B.)

### PUCK SHAPE
Two round disks connected by the axle create the puck shape. Some allowance is given for a slight beveling of the rim margin, but not enough to call this a butterfly.

### SATELLITE
This is a very distinctive shape used for the Duncan Satellite series in the '60s. The satellite shape has dome-like bulged faces. Other satellite models have a flat-top variation in the dome-like shape.

### COASTER SHAPE
These yo-yos all have concavity in the face of the yo-yo. The shape resembles a small drinking glass coaster. Unlike the flywheel, this yo-yo shape has flat rims.

### RIVETED DISK
Riveted disks do not have true rims. They are two flat disks or concave disks, usually metal, that are riveted together. These yo-yos do not have a true axle.

### BUTTERFLY
This is a distinctive yo-yo with beveled rims sharply tapering inward towards the axle.

## COLLECTING AND HISTORY

### ● YO-YO SIZES

Much confusion exists regarding size terminology. Many collectors consider yo-yos with the same seals in two different size categories different yo-yos. Originally, Duncan used the stocking numbers 33, 44 and 77 to indicate increasing sizes. Unfortunately, a Duncan Yo-Yo with the stocking number 22 is not smaller than a 33 and an 88 is not necessarily larger than a 77. Using Duncan stocking numbers to indicate size can be confusing and should not be done. Trying to use terms such as junior, beginner, and tournament to indicate increasingly larger yo-yo sizes is also flawed as some junior models are larger than some beginners and vice versa.

To clarify this sizing issue, this simplified measurement guide should be used, when describing the size of yo-yos. All yo-yo sizes listed in this guide are standard size unless noted otherwise in the yo-yos description.

**Midget:** 1-1/4 inches in diameter or smaller
**Miniature:** Between 1-1/4 inches and 1-7/8 inches in diameter
**Standard:** 1-7/8 inches to 2-3/4 inches in diameter
**Jumbo:** Larger than 2-3/4 inches in diameter

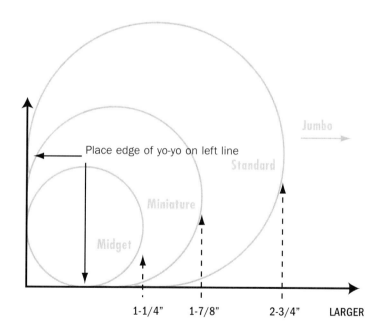

\* *For collectors who previously referred to "Lucky's Size and Shape Guide," the transition size has been merged into the standard for simplicity.*

## COLLECTING AND HISTORY

### ● FREQUENTLY ASKED QUESTIONS

*If a yo-yo does not have its original string, is its value decreased?*
A mint yo-yo should have its original untied string, but replacement of the string is acceptable. The one exception would be pegged string yo-yos which should be collected with original string intact. Loss of an original pegged string does slightly decrease its value.

*Do individual colors make a difference in value?*
Generally no, but there seems to be a new interest developing for specific colors that are more desirable for some particular models. In the future, this demand for certain colors will ultimately influence the value of these yo-yos. If color preferences appear to be developing, I have attempted to address those influences in this guide.

*Should yo-yos be kept in original packaging?*
Non-packaged yo-yos are certainly easier to display, and although not recommended, some collectors will actually play with yo-yos in their collection. This, coupled with the fact that many pre-1960s yo-yos were sold loose (non-packaged), has resulted in less demand for the original packaging. Although many yo-yo collectors do not value original packaging, this trend is slowly changing as more collectors enter the hobby. A general rule is, if the packaging graphics are interesting or unique in any way, keep the packaging. In other toy collecting fields, original packaging can add as much as 20% - 100% to the value of the toy. This is currently not the case with yo-yos, but over the next five years the hobby will likely mature to be more consistent with the rest of the toy market.

*What yo-yos should I collect?*
There are no set rules on collecting. Consideration should be given to storage and display space available and your budget. You can always limit your collection to a specific category that will fit your needs. For example, it would be foolish for a beginning pre-teen collector to specialize in expensive yo-yos, such as early Duncan tin lithographs. New collectors can get their feet wet by collecting recently produced yo-yos, such as Tom Kuhn or the Hummingbird series. If you have an extremely limited budget, you may consider Spectra Star's character yo-yo series. Buy because you like the look of the yo-yo, its history, or its uniqueness.

*Are yo-yos a good investment?*
Who would believe that a PEZ candy dispenser or an old cereal box would sell for thousands of dollars? Many toy collectors believe yo-yo prices will rise dramatically in the future. As with any type of collecting, yo-yo prices will respond to the law of supply and demand. As more collectors enter the hobby and add to the demand, yo-yo prices will increase. Yo-Yos are likely to hold their value because they continue to be produced and new generations of collectors will be exposed to the toy. Future yo-yo collectors will undoubtedly arise from the current generation. Undoubtedly, rare pieces will ultimately command prices similar to other rare toy collectibles, but standard line yo-yos that were produced in large quantities should stay within the price range of the average collector. Yo-Yos are a very durable toy. Although many were thrown in the trash over the years, many were also boxed up and are presently hiding in someone's attic. My personal opinion is that less than one percent of collectible yo-yos have been discovered, and many are still packed away. Many owners do not realize the value of their yo-yos. As soon as they are aware of the collectible value of these yo-yos, they will begin to enter the market. This new supply of yo-yos will offset the additional demand by new collectors. Other factors can affect individual prices of yo-yos. Be aware that a "warehouse find" of a "rare" yo-yo will dramatically decrease its value. The few yo-yo collections that have been sold have rarely equaled the return on their dollar investment once the time investment costs are added. I would discourage anyone who is planning on using their yo-yo collection as an investment for retirement or a college fund. Remember, dealers only pay 40% - 50% of book value for toys, and there are very few individual yo-yo collectors who are interested in purchasing an entire collection. If you want an investment, stick with the stock market.

*Will currently produced yo-yos go up in value?*
Demand, not age, determines the value of a collectible. The reason old yo-yos have value is that nobody thought of them as a collectible, and many were thrown out or packed away over time. Presently, demand exceeds the supply. Recently produced yo-yos may have the same experience in the future. For example, some recently produced Spectra Star character yo-yos now sell on the secondary market for more than ten times their original retail price. I recommend buying newer models for the current needs of your collection, not for long term speculation. If you feel these particular pieces fit well with your future collecting goals, then buy and enjoy.

*Where can I find the best deals on yo-yos?*
As with any collecting hobby, you are going to spend either time or money to make your collection grow. If you are time rich, then search garage sales, flea markets, estate sales, auctions, grandma's attic, and anyone who may have old yo-yos packed away. On the other hand, if you are time poor and have the money to spend, check out toy shows, toy publications, and internet auctions such as eBay. Buying is not the only way to make a collection grow. Most serious collectors trade among themselves. To get in touch with other collectors, consider joining the American Yo-Yo Association listed in the resources section in the back of this guide. You can trade yo-yos with other toy collectors if you have something they desire. I have

*Continued ...* ▶

traded comic books, sports cards, Star Wars figures, and even baseballs for yo-yos. If you have a hard time finding older yo-yos, consider collecting newer models. There are large numbers of yo-yos currently on sale in the retail market and new models are introduced every year.

*What is a beginners yo-yo?*
The manufacturer can label any yo-yo a beginners model. In the "wooden years," beginners models were three-piece wooden yo-yos with pegged strings. They were typically two colored, the most common combination being one half red and the other black. The three-piece models were cheaper to make and were generally smaller then the standard tournament models.

*How do I factor age into condition?*
Age is not a factor in grading a yo-yo's condition. A 50-year-old yo-yo in mint condition should look as new as if it was produced yesterday. A dealer that says he has a mint yo-yo with some aging wear is really saying he is not selling a yo-yo in mint condition. When grading a yo-yo there are no allowances for age.

*How does carving affect the value of the yo-yo?*
Carving yo-yos was a popular sideline activity of early yo-yo demonstrators. The demonstrator, as a reward or to make a sale, would carve a child's name or a scene on a yo-yo. Collectors differ on whether or not this carving adds or detracts from the yo-yo's value. A name scratched by knife or pen by the owner (not the demonstrator) does decrease the value. Elegantly carved scenes of palm trees, sail boats, and flying birds done by a demonstrator can enhance the yo-yo's value. Carvings done by a demonstrator that have resulted in considerable paint loss and chipping will subtract from the value.

*How can I be sure the collectible yo-yo I purchase won't have another production run?*
You can't! In any collecting field there is always the chance that a company will decide to re-release a previously produced model. The good news is that this rarely happens. Many times, if a classic yo-yo is re-released, the seals are different, allowing collectors to tell originals from recent models. Be aware that some plastic yo-yos have been produced from the same molds for decades. There are some 20-year-old yo-yos that have an identical twin for sale today in retail stores.

*Why is the seal so important?*
The seal to a yo-yo is like the date and mintmark on a coin. A rare Lincoln cent that has its date and mintmark worn off is worth just a penny. A yo-yo without the seal has lost its collectible identity and value.

*I have a wood Duncan Yo-Yo that has one half of its face sprayed with a contrasting color. The paint looks like the typical stripe was to be made, but it is much too wide. Does this have any significance?*
Sometime in the '50s the airbrush paint machine at the Duncan plant broke down and could not make the typical stripe, so half the yo-yo was sprayed. Collectors call these yo-yos "half and half" yo-yos. Do not confuse these yo-yos with three-piece beginners yo-yos where one entire half was one color and the other half a different color. Rumor is the paint machine stayed broken for quite awhile so many of these "half-and-half" yo-yos exist. See #449.1 for an example of a half-and-half.

*An old player said he had to "sand the slot and wax the string" to get his original Duncans to work well. What does this mean?*
Original Duncan tournaments were one-piece yo-yos, meaning the lathe turned the entire yo-yo out of one-piece of wood. Unlike modern wooden yo-yos, which are sanded and finished before piecing together, these tended to have rough string slots and axles. The yo-yos frequently had to be "broken in" before they played their best. Sanding the string slot helped smooth the sides, reducing friction. Waxing the string also helped reduce friction. Working the axle with the string also helped smooth the axle for better play. It took a lot of effort to fine tune early yo-yos into smooth playing models.

*What is the highest price ever paid for a yo-yo?*
The highest amount paid for a yo-yo occurred in 1993 at the Roy Acuff estate auction in Nashville, Tennessee. Roy Acuff was a legend at the Grand Ole Opry, having preformed there for over 50 years. Acuff's trademark was doing yo-yo tricks on stage. His affection for the yo-yo was known by all. When then President Richard Nixon visited Acuff at the Grand Ole Opry in March of 1974, he presented Acuff with an autographed yo-yo. Nixon took a couple of throws on stage with the yo-yo and a photo of this was released by the Associated Press making this one of the most famous yo-yos. The yo-yo and a film of the presentation were auctioned off for $16,029.00. The yo-yo now resides in the Country Music Hall of Fame.

*How do I calculate the value of a complete display box with all yo-yos?*
Values of complete boxes are actually less than the value of the individual pieces plus the box value. The reason is, unless the yo-yo is rare, it might take a dealer years to sell off the individual yo-yos loose for full value. A full box is generally 50% to 100% of the value of the individual yo-yos multiplied by the number of pieces in the box. There is no extra charge for the box. A good formula would be: Yo-Yos valued under $20.00 = 50%, $20.00-50.00 =75%, over $50.00 = 100%. For example, take a complete box of 12 Duncan Butterflies #331. Assuming all yo-yos were near mint, (be aware just because yo-yos are in the original box does not make them all near mint) you would take the near mint value of $25.00 times 12, which is 300, and multiply this by 75% for a value of $225.00.

## THE HISTORY OF THE YO-YO

There is magic in the yo-yo that does not exist in any other plaything. Some consider it a toy while others consider it a sport. The yo-yo's staying power as a toy has survived the centuries. A six-year-old can rapidly learn the yo-yo's simple operation, yet an adult can spend a lifetime and never fully succeed in mastering all known skills.

Collectors are familiar with the name Pedro Flores who is credited with introducing the word "yo-yo" to the United States in the late 1920s. Non-collectors recognize Don Duncan based on his popularization of the toy in the United States and around the world. What is not commonly known is that yo-yos were around long before Duncan ever began his first promotion.

*Ornate Greek ceramic disks, representing toys of the youth, may have been used as gifts to the gods as part of the "rites of passage" from childhood to manhood.*

The yo-yo existed centuries before Duncan, but was recognized by other names. Yo-Yo enthusiasts have always debated the country of origin. In the 1960s, Dr. Henry Lee Smith, a professor of Anthropology and Linguistics, testified in the Duncan vs Royal trial, that his research indicated the name "yo-yo" was oriental in origin. Adding credence to this claim is that a similar toy, the Diabolo, is well documented as having existed in China 1000 BC. It is also known that highly ornate Greek ceramic disks have the same design of a modern yo-yo. Archaeologists argue whether these 500 BC Greek disks represent yo-yos, spools for thread, or ornamental supports on which drapery cords were hung. Proponents of the yo-yo theory of these disks cite a Greek vase with a decoration of a young boy playing with what appears to be a yo-yo.

Another popular but fabricated origin of the yo-yo is that of a centuries old Filipino weapon. The story is about an assailant in a tree with a heavy oversized yo-yo waiting for a victim to pass below. At the correct moment, the yo-yo would be hurled at the victim's head presumably rendering him unconscious. A near miss would still allow the assailant a second opportunity. The physics of the yo-yo make this story highly improbable. The transfer of linear motion to rotary motion diminishes the yo-yo's striking force making it a much less effective weapon than that of a free missile. Although there is no question that the toy has existed in the Philippines for centuries, there is no documented evidence that it was ever used as a jungle weapon.

Prior to Pedro Flores, the most common name for the toy in the United States was Bandalore. The word "bandalore" is French in origin. The English also used "bandalore" as well as the word "Quiz" to identify the toy. Other French words used include *l'emigrette* (leave the country), *de Coblenz* (a city with a large number of noble French refugees) and *incroyable* (a French dandy). These terms all have an important historical connection with the French revolution. The yo-yo was a very fashionable toy of the French nobility during the time of the guillotine. When the heads of the nobility started being loped off, many of the nobles wisely emigrated along with their yo-yos. It is this association of the yo-yo and the displaced nobles that resulted in these colorful French names for the toy.

In 1866, James L. Haven and Charles Hettrick patented the first Bandalore in the United States. Their improved construction of the toy included a central rivet to hold the two halves together which allowed the toy to be made out of metal. Weight was distributed to the outer rim which allowed the toy to better function as a fly wheel. Interestingly, 100 years later, the high tech long spin yo-yos have a very similar rim design as the original patent. Several bandalore patents were issued prior to Flores and Duncan, such as a butterfly shaped yo-yo in 1879 by Katz, and a sphere shaped yo-yo in 1911 by Floto. In 1928, Pedro Flores introduced the toy as a "yo-yo" to the United States, and the history of the modern yo-yo began.

*Continued ...*

## COLLECTING AND HISTORY

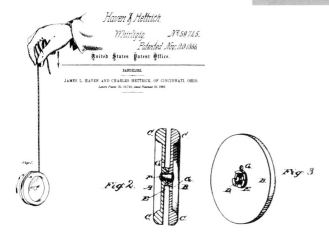

The word "yo-yo" was used in the Philippines for centuries. The first documented use of the word in print appeared in 1860. Although Flores introduced the word "Yo-Yo" to the United States, it was Duncan that popularized the toy and made it a household name. During the "Duncan Golden Years," 1929-1965, it was accepted by many that Duncan held the trademark for the word "yo-yo." Other manufacturers still produced yo-yos, but used other names such as "Swing Top," "Twirler," "Climbing Top," and " Return Top." In 1965, Duncan lost a court battle to maintain the trademark "Yo-Yo." The word "yo-yo" was given generic status. Now all yo-yo manufacturers in the United States can use the word "yo-yo" on the toy.

*The following is a historical time line of references to "yo-yos" prior to Duncan.*

| | |
|---|---|
| Origin | The yo-yo is currently believed to be of oriental origin (1000 BC), but there is no indisputable documentation. |
| 500 BC | Terra-cotta (yo-yo) disk (Greece). |
| 500 BC | Vase with an illustration of a youth playing with a yo-yo (Greece). |
| 1789 | Quizzes could be purchased at the Peckham Fair (England). |
| 1789 | Painting of young King Louis XVII (age 4) with l'emigrette (French). |
| 1780s | Cartoon of General Lafayette leading a procession of soldiers playing with yo-yos (French). |
| 1780s | Cartoon of Mirabeau with troops and yo-yos (French). |
| 1790 | Meyer's Lexikon (pub. 1927) states toy introduced to Paris from the orient this year. Calls it a Kletterkreisel. |
| 1791 | Larousse's Grand Dictionnaire states the yo-yo was invented this year. (French) Published 1866. |
| 1791 | Satirical print of George IV playing with a bandalore (English). |
| 1792 | Edition of Baron Munchausen makes reference to a lady playing with quizzes. |
| 1792 | Booklet (The Constitution in Vaudeville) shows man on cover with large l'emigrette (French). |
| 1824 | Novelist Mary Russell Mitford (Hampshire) makes reference to bygone toy (bandalore). |
| 1833 | Moore makes reference to a fashionable toy called a bandalore in French, a quiz in English. |
| 1852 | Preller refers to the toy as a Jou Jou. |
| 1860 | First documented use of word "yo-yo" in print. Vocabulario de la Lengua Pampanga (Philippines). |
| 1862 | Illustration in Punch shows two boys throwing quizzes at a lady (England). |
| 1866 | Patent for "improved bandalore" was awarded, Haven and Hettrick (United States). |
| 1882 | Patent for securing bells to a bandalore, Wurst and Mooseman (United States). |
| 1887 | Patent for Whirligig, new way to attach string to a bandalore (United States). |
| 1891 | Magazine (Boy's Modern Playmate) makes reference to bandalore being called a "quiz" around the turn of the century, 1800 (English). |
| 1901 | Patent for metal rim protector and circle of fire effect, Boehme (United States). |
| 1904 | An iron yo-yo is sold as a souvenir at the St. Louis World's Fair. |
| 1907 | Patent for new way to attach the two yo-yo halves together, Temple (United States). |
| 1911 | Patent for round yo-yo and new type of axle, Floto (United States). |
| 1916 | Scientific American, supplement, article "Filipino Toys," shows and names a yo-yo. |
| 1923 | Combination cup and ball yo-yo, with ball being a yo-yo, described by Bouasse (French). |
| 1928 | Pedro Flores opens the Yo-Yo Manufacturing Company in Santa Barbara, California. |
| 1929 | Pedro Flores opens the Yo-Yo Manufacturing Company in, Los Angeles, California. |
| 1929 | Popular Mechanics, July 29 issue, publishes "Make a Filipino Yo-Yo." |
| 1930 | Flores receives registration of trademark "Flores Yo-Yo" (July 22). |
| 1930 | Duncan files for trademark "Genuine Duncan Yo-Yo" (November 28). |
| 1931 | Duncan registration denied by patent office examiner citing "Flores Yo-Yo" trademark. |
| 1932 | Trademark "Flores Yo-Yo" assigned to Duncan. |
| 1932 | Duncan files for trademark "Yo-Yo" (October 4). |
| 1932 | Duncan receives registration "Genuine Duncan Yo-Yo" (November 1). |
| 1933 | Duncan receives registration "Yo-Yo" (January 24). |

## HISTORY OF YO-YO CONTESTS

Most yo-yo collectors view the yo-yo as an important piece of Americana from the 20th century. America's love for the yo-yo is intimately linked with the yo-yo contest. From the late '20s through the '60s, the annual appearance of the yo-yo demonstrators and their contests were an expected ritual in towns and cities across the country. Not surprisingly, many yo-yo collectors have fond memories of these contests. This fuels the fire for the preservation of the history of the yo-yo.

Yo-Yo contests originated in California, the first one being held in 1928 in Santa Barbara. Pedro Flores ran the contest. These contests made the "yo-yo" name wildly popular. The early contests were very simple, the goal being who could make the yo-yo go up and down the most times without stopping.

Flores held the trademark for the word "yo-yo" and had the first yo-yo factory, which was based in Santa Barbara. In 1930, Flores sold the trademark rights to Duncan. Flores continued to run the contests for Duncan through the early '30s. It was the Duncan contest promotions that made the word "yo-yo" a household name. Contests had also dramatically changed from the earliest Flores promotions. The use of a slip string opened a whole new universe of tricks for yo-yo players and the introduction of the contest trick list was started. Score cards were developed by Barney Akers, a famed yo-yo demonstrator and cousin to Don Duncan Jr.

Duncan had yo-yo demonstrators, usually young Filipinos, travel from city to city across the United States, teaching yo-yo skills and setting up contests. These contests were not limited to just the United States. International campaigns were run successfully as well.

In the 1930s, the Duncan Company dominated the yo-yo contest market in the United States. This was largely due to the innovative collaboration of yo-yo promotions and local newspapers. Newspapers gave free publicity to contests in exchange for new subscriptions. These subscriptions were a requirement for contest entry. These city wide promotions were hugely successful with some major metropolitan areas generating as many as 50,000 new subscribers for the sponsoring paper and increasing yo-yo sales up to three million in a thirty-day period.

The contest rage simmered in the early '40s due to World War II (WWII). Yo-Yo demonstrators were off to war and raw materials for making yo-yos were difficult to acquire. Factories that had normally produced yo-yos had converted production and equipment to support the war effort.

*School teacher demonstrates "Man on the Flying Trapeze" with a first series Duncan Tin Whistler, circa 1934.*

*Children gather for a Duncan contest in 1931. Notice the Duncan demonstrator in the foreground.*

Continued ...

## COLLECTING AND HISTORY

37

*Duncan contestants, Miami, Florida, Bayfront Park, Jan. 27, 1934.*

Following the War, Duncan decided to manufacture the yo-yos in house and set up a factory in Luck, Wisconsin. During 1946, Cheerio hit the contest road, and with the absence of Duncan, who was still setting up shop, developed some serious market penetration. In 1947, Duncan re-entered the contest trail in a big way, but other companies such as Royal and Goody now had noticeable presence.

The '50s would have to be considered the glory years of the yo-yo contests. Duncan had regained its ground as the market leader. The other companies' strategies included either shadowing the Duncan promotions with their own limited budget promotions or completely avoiding the Duncan campaign areas. Competition between companies for contest recognition resulted in an increasingly vast array of contest awards and prizes, from patches and ribbons to motor scooters and bicycles.

In the early 1960s, Duncan changed its marketing approach from demonstrator based promotions to television based. Some demonstrators used the television to maximum advantage by offering live performances. A week before the contest, Duncan would usually buy three to five 60-second advertising spots during the local afternoon kids' show. Demonstrators would then contact the show and ask if they would like a free live performance. This was usually accepted and the yo-yo demonstrator would get five to ten minutes of free airtime on the show. Bob Baab, a former demonstrator, remembers appearing on kids' shows such as "Salty Brine" in Providence, RI, "Bob Brandy the Cowboy with a Guitar," Chattanooga, Tenn., and "Sailor Bob," Indianapolis, Ind. Many other children's shows also hosted yo-yo demonstrators in the 1960s.

Duncan initially had wild success with television, but with a decreasing number of demonstrators to teach skills and set up contests, interest faded. In 1965, Duncan lost a court battle to retain the word "yo-yo" as a registered trademark and the word became generic. Later, in the same year, Duncan was forced into bankruptcy, and in 1968 was bought by Flambeau Plastics. Flambeau continued Duncan's tradition of contests with traveling demonstrators, but they were never of the scale or had the success of the early years.

During the late 1980s, state championships were reintroduced in Chico, California, mainly due to the efforts of Bob Malowney. In 1993, the first modern National Yo-Yo Championships were held in Chico, California. The championships were sponsored by the Bird-in-Hand store, which is also the site of the National Yo-Yo Museum. The National Yo-Yo Championships have been hosted in Chico ever since.

The '90s have seen the formation of the American Yo-Yo Association (AYYA) whose mission is to promote yo-yo play, contests, and collecting. The AYYA sanctions yo-yo contests around the country and has developed rules and guidelines for competition play. The introduction of high tech transaxle yo-yos, which spawned freestyle play, has changed the way we view yo-yoing expertise. Manufacturers and retailers are increasing the number of competitions held each year and the future is bright for yo-yo contests to once again regain prominence in American culture.

## COLLECTING AND HISTORY

### AN EARLY CONTEST

The following is information from a yo-yo contest run by the Duncan Yo-Yo Company in April 1931. The contest was held in Akron, Ohio at the RKO Theater and organized by Pedro Flores. Theaters were a common site for yo-yo contests in the '30s. Contestants had to be between 7 and 25 years old and were required to use a Genuine Gold Seal Duncan Yo-Yo. Training camps were held daily for a few weeks prior to the contest and a preliminary tournament was held one week prior to the finals. News releases indicated all demonstrators were native Filipino boys and called the contest an "invasion" of Akron. This was obviously a big happening in Akron.

Yo-Yos were supplied on consignment to dealers throughout Akron. Dealers could sell two models, the 10 cent version, which was most likely the O-Boy Beginner model, or the 25 cent model, the Gold Seal Tournament. Dealer cost was 80 cents a dozen for the 10 cent model and $2.00 a dozen for the 25 cent models. Retailers could return unsold yo-yos for a full refund at any time.

One hundred dollars in cash prizes were given. The winner received a trip to Cleveland to see a professional baseball game. Prizes were as follows: 1st = $30.00, 2nd = $20.00, 3rd = $15, 4th = $10, 5th = $8, 6th = $6, 7th = $5, 8th = $3, 9th = $2, 10th = $1. This was a lot of money during The Depression and it brought out hundreds of contestants. A total of eighteen people qualified for the finals. They were required to do 12 tricks in the championship competition. The most Loop the Loops completed decided ties.

Below is a description of the trick list from this 1931 contest. (Most of the tricks did not have names in 1931.)

1. Snap the Yo-Yo with the left hand and wind the string in full length with the right hand.
2. Throw the Yo-Yo over hand, five times.
3. Throw the Yo-Yo straight forward, five times.
4. Throw the Yo-Yo straight up into the air, five times.
5. Throw the Yo-Yo downward and make it spin at the end of the string, two times.
6. Throw the Yo-Yo downward. Make it spin at the end of the string. Roll it on the floor, then bring it back to the hand without winding.
7. Throw the Yo-Yo downward. Make it spin at the end of the string. Fold the string three times and then let the Yo-Yo drop, causing the Yo-Yo to come back to the hand.
8. **THE WATERFALL.** Throw the Yo-Yo straight up, then let it fall down close to the body, catching the Yo-Yo as it comes back up.
9. Throw the Yo-Yo straight down, making it spin at the end of the string, then bounce it on the floor, causing the Yo-Yo to jump back to the hand.
10. **AROUND THE WORLD.** Two times. Throw the Yo-Yo in a full circle at the right hand side of the body, letting the Yo-Yo rewind itself after completing the circle.
11. Wind the Yo-Yo by means of rolling it on the floor.
12. **LOOP the LOOP.** Throw the Yo-Yo straight forward from the body in rapid succession without catching it until the Yo-Yo fails to wind.

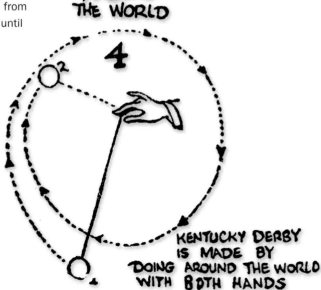

*This 1930s illustration demonstrates how to perform these two yo-yo tricks, Around The World and the Kentucky Derby.*

# Yo•Yo Listings

## YO-YO LISTINGS

 **ADULT**

This is a tin photo lithograph yo-yo series from the '70s. These yo-yos were retailed either loose out of a display box or on blister cards. The blister card had photos of nude models that were different from those displayed on the yo-yo. The blister card referred to this as an "Executive Yo-Yo."

| | | |
|---|---|---|
| 1 | 19 | **Photo lithograph seal, tin, tournament shape, '70s** |
| | | There are 12 photo lithographed variations of this yo-yo. For the display box, see Plate 44, #1 |

 **ADVERTISING METAL**

Over the decades, yo-yos of all types have been used for advertising promotions. Using metal for advertising yo-yos has been somewhat limited due to the cost of production. Wood or plastic models are much less expensive. Regardless of the metal used in its manufacturing, collectors usually refer to metal yo-yos as "tins." The value of tin ad yo-yos depends on several factors: cross-collectibility of the ad, design, age, condition and manufacturer. Recently produced tin advertising yo-yos carry little value. However, some early metal ad yo-yos can have high value. Collectors consider pre-1960 metal ad yo-yos very desirable. Be on the lookout for unusually shaped yo-yos or those that produce sounds. For other metal yo-yos, see Duncan Tin, Metal Miscellaneous Manufacturers, Souvenir, Clown, and Character sections.

| | | |
|---|---|---|
| 2 | 19 | **BOSCO 3-FOOD DRINK FOR HEALTH, ink stamped seal, concave riveted disk axle** |
| | | This was a promotion for the popular chocolate drink mix. The date of production for this yo-yo has not been documented. |
| 3 | 01 | **BUSTER BROWN, lithograph seal, tournament shape, metal axle** |
| | | This promotional yo-yo was for Buster Brown Shoes. It is believed to have been produced in the late '60s or early '70s. Several color variations of this tin model exist. Although this is the most recognized Buster Brown yo-yo, it was only one of many Buster Brown yo-yos produced. Many of the earlier versions from the '50s were wood. See also #75. |
| 4 | 01 | **CHUCK E CHEESE, lithograph seal, tournament shape, '90s** |
| | | Chuck E Cheese is the mascot for Showbiz Pizza. This tin model is considered the most collectible of the Showbiz Pizza yo-yos. The other Chuck E Cheese yo-yos are plastic. See #19, #20, #21, and #22. |
| 5 | 19 | **HILL TOYS AND BELLS, paper seal, bulge face shape, wood axle** |
| | | This yo-yo has metal balls inside that produce sounds when played. It also has a small wooden disk tied to the string end. |
| 6 | 19 | **JOSEPH STRAUSS COMPANY, inked seal over gradient paint, convex disks, metal axle, '50s** |
| | | This is a very unusual shaped yo-yo. Note the four digit phone number. |
| 7 | 23 | **PEZ (with peppermint package), lithograph seal, riveted disk axle, '50s** |
| | | The PEZ ad yo-yo carries one of the highest values of any ad yo-yo. It is valued more by PEZ collectors than by yo-yo collectors. Two variations of this yo-yo were produced as PEZ premiums in the '50s. Although both tin PEZ yo-yos sell in a similar price range, this style is slightly easier to find. Both tin PEZ are riveted disk yo-yos. There is also a plastic version of the PEZ yo-yo. See #54. |
| 8 | | **PEZ (without PEZ peppermint package), lithograph seal, riveted disk axle, '50s** |
| | | This is the least common of the two styles of tin PEZ premium yo-yos. It does not have a PEZ package lithographed on the face. See also #7. |
| 9 | 19 | **TOASTED WHEAT FLAKES • KEEP GOING ON FORCE, paint seal, coaster shape, riveted disk axle, '40s** |
| | | This yo-yo was a cereal premium from WWII. These war years yo-yos had very low production numbers due to limited resources and manpower. |
| 10 | | **VICTOR TOY OATS, paper seal, tin, coaster shape, riveted disk axle** |
| | | The date of production for this yo-yo has not been documented, but many collectors feel this is from the '30s or earlier. The reverse face gives a description on how to use the toy and refers to it as a "top." An example of this yo-yo is shown next to this section's title. |

## YO-YO LISTINGS

### ADVERTISING PLASTIC

This section includes a variety of plastic advertising yo-yos. The majority are of inexpensive design and play poorly. The value of an advertising yo-yo depends on several factors: how cross-collectible it is, the ad design, age, and maker of the yo-yo. The most common plastic advertising yo-yos are manufactured by Humphrey. (See All Pro #762.) Because Humphreys are more common than any other plastic ad yo-yos, they are listed in their own section.

Thousands of different plastic ad yo-yos have been produced over the years. It would be impossible to collect them all. Collecting plastic advertising yo-yos is a great way to get a collection started, especially for the collector on a limited budget. This section contains a small sampling of plastic ad yo-yo styles that may be found. For other advertising yo-yos, see Duncan Advertising, Advertising Wood, Advertising Metal, Humphrey, Sports Related, Coca-Cola and Souvenir sections.

| | | |
|---|---|---|
| 11 | | AMDAHL, paper sticker seal, coaster shape, '90s |
| 12 | | AMDAHL, imprint seal, bulge face F-210 shape, '80s |
| 13 | | BARCALOUNGER, paper sticker seal, coaster shape, '90s |
| 14 | 04 | B.F. GOODRICH TIRES, embossed seal, sculpted, miniature, plastic axle, pegged string, '70s |

This yo-yo is sculpted in the shape of a tire. Sculpted yo-yos are favorites of ad yo-yo collectors and carry a higher value than non-sculpted varieties.

| | | |
|---|---|---|
| 15 | 01 | BURGER CHEF © 1972, imprint seal, tournament shape, early '70s |

The seal shows Burger Chef and Jeff. This yo-yo appears to have been produced from a Festival mold.

| | | |
|---|---|---|
| 16 | | BURGER THING, paper sticker seal, puck shape, 1979 |

This is a snap together premium yo-yo given out by Burger King in 1979. The seal shows a funny faced hamburger called "The Burger Thing" and carries a copyright date of 1979. An example of this yo-yo is shown next to this section's title.

| | | |
|---|---|---|
| 17 | | CARLISLE PRODUCTIONS, large disk insert seal, Humphrey shape, '80s |
| 18 | | (Chevrolet logo), paper sticker seal, coaster shape, '90s |
| 19 | 01 | (Chuck E Cheese) 1991 SHOWBIZ PIZZA TIME, large disk insert seal, tournament shape, '90s |

Chuck E Cheese is the character mascot of Showbiz Pizza. This restaurant chain caters to children and offers games and prizes for patrons. Several different styles of inexpensive yo-yos have been offered as prizes.

| | | |
|---|---|---|
| 20 | | CHUCK E CHEESE PIZZA TIME THEATER, imprint seal, tournament shape, '90s |
| 21 | 01 | CHUCK E CHEESE, paper sticker seal, tournament shape, midget, '90s |
| 22 | 01 | CHUCK E CHEESE, paper sticker seal, tournament shape, miniature, '90s |

This is a soft plastic yo-yo.

| | | |
|---|---|---|
| 23 | 01 | COUNTRY KITCHEN, paper insert seal, tournament shape, view lens, '70s |

This restaurant promotional yo-yo has a pinball game on one face and the Country Kitchen characters on the reverse. This yo-yo appears to have been made from a Festival mold.

| | | |
|---|---|---|
| 24 | 01 | (Country Kitchen characters), paper sticker seal, tournament shape, '70s |

The Country Kitchen girl is on the reverse face.

| | | |
|---|---|---|
| | • | **Cracker Jack** - *See Character section #162.* |
| 25 | | DOLE, paper sticker seal, butterfly shape, '80s |

This is from a Dole Pineapple promotion.

| | | |
|---|---|---|
| 26 | | DRUGS "JUST SAY NO," imprint seal, bulge face F-210 shape, '80s |
| 27 | | FIRESTONE TIRE AND RUBBER CO., paper sticker seal, tournament shape, '60s |

This is a heavier, better playing yo-yo than most plastic ad yo-yos.

| | | |
|---|---|---|
| 28 | | FRIENDS ARE SPECIAL, imprint seal, tournament shape, metal string slot axle, '90s |
| 29 | | GIRL SCOUTS GROW AT CAMP, paper sticker seal, coaster shape, '90s |
| 30 | | GOODYEAR #1 IN RACING, imprint seal, coaster shape, '90s |

A paper sticker seal of a hubcap is in the concavity of the face. This model was marketed by Leading Eagle Technologies as a "Go-Yo." It includes a trick sheet showing three tricks with car racing names: Around the Speedway (Around the World), Checking Out (Walk the Dog), and Spinning Wheels (Sleeper). Similar yo-yos were retailed with photo sticker seals of race cars and drivers. See #1496.

| | | |
|---|---|---|
| 31 | | GRISHAM STEEL, imprint seal, bulge face F-210 shape, '80s |
| 32 | 01 | HAWAIIAN PUNCH, paper sticker seal, Imperial Toy Co., butterfly shape, 1996 |

This was retailed on a blister card, Stock No. 7120. The yo-yo features "Punchy," the Hawaiian Punch character, on the seal.

| | | |
|---|---|---|
| 33 | 03 | HBO (Independence Day), embossed seal, sculpted, late '90s |

This yo-yo is sculpted in the shape of the alien space ship featured in the 1996 blockbuster movie "Independence Day."

| | | |
|---|---|---|
| 34 | 34 | HERTZ, imprint seal, sculpted, '90s |

This is a globe shaped yo-yo made by the Humphrey Co. In addition to Hertz, other advertisers have used this globe mold for premiums. Two sculpted caps snap onto the faces to make this yo-yo.

*Continued ...* ▶

*The History and Values of Yo-Yos*

## YO-YO LISTINGS

● ADVERTISING PLASTIC *Continued ...*

| | | |
|---|---|---|
| 35 | | **HOME FEDERAL FLOPPY SAVERS CLUB**, paper sticker seal, tournament shape, '80s |

The seal has a cartoon character of a floppy-eared dog. This is a savings and loan promotion.

36          **I MADE A PIG OF MYSELF AT FARRELL'S**, large disk insert seal, Humphrey shape, '80s

37          **JACK IN THE BOX**, paper sticker seal, tournament shape, miniature, '90s

38          **KEDS**, imprint seal, tournament shape, '80s

39      *03*     **KEDS**, molded seal, miniature, view lens, '70s

This is a snap together premium with a pinball game on one face and a decoder on the opposite that was packaged in cellophane. This yo-yo has a space ship and Saturn as graphics and was a promotional give-away from the giant shoe store chain.

40          **LEE**, imprint seal, slimline shape, '90s

This is a promotional yo-yo from the Lee Jeans Company.

41          **LEE'S FAMOUS RECIPE COUNTRY CHICKEN**, embossed seal, Humphrey shape, '80s

The yo-yo has internal fins and a pegged string.

43      *01*     **(Long John Silver's • octopus)**, imprint seal, tournament shape, '80s

This is a premium yo-yo produced for the fast food chain. Three different animal logos were used on this yo-yo. Of the three logos, the octopus is slightly harder to find.

44      *01*     **(Long John Silver's • dolphin)**, imprint seal, tournament shape, '80s

45      *01*     **(Long John Silver's • sea horse)**, imprint seal, tournament shape, '80s

46          **MATH BLASTER MYSTERY**, imprint seal, tournament shape, '80s

The reverse face reads "Davidson Teaching Tools for Teachers."

47          **MEDTOX**, imprint seal, tournament shape, '90s

48          **MILLER GENUINE DRAFT**, removable disk seal, flywheel shape, '80s

This Pro Yo has removable ad inserts. See the Playmaxx section for other yo-yos of this type.

49          **MOLINE NATIONAL BANK**, ink stamped seal, tournament shape, '50s

"Big Chief" is hot stamped on the reverse face. The yo-yo has internal fins and is made from marbleized plastic. Manufacturers used marbleized (variegated colors) plastic in the '40s and '50s to economize. Molding different colors of scrap plastic together gives this plastic a variegated appearance.

50      *28*     **MONARCH BUTTERFLY • LARGEST IMPRINT IN THE INDUSTRY**, imprint seal, butterfly shape, '70s

This is a butterfly yo-yo briefly produced by the Humphrey Corp. in the late '70s and was used in marketing Humphrey's own Monarch advertising line. In addition to the All Pro, known as the "Humphrey" style to collectors, Humphrey produced both a butterfly shape (Monarch) and tournament shape (Classic).

51      *01*     **MORTON YO-NUT**, molded seal, sculpted, '80s

This was a Morton Donut promotional yo-yo. It was sculpted into the shape of a chocolate covered donut and packaged with a small trick book.

52      *01*     **MR. PEANUT**, imprint seal, tournament shape, miniature, '80s

For other Mr. Peanut yo-yos, see also #307.34 and #872.

53      *01*     **OREO**, embossed seal, sculpted, metal axle, '80s

This cookie shaped yo-yo is one of two versions of the Oreo Yo-Yo. See also #1038.

54      *23*     **PEZ**, imprint seal, slimline shape, '70s

It is unclear whether this yo-yo was ever licensed by PEZ. The date and maker are currently not documented, but the yo-yo is believed to be from the '70s. Several thousand of them turned up in a warehouse in Europe in 1997. For other tin PEZ yo-yos, see #7 and #8.

55          **SCOTCHGARD**, imprint seal, bulge face F-210 shape, '80s

56      *01*     **SHOWBIZ PIZZA TIME**, large disk insert seal, tournament shape, '80s

Showbiz Pizza, a restaurant chain, has produced a variety of inexpensive plastic yo-yos that were used as prizes for games. These yo-yos are from the late '80s and '90s. See also Chuck E Cheese #4 and #19.

57          **SHOWBIZ PIZZA PLACE**, large disk insert seal, tournament shape, '80s

The yo-yo pictures a bear playing the banjo, copyright 1986. See also #56.

58          **STAR INSURANCE**, imprint seal, ball shape, '90s

Sports ball models, such as this bowling ball shaped yo-yo, are frequently used for promotional advertising. These carry a slightly higher value than standard yo-yos for the same promotion. For other sports balls, see the Sports Balls section.

59          **SUN BEAM BREAD**, paper sticker seal, coaster shape, '80s

60          **TASTY TACOS**, imprint seal, bulge face F-210 shape, '80s

61      *28*     **THE CLASSIC "ANOTHER WINNER FROM THE HUMPHREY LINE,"** imprint seal, tournament shape, '70s

This is a promotional yo-yo for Humphrey advertising yo-yos. The Humphrey "Classic" was tournament shaped, but the mold is now discontinued. Humphrey opted to eliminate this style of yo-yo because the "All Pro" (standard Humphrey shaped yo-yo) had a much larger imprint area for advertising. These early promotional yo-yos from Humphrey are uncommon.

62      *01*     **THE LAUGHING COW**, imprint seal, tournament shape, miniature, metal string slot axle

63          **VARIAN**, imprint seal, tournament shape, '90s

"Inventive People Making Technology Work" is imprinted on the reverse face.

64          **WAYNE WOODRUM INC.**, paper sticker seal, coaster shape, '80s

The Chevrolet symbol is on the reverse face.

## YO-YO LISTINGS

## ADVERTISING WOOD

The practice of using yo-yos for advertising has been around since the 1930s. The use of wood advertising yo-yos peaked during the '50s and early '60s. Although wood advertising yo-yos are still being produced, their numbers are small in comparison to plastic. Many of the wood ad yo-yos in the '50s and '60s were miniature, three-piece models with pegged strings, but there are exceptions.

The value of a wood advertising yo-yo depends on several factors, the most important is whether the ad is cross-collectible. Other factors that can affect the value include design, seal type, date of production, and manufacturer of the yo-yo. For example, a black ink stamped seal carries less value than a decal with an ornate logo. Older advertising yo-yos with stickers may carry a higher value, especially if the seal is intact. Advertising collectors, as well as yo-yo collectors, compete to collect items from advertisers such as Coca-Cola, Shoney's, Buster Brown Shoes, etc. This can increase the value of these yo-yos. The company that manufactured the yo-yo may also add to the yo-yo's value. Duncan wooden advertising yo-yos, produced before 1965, are favorites of collectors.

Hundreds, possibly thousands, of different wood ad yo-yos have been produced and new finds are constantly turning up. This list is only a sampling of wood advertising yo-yos that may be found. (See also Duncan Advertising section.)

| # | | Description |
|---|---|---|
| 65 | | **ACTION OLDS**, paint seal, tournament shape, '70s |
| 66 | *46* | **AVON**, gold leaf stamped seal, wood, tournament shape, three-piece, '70s |
| | | This yo-yo came in a two compartment box called the "Avon Yo-Yo Set." One compartment held the yo-yo; the other held a yo-yo shaped pink cake of soap. The yo-yo is not uncommon, but the complete set shown is hard to find. |
| 67 | | **BANK AMERICARD**, gold leaf stamped seal, tournament shape, pegged string, '60s |
| | | This yo-yo reads "The Bank of Indiana National" on the reverse face. |
| 68 | *02* | **(Big boy) SHONEY'S BIG BOY**, die stamped seal, tournament shape, '60s |
| | | This yo-yo has the Big Boy seal #69 on the reverse face. |
| 69 | *02* | **(Big boy)**, die stamped seal, tournament shape, '60s |
| | | An example of this yo-yo is shown next to this section's title. |
| 70 | | **BOISE CASCADE PAPER GROUP**, paint seal, jumbo, '80s |
| 71 | | **BRADLEY'S GENERAL STORE**, ink stamped seal, tournament shape, '90s |
| | | This yo-yo is a natural wood yo-yo without a finish. |
| 72 | *02* | **BROCK CANDY BAR**, paper sticker seal, tournament shape, pegged string, '50s |
| 73 | | **BROOKSTONE**, gold leaf stamped seal, tournament shape, '90s |
| | | This yo-yo was manufactured by Hummingbird and retailed in Brookstone stores. Most Hummingbird advertising yo-yos were maple. The Brookstone yo-yo was made of exotic woods. |
| 74 | *02* | **BURGER KING**, imprint seal, tournament shape, pegged string, '70s |
| | | The reverse face reads "Home of the Whopper." |
| 75 | *02* | **BUSTER BROWN SHOES**, ink stamped seal, tournament shape, pegged string, '50s |
| | | Yo-Yo give-aways were very common with this shoe store chain. Many varieties of ink stamped seals were used for Buster Brown shoes. See also Advertising Metal #3. |
| 76 | | **CHIP'S CLOTHES FOR BOYS**, ink stamped seal, tournament shape, three-piece, pegged string, '50s |
| 77 | *02* | **CHEVROLET '56**, gold leaf stamped seal, tournament shape, '50s |
| | | This is a favorite among ad collectors, but the yo-yo is not that rare. |
| 78 | *02* | **CLACKAMAS CO. FAIR**, ink stamped seal, tournament shape, three-piece, '60s |
| 79 | | **CRACKER BARREL**, die stamped seal, Hummingbird, tournament shape, '90s |
| | | This yo-yo was a promotional item for the Cracker Barrel restaurant chain. It retailed in a polybag with a saddle header card. |
| 80 | | **DAN DEE POTATO CHIPS**, gold leaf stamped seal, three-piece, '60s |
| 81 | | **DAVE HOFER IMPLEMENT CO.**, gold leaf stamped seal, tournament shape, miniature, '50s |
| | | The yo-yo is a three-piece model with a pegged string. The logo also has a two digit phone number, 13, from Menno, South Dakota. |
| 82 | | **DODGE TRUCKS**, ink stamp seal, tournament shape, three-piece, pegged string, '60s |
| 83 | *28* | **FAO SCHWARZ FIFTH AVE.**, gold leaf stamped seal, tournament shape, '90s |
| | | Hummingbird manufactured this yo-yo for the famous New York City toy store. It was one of the few painted yo-yos Hummingbird produced. Hummingbird is better known for its natural wood yo-yos. See also Duncan Advertising #307.21. |
| 84 | *02* | **G.E.**, stamped seal, tournament shape, pegged string, '80s |
| | | "New Harvest" is die stamped on the reverse face. |
| 85 | | **FIREDOME V8 DESOTO**, gold leaf stamped seal, tournament shape, three-piece, '60s |
| | | The seal on this model also reads "Powermaster 6." |

*Continued ...*

The History and Values of Yo-Yos

## YO-YO LISTINGS

● **ADVERTISING WOOD** *Continued ...*

| | | |
|---|---|---|
| 86 | *02* | **GLOVE KID PEANUT BUTTER**, die stamped seal, tournament shape, three-piece, '50s |
| 87 | *02* | **GULF**, paint seal, tournament shape, three-piece, pegged string, '60s |
| 88 | *02* | **HENRY'S**, gold leaf stamped seal, tournament shape, three-piece, '70s |

This yo-yo advertises Henry's Department Store. There is a gold leaf stamped seal of a propeller on the reverse face (shown). This style of yo-yo, with the propeller imprint, has been found both with and without advertising. The maker is unknown.

| | | |
|---|---|---|
| 89 | | **HOLMS BREAD**, ink stamped seal, tournament shape, '60s |
| 90 | *02* | **HOT POINT**, gold leaf stamped seal, tournament shape, pegged string, '70s |
| 91 | *02* | **JOHN'S SHOE STORE X-RAY FITTINGS**, ink stamped seal, tournament shape, '50s |

Shoe stores frequently gave away yo-yos as promotions in the '50s and '60s. These x-ray fittings were discontinued in the very early '60s for now obvious reasons.

| | | |
|---|---|---|
| 92 | *02* | **KENT**, gold leaf stamped seal, tournament shape, miniature |

"Famous for quality and performance" is inscribed on the seal.

| | | |
|---|---|---|
| 93 | | **LEARNING SMITH**, paint seal, wood, puck shape, 1996 |

This yo-yo is a three-piece model with a pegged string. The wood has a colored stain with a polyurethane coating.

| | | |
|---|---|---|
| 94 | *01* | **LIFE SAVER**, paint seal, sculpted, '60s |

This Lifesaver shaped yo-yo is popular with ad collectors. It was reportedly a premium given with the purchase of Beechnut Gum. The yo-yos offered were in the five colors found in a Lifesaver pack.

| | | |
|---|---|---|
| 95 | | **L.L. BEAN**, die stamped seal, Hummingbird, tournament shape, '80s |
| 96 | | **MERCEDES BENZ**, plastic medallion seal, tournament shape, '90s |
| 97 | | **MINWAX**, paint seal, tournament shape, '80s |

This yo-yo is made from cedar wood, a common material for souvenir yo-yos.

| | | |
|---|---|---|
| 98 | | **MODERN WOODSMAN LIFE INSURANCE**, die stamped seal, tournament shape, '60s |
| 99 | | **MR. DOUGHNUT**, die stamped seal, tournament shape, '90s |
| 100 | *02* | **MISTER TEE'S POTATO CHIPS**, gold leaf stamped seal, tournament shape, pegged string, '50s |
| 101 | | **NAPA ASSURANCE IN QUALITY**, paint seal, tournament shape, miniature, three-piece, pegged string, '70s |
| 102 | | **NATURAL WONDERS**, paint seal, tournament shape, '90s |
| 103 | | **NEW DAY FLOWER**, ink stamped seal, tournament shape, '50s |
| 104 | *02* | **NOPCO CTC THE FULL POTENCY ANTIBIOTIC**, paint seal, tournament shape, pegged string, '70s |
| 105 | | **PENNY'S**, paint seal, tournament shape, miniature, pegged nylon string, '70s |
| 106 | | **PLAY ALONG WITH SHEARER • FISHER • HACKBACK**, ink stamped seal, tournament shape, miniature, '60s |
| 107 | | **POLK COUNTY FEDERAL**, imprint seal, tournament shape, miniature, three-piece, pegged string, '60s |
| 108 | *02* | **POLL-PARROT SHOES FOR BOYS AND GIRLS**, paper sticker seal, wood, tournament shape, '50s |
| 109 | *02* | **POTTER'S JUNIOR SHOES**, ink stamped seal, ball shape, miniature, pegged string, '50s |

This yo-yo is almost spherical in shape. It also has a metal finger ring at the end of the string.

| | | |
|---|---|---|
| 110 | *02* | **RED BARN**, gold leaf stamped seal, tournament shape, '70s |

There are three variations of the red barn seal and there may be others. All versions are solid red and have pegged strings.

| | | |
|---|---|---|
| 111 | *02* | **ROYAL NEIGHBORS OF AMERICA**, gold leaf stamped seal, tournament shape, '60s |

Royal Neighbors of America is an insurance company. More than one style of logo was produced for this company.

| | | |
|---|---|---|
| 112 | | **SEALTEST**, paint seal, wood, tournament shape, jumbo, 3 piece, pegged string, '50s |

This is a 4-7/8" diameter yo-yo used in ice cream store window displays. In the '50s, Duncan ran many Sealtest ice cream promotions. Many ice cream shops also retailed yo-yos during these campaigns. See also #1727 for a Sealtest Yo-Yo Award coupon.

| | | |
|---|---|---|
| 113 | *02* | **SERVE MORRELL MEATS**, gold leaf stamped seal, Alox, tournament shape, pegged string, '70s |

This is an example of one of the many wood advertising yo-yos produced by Alox. A few Alox ad yo-yos have an Alox ink stamp on the face opposite the advertising. These carry a higher value than those without a manufacturer's mark.

| | | |
|---|---|---|
| 114 | | **SOLD ONLY AT KINNEY'S**, ink stamped seal, tournament shape, '60s |
| 115 | | **SOUTH EUCLID CENTRAL NATIONAL BANK**, ink stamped seal, tournament shape, miniature, '60s |
| 116 | *02* | **S & W CAFETERIA**, die stamped seal, tournament shape, '70s |
| 117 | *02* | **TASTY FREEZE**, die stamped seal, tournament shape, '70s |
| 118 | | **THE NATURE COMPANY**, die stamped seal, Hummingbird, tournament shape, '80s |
| 119 | *02* | **TRADE HOME SHOES**, ink stamped seal, tournament shape, '50s |
| 120 | | **UNIVERSITY STATE BANK • HOUSTON • TEXAS**, paint seal, tournament shape, '70s |
| 121 | *02* | **VIKING SNUFF**, paint seal, tournament shape, miniature, pegged nylon string, '70s |
| 122 | *23* | **WAVELAND ELEVATOR INC.**, paint seal, tournament shape, jumbo, '60s |

This is an unusual ad yo-yo for a grain elevator. Jumbo advertising yo-yos are rare, due to the expense of their production.

| | | |
|---|---|---|
| 123 | *02* | **WEATHER BIRD SHOES**, decal seal, Alox, tournament shape, '50s |

Older, wood ad yo-yos with decal seals, such as this, are more prized than ink stamped or die stamped seals.

| | | |
|---|---|---|
| 124 | | **WEATHER BIRD SHOES**, ink stamped seal, tournament shape, '50s |
| 125 | *02* | **"YO-YO" COLUMBIA RECORDS**, gold leaf stamped seal, tournament shape, '60s |

This was a promotional yo-yo from Columbia Records for the song "Yo-Yo."

# YO-YO LISTINGS

 **ALOX**

The Alox Manufacturing Company was started in St. Louis, Missouri, in 1919 by John Frier. Frier ran the company until his death in 1974 when his son John Jr. took over. His son John Frier, Jr., who joined the company in 1950, continued to operate it until 1989.

Alox originally did not make yo-yos. They were known as the first shoelace manufacturer in the West. When corsets and high top laced shoes became obsolete, Alox branched out into other areas. Low cost kites, marbles, and jump ropes were just some of the items Alox manufactured. In the '30s, in the midst of the yo-yo craze, Alox added yo-yos to their product line.

Alox did not have their own wood lathes, so they bought yo-yo halves from suppliers up North. Alox then painted and assembled the yo-yos in their factory. Many of three-piece wooden advertising yo-yos from the last five decades are Alox in origin. Although best known for these three-piece pegged string yo-yos, Alox did produce some one-piece free spinning models. They also produced a jewel model in limited quantities and briefly experimented with a butterfly model that was never released.

Alox yo-yos originally retailed for 10 cents for the pegged string models and 15 cents for the free spinning style. Most were sold loose out of counter display boxes, but in later years some could be found in polybags with saddle header cards. Alox made their counter display boxes in house. The boxes were used for decades without any change in graphics.

Keeping with their ongoing price control strategy, Alox produced yo-yos as low cost, non-promoted items. Their early distributors were tobacco and candy store Jobbers and paper companies. In the later years, they sold to chain stores such as TG&Y and small regional chains. Alox sales were influenced by the large Duncan campaigns. During successful Duncan promotions, Alox would frequently sell out of yo-yos. It was then difficult for Alox to get materials from suppliers during these peak periods of yo-yo sales.

Alox never ran yo-yo contests or promotions, so collectors will not find any memorabilia such as patches, posters, or award yo-yos with the Alox name.

| | | |
|---|---|---|
| 126 | 25 | **ALOX FLYING DISC**, ink stamped seal, tournament shape, three-piece |
| 127 | | **ALOX FLYING DISC**, decal seal, tournament shape, one-piece |

One-piece Alox yo-yos are more rare than three-piece models. One-piece models are the same color on both sides, while many of the three-piece models are two colors. This yo-yo has the same seal design as the other Alox yo-yos, but in decal form.

128    **ALOX FLYING DISC**, foil sticker seal, tournament shape, one-piece

This is considered the rarest of the one-piece Alox Flying Discs. It is the only model known to have been made with an airbrushed stripe.

129    **ALOX FLYING DISC** (jewel), tournament shape

This has the standard seal on the front face and three rhinestones on the reverse face.

130   25   **ALOX FLYING DISC**, gold leaf stamped seal, tournament shape, three-piece

The gold leaf stamped seal was used only on the Alox three-piece models. The press cracked the axle on the one-piece styles. This yo-yo has a deep indented gold leaf stamped seal with each half a different color.

131    **ALOX FLYING DISC**, paint seal, tournament shape, one-piece

Most Alox yo-yos with a silk-screened paint seal are one-piece models, but three-piece versions may exist. The silk-screened seal can easily be identified from a decal seal. An example of this yo-yo is shown next to this section's title.

## YO-YO LISTINGS

## AMERICAN SPINNERS

Bill Cress founded this company in 1988 and began production of one-piece wooden yo-yos with exotic finishes. These yo-yos are still being produced, but have not been advertised in the last few years. Two predominant design styles are a maple, bulge face yo-yo called the "standard," and a Swedish imported, beechwood yo-yo called the "pocket rocket." Over one hundred different paint styles were produced. Most collectors are interested in the high gloss, bulge face models. These were painted by a company that specialized in exotic finishes for pace cars.

Many of these specialty paint versions do not have seals, so they are of little interest to collectors. Several different "jewel" varieties were also produced. These yo-yos had rhinestones added after painting. Jewel versions were hand-made by Cress and produced by special order. Prototypes for a butterfly model were made, but this model was never put into production.

132        (Classic), die stamped seal, bulge face shape, '90s
           This yo-yo has a seal like #135, but without the word "Classic." It was called the "Classic" only in the ad literature. There are four rhinestones on each face.
133        (Diamond Spinner), die stamped seal, bulge face shape, '90s
           This yo-yo has a seal like #135, but without the words "Diamond Spinner." It was only called the "Diamond Spinner" in the ad literature. There are fourteen rhinestones on each face.
134        (Rhinestone Spinner), die stamped seal, bulge face shape, '90s
           This yo-yo has a seal like #135, but without the words "Rhinestone Spinner." It was only called the "Rhinestone Spinner" in the ad literature. There are six rhinestones on each face. An example of this yo-yo is shown next to this section's title.
135    43  (Standard), die stamped seal, bulge face shape, '90s
           This is the standard model without the word "standard" on the seal. It was made with both a high gloss enamel finish and a neon finish.
136        (Standard • pocket rocket), paint seal, tournament shape, '90s
           This is the standard model without the word "standard" on the seal. "Pocket Rocket" was only used in the ad literature. This seal varies in that it has "Made in Sweden" imprinted on the face. The enamel paint is not a high gloss finish like #135. These yo-yos are made of beechwood and imported from Sweden.

## BC

This company appears to be the successor of the Hummingbird Toy Co. Brad Countryman was the driving force behind Hummingbird. Following the closure of Hummingbird in 1995, Countryman founded a new company, "What's Next Mfg., Inc.," and began production of the BC yo-yo line. Although the company began production in September 1995, it did not start selling yo-yos in the United States until 1996. The first yo-yos were retailed without packaging, but started packaging in February 1996. The yo-yos listed below are "BC" yo-yos. All had seal design changes in 1998. All BC yo-yos listed are made of wood.

137    28  APOLLO PRO, gold paint seal, butterfly shape, '90s
           This is a colored wood laminate, similar to the Hummingbird Trickster #747. (Seal redesigned in 1998).
138    28  BLACK BIRD, paint seal, tournament shape, '90s
           This is a jet-black yo-yo. An example of this yo-yo is shown next to this section's title. (Seal redesigned in 1998).
139        COMPETITION 1ST, gold leaf seal, tournament shape, '90s
           This first place award yo-yo is black with a white stripe. A second place yo-yo with a silver leaf seal and third place with bronze leaf seal were also made.
140    28  NATURAL, die stamped seal, tournament shape, '90s
           This yo-yo is a natural wood yo-yo with a clear lacquer coat. (Seal redesigned in 1998).
141    28  PHANTOM, paint seal, tournament shape, '90s
           This yo-yo is made in three colors: magenta, purple, and teal. (Seal redesigned in 1998).
142    28  RAINBOW, gold paint seal, tournament shape, '90s
           This is a colored wood laminate, similar to the Hummingbird Trickster #747. (Seal redesigned in 1998).

## YO-YO LISTINGS

### BIRD IN HAND

Bird in Hand yo-yos were produced from 1988 through 1995. Bob Malowney was the driving force behind the production of the Bird In Hand yo-yos. He was also the director of the California State Yo-Yo Championships for several years, and has been head of the National Yo-Yo Championships since 1993. In the 1980s, there were few one-piece wooden yo-yos for tournament players. Bird in Hand yo-yos attempted to fill this need with the production of its own line. Steve Smith was the wood turner for the Bird in Hand yo-yos.

The original production runs were laminates without a "Bird In Hand" logo. Shortly thereafter, production of one-piece yo-yos began and these were referred to as "22's" (painted) and "23's" (natural wood). The numbers represent the stock order identification, but are not found on the seal. The original one-piece yo-yos have an unpainted, raw wood, string slot. Later, Bird in Hand began producing three-piece yo-yos called the "32's" (painted) and "33's" (natural wood). These yo-yos had enamel or clear lacquer finishes. These finishes extended through the string slots making these easy to identify from the original one-piece models. All yo-yos that have the logo "Bird in Hand Yo-Yo" were produced in Chico, California. Although the national championship yo-yos have "Bird In Hand" on the reverse face, they were not produced in Chico. They were produced by Hummingbird. These can be distinguished from original Bird in Hand yo-yos by the absence of the word "yo-yo" on the reverse face.

| | | |
|---|---|---|
| 143 | 39 | **AWARD YO-YO**, paint seal, tournament shape, metallic gold finish, three-piece |
| | | This is an award yo-yo used by Bird in Hand for its contests and promotions. Two other award yo-yos were also made, a second place silver finish and a third place metallic blue finish. |
| 144 | 39 | **BIRD IN HAND YO-YO**, paint seal, tournament shape, '80s and early '90s |
| | | The original one-piece yo-yo can be distinguished from the three-piece by lack of paint in the string slot. Early versions of the one piece models are sometimes called "oil cans" because of their flat rims. Later versions had rounded rims. An example of this yo-yo is shown next to this section's title. |
| 145 | 39 | **CHICO TRAVELER**, paint seal, tournament shape, 1995 |
| | | This yo-yo is modeled after the original Goody Master. |
| 146 | | **I WOUND UP IN CHICO**, paint seal, tournament shape, '90s |
| 147 | 39 | **SUPER CHICO TOURNAMENT 77**, paint seal, tournament shape, one-piece, '90s |
| | | This yo-yo is modeled after the original Duncan Super Yo-Yo. Do not confuse this yo-yo with the Chico yo-yos produced by Radovan in the '50s. |
| 148 | | **THE RETURN OF THE YO-YO**, paint seal, tournament shape, '90s |
| | | A gold finish award yo-yo was given to first place finishers in contests sponsored by the traveling "Return of the Yo-Yo" show in 1990. For other "Return of the Yo-Yo" yo-yos, see #1260-1265. |

### BOB ALLEN

Bob Allen was a former Duncan demonstrator. In the '50s, he was one of the most popular yo-yo demonstrators in California. While working for Duncan, he put together a competition kit and worked closely with the Parks and Recreation system promoting contests. During the early 1960s, Duncan changed its marketing plan and laid off many demonstrators. Duncan elected to market their yo-yos through television ads rather than the traditional traveling demonstrator. Believing there was still a role for the traveling demonstrator, Bob Allen began producing his own line of yo-yos in the early 1960s. He continued to run yo-yo contests in California. The Bob Allen yo-yos were called "return tops," because at the time, Duncan was still considered to own the trademark for the word "yo-yo." The Pepsi Cola Company sponsored some of the Bob Allen contests. The distribution of Bob Allen yo-yos was largely limited to California. Four styles of Bob Allen yo-yos were produced and are listed in this section.

*Continued ...*

## YO-YO LISTINGS

● **BOB ALLEN** *Continued ...*

149   *37*   **SIDEWINDER, hot stamped seal, plastic, slimline shape, wood axle, '60s**
This yo-yo was retailed on a blister card with a pre-printed price of 79 cents. An example of this yo-yo is shown next to this section's title.

150       **SIDEWINDER, lithograph seal, tin, tournament shape, whistler holes, '60s**

151   *32*   **SIDEWINDER, hot stamped seal, plastic, tournament shape, '60s**
This was retailed on a blister card with a pre-printed price of 99 cents. The display card is marked Item No. SW 104.

152       **SIDEWINDER, paint seal, wood, tournament shape, '60s**
This model retailed on blister card with a pre-printed price of 49 cents. The display card is marked Item No. SW 102.

## CHARACTER

The production of licensed character yo-yos was not documented prior to WWII. The first character yo-yos appeared in the 1950s when television stimulated massive interest in licensed toys. Childhood heroes of the '50s, such as Roy Rogers and Davy Crockett, led the way for the character yo-yos. Yo-Yo collectors compete with toy collectors and character collectors for these items. For other character yo-yos, see these sections: Disney, Hanna Barbara Characters, Peanuts, Spectra Star, and Duncan Super Hero.

          •   **Alvin** - *See Light-up section #1030.*

153   *04*   **BABAR, imprint seal, vinyl covered wood, puck shape, pegged string, early '90s**
Babar the Elephant is a French based children's cartoon book character. The yo-yo was released after the 1989 animated movie "Babar."

154   *06*   **BARNYARD COMMANDOS • P.O.R.K.S., paper sticker seal, plastic, tournament shape, metal axle, 1990**
This yo-yo was retailed by the Imperial Toy Company on a blister card, Stock No. 8156. Other variations of the Barnyard Commandos were also produced.

155   *18*   **(Batman), photo insert seal, plastic, slimline shape, view lens, late '80s**
This yo-yo has the Batman logo on one face, and a photograph of Michael Keaton and Kim Basinger on the reverse side. Batman is a comic book character introduced in "Detective Comics" in 1939. This yo-yo was based on the blockbuster movie released in 1989. For other Batman yo-yos, see the Duncan Super Heroes and Spectra Star sections.

156       **(Batman), imprint seal, plastic, Canada Games, tournament shape, '90s**
The seal shows Batman and the blister card has a hologram of the Batman logo.

157       **(Batman), paper sticker seal, plastic, tournament shape, midget, '60s**
This appears to be an unlicensed Batman product sold as a gumball machine prize.

158   *03*   **BATTLESTAR GALACTICA, paper sticker seal, plastic, tournament shape, '70s**
This yo-yo is based on the sci-fi TV series which aired in 1978 and '79. This yo-yo was retailed on a blister card without a pre-printed price or bar code.

159       **(Biker mice from mars), imprint seal, plastic, Canada Games, tournament shape, 1996**

          •   **Bozo** - *See Clown section #218, #218.1, and #218.2.*

160       **(Breakfast bears) • Teddy • Eddy • Freddy, embossed seal, plastic, '90s**
This is a premium snap together yo-yo found inside Breakfast Bears Cereal. The yo-yos were individually packaged in a cellophane bag. Each of the three bears is on their own yo-yo.

161   *03*   **(Children's occupation characters), lithograph, tin, tournament shape, 1990**
This is a series of cartoon characters depicted in different occupations. The series includes a doctor, sailor, fireman, nurse, racecar driver, astronaut (shown), and soldier. These yo-yos were retailed individually on blister cards along with two other tin yo-yos without character seals.

          •   **Cobi • 1992 Olympic Mascot** - *See Souvenir section #1394.1.*

162   *01*   **(Cracker Jack and bingo), embossed seal, plastic, midget, '60s**
This is a snap together plastic Cracker Jack prize given out in the '60s. Prizes have been supplied in Cracker Jack boxes since 1912. Several different Cracker Jack tops have been produced, but this is the only known yo-yo. The yo-yo uses a thread for a string and is a poorly functioning toy. Several different color variations of this yo-yo exist.

163   *04*   **DAVY CROCKETT, gold leaf stamped seal, wood, tournament shape, three-piece, pegged string, '60s**
This is a promotional yo-yo which sold advertising on the reverse face. It was believed to have been produced from the middle to late '60s.

*Continued ...* ▶

## YO-YO LISTINGS

164    *04*    **DAVY CROCKETT**, gold leaf stamped seal, wood, tournament shape, three-piece, pegged string, '50s
The Davy Crockett craze began in 1954 when Fess Parker starred in the Disneyland series on ABC. This yo-yo was made by Fli-Back which also made the Davy Crockett Paddleball.

165    *26*    **DENNIS THE MENACE**, paper sticker seal, plastic, tournament shape, '90s
Dennis the Menace began as a cartoon in 1951 and became a TV series in 1959. For another example of Dennis the Menace yo-yos, see #813.

166    *04*    **(Doonesbury • Zonker)**, leather debossed seal, wood, tournament shape, 1990

167    *03*    **DR. WHO**, paper sticker seal, plastic, butterfly shape, '70s
This is from the long running British sci-fi series "Dr. Who," which began in 1963. The yo-yo was retailed on a blister card that pictures a soccer player.

168       **(Duck • Sanrio)**, imprint seal, plastic, slimline shape, 1994
This yo-yo retailed in a polybag with a header card, AHIRU NO PEKKLE.

•    **Dukes of Hazzard** - *See the Duncan Wheels section #514, #514.1, #514.2*

169    *04*    **(Ed Sullivan)**, paint seal, wood, tournament shape, three-piece, pegged string. '60s
The Ed Sullivan TV show was an hour-long variety show which ran from 1955-71. Sullivan was one of the most recognizable TV personalities of the '60s.

170    *03*    **(E.T.)**, paper insert seal, glitter plastic, slimline shape, view lens, early '80s
"E.T. The Extra Terrestrial" was the 1982 sci-fi blockbuster film by Universal Studios.

171    *34*    **FELIX THE CAT**, decal seal, wood, tournament shape, 1997
This yo-yo was retailed with a limited edition "Felix the Cat" watch made by Fossil. Both watch and yo-yo were displayed in a Felix the Cat collectible tin box.

172    *06*    **(Frog • Sanrio)**, paper sticker seal, plastic, bulge face shape, '90s

173    *06*    **FREDDIE THE FROG**, decal seal, plastic, tournament shape
The animated movie "Freddie the Frog" was released in 1992. It is believed that this yo-yo was made several years before the movie.

174    *04*    **GARFIELD**, imprint seal, plastic, miniature, '80s
Garfield is a long running comic strip character created by Jim Davis. There are at least two different variations of this miniature yo-yo. See also the Spectra Star section #1433.

174.1       **(Goosebumps)**, imprint seal, plastic, Canada Games, tournament shape, '90s
The yo-yo is based on the Canadian produced horror show for kids.

175    *18*    **(Joker)**, paper insert seal, plastic, slimline shape, view lens, late '80s
The Joker appears on the front face; the bat logo on the reverse face. The yo-yo is licensed from the 1989 movie "Batman." Jack Nicholson portrays the Joker in this film. See also the Spectra Star section #1443.

176    *03*    **KNIGHT RIDER**, paper sticker seal, plastic, butterfly shape, 1982
The TV action series "Knight Rider" ran from 1982-86 and starred David Hasselhoff. The blister card shows Knight and Kitt, the talking sport car with artificial intelligence.

•    **Louie Anderson** - *See Light-Up section #1035.*

177    *18*    **MADMAN**, paper insert seal, plastic, slimline shape, view lens, 1996
This yo-yo was given out to promote the Madman comic book characters. The yo-yo was packaged in a polybag with a Duncan trick mini booklet.

178    *04*    **(Oliver) HARDY**, lithograph seal, tin, tournament shape, '70s
The yo-yo features the famous '20s and '30s comic film star, Oliver Hardy. See companion yo-yo #187.

•    **Ninja Mutant Turtles** - *See Spectra Star section #1450, #1450.1, and #1451.*

•    **Pac-Man** - *See Duncan Miscellaneous Plastic section #427.*

179       **PINK PANTHER**, imprint seal, plastic, Humphrey shape, '90s
There is a tri-colored Pink Panther cartoon face on one face and the words "Pink Panther" on the reverse. It is dated 1994.

180    *35*    **POWER RANGERS**, lithograph seal, tin, tournament shape, 1994
"Mighty Morphin Power Rangers" is a Japanese produced super hero series that first aired on FOX in 1993. The name changed to Power Rangers Zeo in 1996. For other Power Ranger yo-yos, see the Light-Up section #1036.

181       **(Power Rangers)**, imprint seal, plastic, Canada Games, tournament shape, '90s

182    *31*    **(R.X. Freud)**, imprint sticker seal, plastic, tournament shape, early '90s
This is a bogus knock off of the Yomega "Brain" type yo-yo with a picture of Sigmund Freud on the seal. The yo-yo was briefly produced in Taiwan during 1990. This yo-yo retailed in a display box and was apparently withdrawn from the market when legal action was threatened.

183    *04*    **(Raggedy Andy at Christmas)**, lithograph seal, tin, tournament shape, '80s
This yo-yo carries a copyright date of 1980 by the Bobb Merrill Company. The yo-yo is based on the best selling series of children's stories.

183.1       **(Raggedy Ann and Andy)**, die stamped seal, wood, tournament shape, 1997
This yo-yo was made for the 1997 Toy Fest.

183.2       **(Raggedy Ann and Raggedy Arthur)**, lithograph seal, tin, tournament shape, '80s
The seal shows Raggedy Ann reading a book to Raggedy Arthur. This yo-yo was retailed in a polybag with a header card from Santa's World.

*Continued ...*

The History and Values of Yo-Yos

## YO-YO LISTINGS

● **CHARACTER** *Continued ...*

183.3    *04*    **(Raggedy Ann), imprint seal, plastic, tournament shape, '70s**
The copyright on the reverse face of this model reads "1974 Bobb Merrill Company." This yo-yo was retailed by Hallmark on a blister card as a "Raggedy Ann Champion Yo-Yo" with a pre-printed price of $1.35.

       •    **Roy Acuff** - *See Hummingbird and Souvenir section #742 and #1395.*

184    *04*    **ROY ROGERS AND TRIGGER, paper insert photo seal, plastic, slimline shape, view lens, '50s**
The Roy Rogers "Round Up King" yo-yo was produced by All Western Plastics. A national contest is advertised on the reverse face of the yo-yo. A large warehouse find of these Roy Rogers yo-yos in the '80s allowed thousands to be distributed to antique dealers throughout the country. Each yo-yo was individually wrapped in a cellophane wrapper with extra strings. They retailed out of a counter display box. In the '50s, the "Roy Rogers" mold was sold to Duncan. Yo-Yos made from this mold were called the "eagle style." Duncan used the mold largely for foreign promotions. Later, Jack Russell used the same mold for his foreign Coca-Cola promotions.

185          **(Ren and Stimpy), paper sticker seal, tournament shape, 1996**
This was a Kentucky Fried Chicken premium release in Australia. The yo-yo is made of glow plastic.

186          **(Sailor moon), imprint seal, plastic, Canada Games, tournament shape, '90s**
This yo-yo is based on the 1995 Japanese animated TV series. An example of this yo-yo is shown next to this section's title.

187    *04*    **(Stan) LAUREL, lithograph seal, tin, tournament shape, '70s**
This yo-yo features Laurel from the famous '20s and '30s comic team Laurel and Hardy. See companion yo-yo #178.

188    *03*    **STAR TREK, plastic sticker seal, glitter plastic, tournament shape, metal string slot axle, 1979**
This yo-yo is from "Star Trek: The Motion Picture." The Enterprise is on one face; Kirk and Spock are on the reverse. The yo-yo was released by Ariva Enterprises on a blister card, Stock No. 847. The blister card shows an illustration of Spock.

188.1        **(Star Trek The Next Generation), imprint seal, plastic, Canada Games, tournament shape, '90s**

189    *01*    **(Sugar bear), paper insert seal, plastic, miniature, view lens, '70s**
This is a Sugar Crisp cereal snap together premium. The yo-yo has a small view lens with a pinball game on one face and a decoder on the reverse face. The yo-yo was sealed in a cellophane bag. For other Sugar Crisp yo-yos, see Duncan Advertising #307.54 and Duncan Yo-Yolympics #447.2 and #447.4.

190          **THE CLINTON, paper sticker seal, plastic, tournament shape, '90s**
A cartoon face of President Clinton is shown on the seal and reads, "spins on everything." The yo-yo retailed in a display box with humorous jabs at the president written on the inside lid.

191          **(Troll), photo sticker seal, plastic, tournament shape, miniature, '90s**
This yo-yo was retailed with a gold sticker seal over the string slot that reads "Papel Freelance." There were several different troll picture seals in this series.

192          **UNDERDOG, paper sticker seal, plastic, tournament shape, '90s**
This yo-yo was retailed on a blister card with a copyright date of 1992.

193    *04*    **WILL ROGERS, ink stamped seal, wood, tournament shape, '90s**

## CHEERIO

The Cheerio brand was started by Wilfred Schlee, Sr., in Kitchener, Ontario, Canada. Schlee took over the Kitchener Buttons Limited factory founded by his father in 1907. A local printer brought Schlee a yo-yo in 1931, and told Schlee he should make them because they were selling in large numbers in the United States. Schlee made several thousand, but sales were very poor. Schlee quickly realized that the toy required considerable promotion. Once promotions began, he could not make them fast enough. Promotions during the '30s were limited to Canada and England.

Schlee's original yo-yos were called Hi-Ker, but a new name, "Cheerio," was introduced to broaden their appeal in the British market. The popularity of the Cheerio brand name rapidly surpassed the Hi-Ker Yo-Yo. By 1932, the factory, working around the clock, could not meet the demand. Although the Cheerio Yo-Yo was being sold in both Canada and England, it was still not being marketed in the United States. They sold so well in England that a large shipment was sent each week and Schlee himself went to England six times that year. In England, Edward the Duke of Wales was photographed playing a Cheerio Yo-Yo. During this period, Joe Young was the most popular Cheerio demonstrator. He was considered the Cheerio "world champion." Currently, the earliest known film record of yo-yo playing is that of Joe Young demonstrating the Cheerio Yo-Yo in England, circa 1934.

*Continued ...* ▶

## YO-YO LISTINGS

Sam Dubiner, a former Duncan promoter, joined Schlee in the early '30s. Dubiner ran promotions and Schlee produced the yo-yos out of the Kitchener factory. Dubiner registered the "Yo-Yo" trademark as well as the Cheerio name in 1938. Dubiner held the yo-yo trademark in Canada until the late '60s. Although the yo-yos were being produced by Schlee, the "yo-yo" trademark belonged to Dubiner.

Al Gallo began working as the head field promoter for Dubiner in 1938. During this period, promotions and contests were still limited to Canada and England. Demonstrators would often drive up to schools on flat bed trucks, do tricks and hand out yo-yos and shirts. Trick cards were included in string packs during this period and if a child collected all 48 trick cards, he could redeem them for a box of Cheerio 99s. With the onset of WWII, the production of yo-yos in Canada stopped. All resources were going to the war effort.

In 1946, following the War, Cheerio restarted production of yo-yos in Canada and finally entered the U.S. market. Sam Dubiner set up Cheerio operations in the United States. Many of the promotions were run by Al Gallo. Wilf Schlee Jr. accompanied Dubiner to the United States and set up production of the yo-yos while Dubiner ran operations.

In 1946, Duncan was not running promotions in the United States because they were busy setting up their factory in Luck, Wisconsin. Cheerio rapidly became Duncan's major competitor. Since Cheerio could not use the word "yo-yo" in the United States, Duncan owned the trademark, they marketed their product as a Cheerio Top. They had so much market penetration in some areas that kids began calling the toy a "Cheerio" instead of a "yo-yo."

The rivalry between Cheerio and Duncan was fierce. In the late '40s and early '50s, Cheerio had a team of approximately fifteen Canadian demonstrators working in the United States. Duncan field promoters frequently called immigration officials to try and disrupt the Cheerio promotions. On more than one occasion, the Cheerio demonstrators were hauled in by police to have their paperwork checked. Although the Canadians had appropriate paperwork, the tactic did work. During this period, Duncan used the American News Company for their distribution and Cheerio used the Independent News Company. When one company would hear of a promotion being run in a city by the other company, they would flood that city with their own product.

During this time, Cheerio yo-yos were still being made in Canada. The Canadian versions could use the word "yo-yo" on their seals. The Kitchener plant, run by Wilf Schlee, Sr., still produced the yo-yos and the operation of the Cheerio company was run by Dubiner. In 1950, Schlee sold the plant, but his son John wanted to continue to make yo-yos. John Schlee started the Mastercraft company and produced yo-yos for Dubiner. In 1952, financial problems forced the sale of Mastercraft to Frank Neibert, but yo-yo production continued for Cheerio.

In 1954, Sam Dubiner decided to get out of the yo-yo business. He sold the American Cheerio operation to the Donald F. Duncan Company. The deal also included the transfer of the demonstrators to Duncan. Many demonstrators were not told of the deal and only found out when they noticed their checks were now from Duncan. Although some continued with Duncan, others went back to Canada and several joined Wilf Schlee Jr. for the U.S. promotion of the Hi-Ker line. The Duncan company sold off the old Cheerio stock and continued to produce the Cheerio line as a non-promoted line until they dropped it around 1963.

Dubiner also sold the Canadian Cheerio operation to Albert Krangle and Al Gallo in 1955. Krangle owned the majority and handled the operations while Gallo ran the promotions. Dubiner gave them a non-exclusive registered user license, but retained the Canadian "yo-yo" trademark and received royalties for its use. During this period, Krangle and Gallo purchased most of their yo-yos from the Mastercraft Co. Supply problems with Mastercraft forced Cheerio to look for additional sources for yo-yos. A supplier was found with a company in Kalamar, Sweden. Some Cheerio Yo-Yos from this period can be found with a Sweden die stamp on the reverse face.

In 1962, a dispute between Krangle, Gallo, and Dubiner resulted in Gallo leaving Cheerio and a withdrawal of the "yo-yo" trademark license by Dubiner. A landmark trademark suit followed, very similar to the Royal vs. Duncan suit in the United States. Unlike the suit in the USA where the word "yo-yo" lost its trademark protection due to popular usage, the courts declared "yo-yo" a registered trademark in Canada. During the '60s, Krangle tried to get around the yo-yo trademark issue by making his own set of Cheerio seals which said "Big C," but did not have the word "yo-yo" on the seal. These yo-yos were only produced briefly in the '60s for Cheerio by Mastercraft.

| | | |
|---|---|---|
| 194 | 25 | **25 TOURNAMENT PRACTICE RETURN TOP, foil sticker seal, wood, tournament shape, '50s and '60s** |
| | | This yo-yo was only marketed in the United States. Some of these yo-yos may have airbrushed stripes. |
| 195 | 56 | **33 FAMOUS, die stamped seal, wood** |
| | | The item shown is the original die used to stamp the seal of this yo-yo. The die is from the Mastercraft Company, a wood product company that made the Canadian Cheerio yo-yos in the '50s and '60s. |
| 195.1 | 25 | **33 OFFICIAL JUNIOR YO-YO, gold leaf stamped seal, wood, tournament shape, '50s** |
| | | This yo-yo was retailed in Canada, but not in the United States. |

*Continued ...*

## YO-YO LISTINGS

● **CHEERIO** *Continued ...*

| | | |
|---|---|---|
| 196 | 25 | **55 BEGINNERS**, foil sticker seal, wood, tournament shape, '50s and '60s |

This yo-yo was marketed only in the United States.

| | | |
|---|---|---|
| 197 | | **55 BEGINNERS**, die stamped seal, wood, tournament shape, '50s |

Production of this yo-yo began in 1946. It continued to be manufactured after Duncan bought out Cheerio in the '50s.

| | | |
|---|---|---|
| 198 | | **99 GENUINE PRO**, die stamped seal, wood, tournament shape, late '50s |

This was the most popular of the tournament models sold in the USA. The seal says "Made in USA." This model was released both before and after Duncan bought out Cheerio.

| | | |
|---|---|---|
| 199 | | **99 OFFICIAL TOURNAMENT YO-YO (no maple leaf)**, foil sticker seal, wood, tournament shape, '30s |
| 200 | | **99 TOURNAMENT PRACTICE**, foil sticker seal, wood, tournament shape, late '40s - early '50s |

This yo-yo was marketed only in the United States.

| | | |
|---|---|---|
| 201 | 25 | **99 PRO (maple leaf)**, foil sticker seal, wood, tournament shape, '50s and '60s |

This yo-yo model was marketed only in the United States and was produced with either a glitter enamel finish or a regular enamel finish. Duncan continued to release this model even after they bought out Cheerio.

| | | |
|---|---|---|
| 202 | 25 | **(Beaver) MEDALIST**, foil sticker seal, wood, tournament shape |

This model is rare and very desirable. The seal has a beaver on the maple leaf. Although the yo-yo is named the "Medalist," collectors call it the "Cheerio Beaver."

| | | |
|---|---|---|
| 203 | 25 | **BIG CHIEF BEGINNER**, die stamped seal, wood, tournament shape, three-piece, '50s |

This yo-yo was only marketed in Canada.

| | | |
|---|---|---|
| 203.1 | | **BIG CHIEF BEGINNER (Big "C")**, die stamped seal, wood, tournament shape, '60s |

There is a large "C" on the seal of this yo-yo. It was produced very briefly in the early '60s. This version replaced #203 and although more recently produced, is considered more rare. Like the original Big Chief, this was marketed only in Canada.

| | | |
|---|---|---|
| 204 | 25 | **CHAMPION**, plastic medallion seal, wood, tournament shape, '50s |

This yo-yo is considered one of the most collectible of the Cheerio yo-yos. First produced in 1953, it is an unusual yo-yo because it has a medallion seal. The medallion is made of molded plastic with a metallic paint finish. The medallions were molded by Sam Dubiner's plastic factory in Toronto. This model has a United States trademark, and was released in both Canada and the United States. The Champion should not be confused as an award yo-yo, it was produced solely for retail sale for at least five years. Following the Cheerio buy-out, Duncan continued to retail the Champion yo-yo in the United States. Most demonstrators did not like using the yo-yo, because the elevated medallion was hard on the hand when the yo-yo returned. Some collectors refer to this yo-yo as the "Eagle Medallion Cheerio."

| | | |
|---|---|---|
| 204.1 | 56 | **FAMOUS**, die stamped seal, wood |

The item shown is the original die used to stamp the seal of this yo-yo. The die is from the Mastercraft Company, a wood product company that manufactured the Canadian Cheerio yo-yos in the '50s.

| | | |
|---|---|---|
| 205 | 25 | **GLITTER SPIN, (4 jewel)**, foil sticker seal, wood, tournament shape, '50s |

This is the most common of the four Cheerio Glitter Spin yo-yos. The jewels are in a square pattern, unlike jewel yo-yos from other companies. The jewels had to be set in this pattern because of the foil sticker. The sticker would have covered the jewels in a linear pattern. The four jewels were set on both faces of this yo-yo. A large red dot starting the letter "G" gives this yo-yo the name "Red Dot Glitter Spin." The Glitter Spin Jewel Yo-Yo went into production in 1947 and continued until the late '50s. This model was retailed in both Canada and the United States. Duncan continued to market this model following the Cheerio buy-out.

| | | |
|---|---|---|
| 205.1 | 25 | **GLITTER SPIN, (4 jewel)**, foil sticker seal, wood, tournament shape, '40s and '50s |

Unlike the "Red Dot" Glitter Spin #205, this version has the word "yo-yo" under Cheerio. Al Gallo, Cheerio's head promotion man, during the '40s and early '50s, named the Glitter Spin. Some collectors refer to this as the "Black Label Glitter Spin." It is considered more rare than the "Red Dot Glitter Spin." This model precedes the "Red Dot Glitter Spin" #205 and was only retailed in Canada. An example of this yo-yo is shown next to this section's title.

| | | |
|---|---|---|
| 205.2 | 56 | **GLITTERSPIN, (4 jewels)**, die stamped seal, wood, tournament shape, '60s |

The original die for this model is shown. This yo-yo does not have the aesthetic value of the foil seal Glitterspins, but its value is higher. This was a Canadian produced model not marketed in the United States. In the die stamped versions, "Glitterspin" is one word; on the foil seal versions it is two words.

| | | |
|---|---|---|
| 205.3 | 56 | **GLITTERSPIN CONTEST MODEL**, die stamped seal, wood, tournament shape, '60s |

The item shown is the original die used to stamp the seal of this yo-yo. Of the two die stamped Glitterspin variations, this is considered the more rare yo-yo. The die is from the Mastercraft Company, a wood product company that manufactured the Canadian Cheerio yo-yos in the '50s. This is a Canadian model not marketed in the United States.

| | | |
|---|---|---|
| 206 | 25 | **KITCHENER BUTTONS LIMITED**, foil sticker seal, wood, tournament shape, one-piece, '30s |

The seal also reads "Buttons and Radio Knobs." Kitchener yo-yos, labeled as Cheerio, started production in 1931 and were produced into the '40s. These yo-yos are the earliest of the Cheerio line and considered highly collectible. Some of the Kitchener Buttons yo-yos may have a black ink stamp on the reverse that reads "Pat. Pend. Reg. Made in Canada" with the number 99. These yo-yos are believed to have been marketed in England. The "Made in Canada" seal was required for import regulations. Kitchener yo-yos can be found both with and without airbrushed decorative stripes.

| | | |
|---|---|---|
| 206.1 | | **MEDALIST**, die stamped seal, wood, tournament shape, '60s |

The Big C Medalist was produced by Cheerio in the '60s. The word "Yo-Yo" is not found on the seal, as Cheerio had lost the Canadian yo-yo trademark at the time.

*Continued ...* ▶

## YO-YO LISTINGS

207    25    **OFFICIAL CHAMPION, foil sticker seal, wood, tournament shape, '50s**
              This was the first American promoted Cheerio yo-yo produced after WWII. Production began in 1947 and went through the middle '50s.
207.1  56    **PEE-WEE, die stamped seal, wood**
              The item shown is the original die used to stamp the seal of this yo-yo. The die is from the Mastercraft Company, a wood product company that made the Canadian Cheerio yo-yos in the '50s.
208    56    **RAINBOW CONTEST MODEL, die stamped seal, wood, '50s**
              The item shown is the original die used to stamp the seal of this yo-yo. The die is from the Mastercraft Company, a wood product company that manufactured the Canadian Cheerio yo-yos in the '50s.
208.1  56    **RAINBOW CONTEST MODEL, die stamped seal, wood, '50s**
              The item shown is the original die used to stamp the seal of this yo-yo. The die is from the Mastercraft Company, a wood product company that manufactured the Canadian Cheerio yo-yos in the '50s.
208.2  56    **RAINBOW PRO MODEL, die stamped seal, wood, tournament shape, '50s**
              The item shown is the original die used to stamp the seal of this yo-yo. The die is from the Mastercraft Company, a wood product company that manufactured the Canadian Cheerio yo-yos in the '50s.
   •          **Tournament Duncan Cheerio** - *See Duncan Tournament section #498.*
209    56    **WHISTLER (big) BUTTERFLY, die stamped seal, wood, butterfly shape, '50s**
              The item shown is the original die used to stamp the seal of this yo-yo. The die is from the Mastercraft Company, a wood product company that manufactured the Canadian Cheerio yo-yos in the '50s.
209.1  56    **WHISTLER (little) BUTTERFLY, die stamped seal, wood, '50s**
              The item shown is the original die used to stamp the seal of this yo-yo. Cheerio Toys and Games Ltd., which is also on the die, was the name of the Canadian Company that marketed Cheerio yo-yos. The die is from the Mastercraft Company, a wood product company that manufactured the Canadian Cheerio yo-yos in the '50s.
209.2  56    **WHISTLER (little) BUTTERFLY MODEL, die stamped seal, wood, '50s**
              The item shown is the original die used to stamp the seal of this yo-yo. The die is from the Mastercraft Company, a wood product company that manufactured the Canadian Cheerio yo-yos in the '50s.

## CHEMTOY

Chemtoy, based out of Cicero, Illinois, produced a series of three plastic, non-promoted yo-yos in the '70s. Blister carded yo-yos carry a copyright date of 1972.

210    06    **MARK V, imprint seal, tournament shape, '70s**
              This was retailed on 4" x 5" blister card, Stock No. 2071.
210.1  06    **CROWN ROYAL, hot stamped seal, tournament shape, '70s**
              This was retailed on 4" x 5" blister card, Stock No. 2072.
210.2  06    **NUMBER 1 YO-YO, embossed seal, puck shape, '70s**
              This was retailed on 4" x 5" blister card, Stock No. 2070, with a pre-printed price of 69 cents. This was Chemtoy's equivalent of a beginners model. An example of this yo-yo is shown next to this section's title. For an example of the blister card, see Plate #32, #210.

YO-YO LISTINGS

## CHICO

Joe Radovan started the Chico yo-yo line in 1949. During the introduction of this yo-yo line, Radovan had already been the owner of Royal for over ten years. It is often thought that Chico was Radovan's new company, but in reality it was just a different label used by Royal to market yo-yos. Both Royal and Chico yo-yos were marketed at the same time.

In the early '30s, some of the yo-yo demonstrations were run like vaudeville stage shows, so many of the demonstrators had stage names. Radovan would frequently wear a large sombrero hat, so he was given the Mexican nickname "Chico." This is where the Chico line got its name.

Pedro Flores helped Radovan in the promotion of the Chico yo-yo during the early years. Several different styles were produced during the '50s and into the '60s. Most were made of wood, but there have been a few plastic Chico yo-yos found. It is unclear whether these were prototypes or were actually retailed.

| | | |
|---|---|---|
| 211 | | **CHICO OLYMPIC TOURNAMENT YO-YO TOP**, foil sticker seal, tournament shape, '50s and '60s |
| | | This is the only Chico model with the word "Yo-Yo" on the seal. |
| 212 | 22 | **SUPERB CHICO JUNIOR TOP**, gold leaf stamped seal, wood, three-piece, pegged string, '50s and '60s |
| 213 | 22 | **SUPERB CHICO STANDARD TOP**, gold leaf stamped seal, wood, tournament shape, '50s and '60s |
| 214 | 22 | **SUPERB CHICO TOURNAMENT TOP**, decal seal, wood, tournament shape, one-piece, '50s and '60s |
| | | This model was also made with jewels. Three jewels in a row are located on the reverse face. An example of this yo-yo is shown next to this section's title. |
| 215 | 22 | **SUPER CHICO**, decal seal, wood, tournament shape, '50s and '60s |
| | | This yo-yo was retailed on a Chico blister card reading "Ola! Chico." The card had a pre-printed price of 49 cents. |
| 216 | | **SUPER CHICO**, foil sticker seal, wood, tournament shape, one-piece, '50s and '60s |
| | | Foil sticker seals are harder to find than decal seals of this same design. This seal is almost identical to #215 except the ® mark is below the word "Chico." |
| 217 | | **SUPER DELUXE CHICO**, decal seal, wood, tournament shape, '50s and '60s |
| | | These banner shaped decals were sometimes placed over Royal yo-yos that already had die stamped seals. These "Chico" yo-yos can be found on Royal blister cards. |
| • | | **Super Chico 77** - *See Bird in Hand section #147.* |

## CLOWN

Most clown related yo-yos are inexpensive tin lithographed models. The most collectible of the clown related yo-yos is the rare spinning clown yo-yo. This was an early plastic yo-yo patented in 1954 by J.M. Field and produced very briefly in the middle '50s. It was one of the first yo-yos made with a view lens. For other clown related yo-yos, see the McDonald's section.

| | | |
|---|---|---|
| 218 | 34 | **BOZO**, lithograph seal, tin, ball shape, miniature, '70s |
| | | This is an unusual round tin yo-yo. Bozo was based on a comic book character from the early '50s and became popular on his own TV series. |
| 218.1 | | **BOZO THE CLOWN**, lithograph seal, tin, tournament shape, '70s |
| | | This yo-yo was retailed both loose and on a blister card. This version is more common than #218. |
| 218.2 | | **BOZO THE CLOWN**, paper sticker seal, plastic, Roalex Co., tournament shape, '60s |
| 219 | 34 | **CIRCUS WORLD MUSEUM**, hot stamped seal, plastic, Duncan Flambeau, butterfly shape, early '70s |
| | | This was a souvenir of the Circus World Museum in Baraboo, Wisconsin. There is a Duncan psychedelic butterfly seal on the reverse face. See #330. This is one of the few plastic advertising Duncans that had a hot stamped seal. It was retailed on a Duncan butterfly blister card. |

*Continued ...* ▷

## YO-YO LISTINGS

220           **CELEBRATE (clown face), lithograph seal, tin, tournament shape, '90s**
                This yo-yo is part of the Birthday series. See #1141.
221    *34*   **(Clown face), lithograph seal, tin, tournament shape, '90s**
222          **(Clown face), lithograph seal, tin, tournament shape, '90s**
                The U.S. Toy Co. released this model. The yo-yo has © C-P on the rim and is called the "Smile Yo-Yo."
223          **(Clown with umbrella), lithograph seal, tin, tournament shape, '80s**
224    *34*   **(Clown), lithograph seal, tin, tournament shape, '70s**
                This yo-yo has an internal bead for making sounds. It was made in Japan and came in two styles, the one shown and a version with stripes instead of polka dots.
225    *34*   **(Spinning Clown), paper insert seal, plastic, puck shape, jumbo, view lens, fixed string, '50s**
                The Monfield Corporation briefly produced this model in the early '50s. J.M. Field applied for a patent on this yo-yo in 1950 and received it in 1954. (Patent Number 2,676,432.) This yo-yo has the words "Patent Pending" embossed on the view lens. A different style of clown is on each face. Collectors refer to this as the "Spinning Clown" yo-yo. An example of this yo-yo is shown next to this section's title.

*This decorative piece, of a clown playing with a yo-yo, is one of a variety of clown/yo-yo theme products available to the collector.*

 **COCA-COLA**

With the exception of World's Fair yo-yos, Coca-Cola yo-yos are the most cross-collected brand. Yo-Yo collectors compete with Coca-Cola collectors for these yo-yos. Coca-Cola has been producing advertising yo-yos and sponsoring contests for decades. The popularity of Coca-Cola advertising yo-yos has resulted in the production of a better playing, higher quality yo-yo.

Although Coca-Cola did sponsor yo-yo contests in the United States, the number of promotions were limited, but foreign promotions have flourished. Foreign Coca-Cola models are not listed in this guide, but literally hundreds of different styles exist. Since 1960, all foreign Coca-Cola yo-yos were produced by the Russell Yo-Yo Company. (See also Russell Yo-Yos.) Dan Volk is the most knowledgeable authority on foreign Coca-Cola yo-yos. He worked overseas as a demonstrator and campaign manager for the Jack Russell Company from 1976 through 1982.

This section lists Coca-Cola yo-yos that were released in the United States. Allied brands such as Fanta, Sprite, and Fresca can be found in the Soft Drink (Non-Coke) section.

226    *0.5*   **(Always Coca-Cola series), imprint seal, plastic, Russell, bulge face shape, 1994**
                The "Always Coke" series was the second to last series of Russell Coca-Cola yo-yos released in the United States. In all styles the seal is similar to this yo-yo shown. In addition to red translucent rims, two styles of glitter plastic were used, gold glitter (shown) and silver glitter.
227    *0.5*   **COCA-COLA, paper sticker seal, plastic, bulge face shape, metal axle, '80s**
228    *0.5*   **COCA-COLA, paper sticker seal, plastic, midget, metal axle, '80s**

*Continued ...*

## YO-YO LISTINGS

● COCA-COLA Continued ...

229    0.5    **COCA-COLA • COKE, imprint seal, plastic, miniature, metal string slot axle, '80s**

230    0.5    **COCA-COLA GALAXY, imprint seal, plastic, Russell, bulge face shape, view lens, wood axle, 1983**
This yo-yo is unique in that it has foil inserts under the lens. It was a limited release yo-yo used only in 1983 for the El Paso and Salt Lake City promotions. The yo-yo was awarded as a contest prize and also given as a premium for turning in 50 bottle caps. This yo-yo was not sold retail.

231          **COCA-COLA • BOB ROLA • WORLD CHAMPION, paper insert seal, plastic, Duncan, slimline shape, '50s**
This is an eagle style yo-yo, see #414.

232    0.5    **(Coca-Cola series), imprint seal, plastic, Russell, bulge face shape, 1992**
This series of five different Coke yo-yos was retailed by the Jack Russell Company in 1992. Unlike many earlier Russell Coca-Cola yo-yos, the Russell Company name does not appear on the yo-yo, only on the blister card. The five yo-yos in the series include: the Professional (white), Master (black), Super (red or green), Special Spin (white face and red rim), and a Hi-Tech light-up. The graphic design of all the yo-yos in this series are similar. The Master model can be seen on Plate #5.

233          **(Coca-Cola series), imprint seal, plastic, Russell, bulge face shape, 1996**
The last Coke brand series released by the Jack Russell Company in the United States. Unlike many earlier Russell Coca-Cola yo-yos, the Russell name does not appear on the yo-yo, only on the card. The blister card shows the Coca-Cola polar bear playing a yo-yo. Super, Professional and Master models were released. The graphics are similar on all models.

234    0.5    **DRINK COCA-COLA, imprint seal, plastic, sculpted metal axle, '60s**
This was a yo-yo promotion by Coke that was not associated with a contest. One yo-yo was given with the purchase of a six bottle carton of Coke. The yo-yos were marketed as "Kooky Kaps." This sculpted bottle cap shape was one of the earliest of the sculpted yo-yos. The yo-yo came loose in a counter display box that held 48 yo-yos.

235    0.5    **DRINK COCA-COLA, gold leaf stamped seal, wood, Duncan, tournament shape, one-piece, '50s**
The reverse face has the Duncan Super Return Top seal. This yo-yo is a favorite among both Coke and yo-yo collectors. The model shown is a half-and-half paint style. It is considered to be the most collectible advertising yo-yo produced by Duncan.

236    0.5    **DRINK COCA-COLA, gold leaf stamped seal, wood, Duncan, tournament shape, one-piece, '50s**
This is the rarest Coca-Cola Yo-Yo. In the 1950s, a special production creating six prototypes with the new "pearlescence" white paint. Coke rejected the color, opting to keep their yo-yo the traditional red color. Only one of the original six prototypes is currently known to exist. The Duncan Super seal #507 is on the reverse face. An example of this yo-yo is shown next to this section's title.

237          **DRINK COCA-COLA, paint seal, wood, tournament shape, '60s**
This is one of the earliest Jack Russell yo-yos. It has an airbrushed stripe and the "Jack Russell" logo on the reverse face.

238    0.5    **DRINK COCA-COLA, paper sticker seal, plastic, coaster shape, plastic axle, '90s**

239          **DRINK COCA-COLA • GALAXY 200, imprint seal, plastic, Russell, bulge face shape, wood axle, 1981**
This was used in the 1980 and 1981 Canada promotions. The word "yo-yo" does not appear on the seal.

240    0.5    **DRINK COCA-COLA IN BOTTLES, paint seal, wood, three-piece, '30s**
This is an early wooden Coke advertising piece. Although the date has not been documented, many feel this yo-yo was produced in the '30s.

241    0.5    **DRINK COCA-COLA IN BOTTLES, embossed seal, plastic, Bo-Lo, tournament shape, miniature, '30s**
The reverse face reads "Edwards Bo-Lo Cinti.O." This yo-yo is made out of bakelite plastic. It is believed to be the first plastic, Coke advertising yo-yo and one of the first of any plastic yo-yos produced.

242          **DUNCAN • COCA-COLA, paper insert seal, plastic, slimline shape, view lens, late '50s**
Early Duncan Coca-Cola promotions often used the eagle style mold for yo-yos, see #414. This style of yo-yo was only used by Duncan in the '50s. Jack Russell continued to use this style of yo-yo for his Coca-Cola promotions, in the '60s and early '70s.

243    0.5    **(Duncan Coke series), paper insert seal, plastic, Duncan, slimline shape, view lens, 1997**
In 1997, Coke switched licensing rights in the United States from Russell to Duncan. Duncan began making three series of Coke yo-yos. Three sets of six different yo-yos were made. All are slimline style with a view lens and have the Duncan name imprinted somewhere on the seal. The three series include a character set, which features mostly polar bears, a crazy disk set, and a trademark set.

244    0.5    **ENJOY COCA-COLA, imprint seal, plastic, tournament shape, miniature, metal string slot axle, '80s**
The seal also reads "The Official Soft Drink of Summer." This is an unusual color for a Coke yo-yo most are red and white.

245    0.5    **ENJOY COCA-COLA, imprint seal, plastic, Humphrey shape, metal axle, '80s**
More than one style of Humphrey "Enjoy Coca-Cola" exists.

246    0.5    **ENJOY COCA-COLA • CHAMPIONSHIP, molded seal, plastic, Russell, slimline shape, 1977**
Like many of the Russell Coca-Cola yo-yos, this model has a wooden axle. It was used in a 1977 test promotion in Illinois.

247    0.5    **ENJOY COCA-COLA • DUNCAN IMPERIAL, imprint seal, plastic, tournament shape, metal axle, '80s**

248          **ENJOY COCA-COLA • PROFESSIONAL, molded seal, plastic, Russell, bulge face shape, wood axle, 1977**
This was used in a 1977 test promotion in Illinois.

249    0.5    **ENJOY COCA-COLA • PROFESSIONAL, plastic, Russell, bulge face shape, wood axle**

250    0.5    **ENJOY COCA-COLA • SPECIAL SPIN, imprint seal, plastic, Russell, bulge face shape, 1983**
A small diameter metal axle and extra weight made this a longer spinning yo-yo than other Russell Coca-Cola yo-yos. This was retailed during the 1983 promotion in El Paso and Salt Lake City.

251    0.5    **ENJOY COCA-COLA • SUPER, imprint seal, Russell, plastic, bulge face shape, wood axle, 1983**
In 1983, during the El Paso and Salt Lake City promotions, this yo-yo retailed loose out of a Russell counter display box of 12. All yo-yos in the box were marked "Super." The box included six Coke, two Fanta, two Mello Yellow, and two Sprite yo-yos.

Continued ... ▶

## YO-YO LISTINGS

252  0.5  **ENJOY COCA-COLA • SUPER, molded seal, plastic, Russell, slimline shape, wood axle, 1977**
Distinctive red reflectors are on both faces. This was used in a 1977 test promotion in Illinois.
253  **ENJOY COKE, imprint seal, plastic, Humphrey shape, metal axle, '80s**
The reverse face of this yo-yo has a Schwinn logo.
254  **MERRY CHRISTMAS • COCA-COLA • 96, paper insert seal, plastic, slimline shape, view lens**

## DAKIN

This company has produced inexpensive wooden yo-yos as well as other gift type products.

255  **(Animal series), paint seal, wood, tournament shape**
A series of animal faces are on the front face of this yo-yo. The reverse face shows the animal's stripes, tiger, zebra, etc.
256  43  **(Fido Dido series), paint seal, wood, tournament shape, wood axle, '80s**
This is a series of Fido Dido yo-yos which include a clown, a red heart on both black and white, the character playing a yo-yo, and a face with crossbones. An example of the face with crossbones is shown next to this section's title.
257  **(Heart series), paint seal, wood, tournament shape, early '90s**
This is a series of four wood yo-yos with the following sayings: "I Keep Coming Back To You," "Up, Down Anyway You Want Me," "You've Got Me In A Spin," and "You've Got Me On A String." Yo-Yos were retailed loose in a counter display box.

## DAMERT

Damert is a toy manufacturing company that specializes in science and nature toys. They have a product line of over 300 items, two are yo-yos. Both yo-yos can still be purchased retail. The Turbo Sparkler is a unique novelty yo-yo patented by Bill Halon in 1989. Over a half million Turbo Sparklers have been sold. Due to the popularity of the Turbo Sparkler, knock offs were briefly produced. When legal action was threatened these were withdrawn from the market. Damert also produces a glow yo-yo, the Night Roller. Unlike other glow yo-yos, the Night Roller has a 20% glow concentrate in its plastic compared to 10-12% in other glow yo-yos. Since beginning production in 1990, over 200,000 Night Rollers have been sold.

258  33  **TURBO SPARKLER, imprint seal rim, coaster shape, external diffraction fins, '90s**
As the yo-yo is played, centrifugal force spreads thin plastic fins in a star like pattern.
259  **NIGHT ROLLER, imprint seal, glow plastic, coaster shape, '90s**
An example of this yo-yo is shown above next to this section's title.

## YO-YO LISTINGS

### DELL

Dell Plastics Co., Brooklyn, New York, produced Dell yo-yos in the 1960s. Dell, at the time, was best known for their plastic toy soldiers and plastic flowers. Upon seeing the success of one of the Duncan promotions in New York, they decided to jump on the yo-yo bandwagon and began production of yo-yos. They hired a small number of demonstrators and, like many of the other smaller yo-yo makers of the time, tried to shadow Duncan's heavily promoted campaigns. Dell did produce a few television commercials, but did not spend large amounts on advertising. The catch phrase for Dell on the television ads was "Buy a Big D, Buy a Big D, Buy a Big D" with an echo fade.

All yo-yos produced by Dell were plastic and had metal axles. Most of the blister cards carried a copyright date of 1961. Both Dell and Royal were sued by Duncan during the 1960s for trademark violations for the use of the word "yo-yo." Dell did not use the word "yo-yo" on their seals, but did use it on the blister card packaging. Dell, unlike Royal, did not continue with the legal battle and ultimately dropped their yo-yo line.

| # | Plate | Description |
|---|---|---|
| 260 | | (Dell's big "D" astronaut), no seal, satellite shape, '60s |
| | | The astronaut was made of opaque plastic and was a knock off of the Duncan wood Satellite series that was popular during this time period. The translucent version of this yo-yo was called the "Fireball." The blister card was marked with a pre-printed price of 39 cents. |
| 260.1 | | **DELL'S BIG "D" FLYING STAR, hot stamped seal, marbleized plastic, butterfly shape, '60s** |
| | | Dell produced two butterfly shaped yo-yos. This is considered the rarest of the two. |
| 260.2 | 32 | **DELL'S BIG "D" RETURNER, hot stamped seal, tournament shape, '60s** |
| | | This yo-yo was retailed on a blister card with a pre-printed price of 39 cents and was marked Style No. 600. For the display card, see on Plate Number 32, #260.2. |
| 260.3 | | **DELL'S BIG "D" SLEEPER KING (4 stars), hot stamped seal, glow plastic, tournament shape, '60s** |
| | | This yo-yo was one of the first glow plastic yo-yos ever made. The Dell glow precedes the production of the Duncan Flambeau glow plastic line. This yo-yo has the same seal as #260.4. |
| 260.4 | 06 | **DELL'S BIG "D" SLEEPER KING (4 stars), hot stamped seal, tournament shape, '60s** |
| | | This yo-yo was retailed on blister card with a pre-printed price of $1.00 and marked Style No. 602. |
| 260.5 | 06 | **DELL'S BIG "D" SUPER STAR, hot stamped seal, butterfly shape, '60s** |
| | | This yo-yo was retailed on blister card with a pre-printed price of $1.00 and marked Style No. 604. |
| 260.6 | 06 | **(Dell's) BIG "D" TOURNAMENT (twin flags), hot stamped seal, tournament shape, '60s** |
| | | Collectors refer to this yo-yo as the "Twin Flags Tournament." An example of this yo-yo is shown next to this section's title. |
| 260.7 | 06 | **DELL'S BIG "D" TRICKSTER (3 stars), hot stamped seal, tournament shape, '60s** |
| | | This yo-yo was retailed on blister card with a pre-printed price of 59 cents and marked Style No. 601. This seal is similar to #260.9, but has only 3 stars and is missing the quotation marks around the letter "D." The "TM" is located after the word "Trickster," instead of below the D as in #260.9. |
| 260.8 | | **DELL'S BIG "D" TRICKSTER (4 stars), hot stamped seal, marbleized plastic, tournament shape, '60s** |
| | | The Trickster marbleized plastic pattern is similar to the patterns on the Duncan Imperial marbleized plastic models from the early '60s. There are several color variations of this version. This yo-yo has the same seal as #260.9. |
| 260.9 | 06 | **DELL'S BIG "D" TRICKSTER (4 stars), hot stamped seal, tournament shape, '60s** |
| | | This is the most common variation of the four Dell Trickster yo-yos. See #260.7, #260.8 and #260.10 for other variations of the Trickster. |
| 260.10 | 06 | **DELL'S BIG TRICKSTER (no "D"), hot stamped seal, tournament shape, '60s** |
| | | This is an unusual Dell seal. There is no centrally placed "D" hot stamped into the face. This has the same seal as #260.9, but without the "D." |
| 260.11 | 06 | **ROYAL BIG D YO-YO, hot stamped seal, tournament shape, '60s** |
| | | This yo-yo is somewhat of an enigma as it has both Royal and Dell markings. The yo-yo is definitely a Dell design, but the word "Royal" has a ® mark, which was owned by the Royal Manufacturing Co. Royal did use many different plastic companies to make their yo-yos, but why they would have allowed the "D" to be added to the seal is unclear. |

## YO-YO LISTINGS

 **DISNEY**

Disneyana has high cross-collectibility adding value to these yo-yos. Over fifty different Disney yo-yos have been identified and more are likely to be released. Favorites among collectors are the Duncan Imperial Jr. yo-yos and the Wonderful World of Color yo-yos produced in the 1960s. Several companies have been licensed to produce Disney yo-yos: Admiral Toy Co. 1956-1957, Donald F. Duncan, Inc. 1962-1965, Union Wadding Co. Inc., 1970-1982, and Monogram Products Inc., 1995-present. Because it is easy to stick a Disney sticker on a yo-yo, collectors need to be wary of fakes.

- **101 Dalmatians** - *See Spectra Star section #1453 and #1453.1.*
- **Aladdin "Genie"** - *See Spectra Star section #1418.*
- **Autograph Disney Pals Baseball** - *See Festival section #532.*
- **Autograph Disney Pals Football** - *See Festival section #533.*
- **Dick Tracy** - *See Spectra Star section #1430 and #1430.1.*

261  07  **DISNEY MGM STUDIOS**, imprint seal, brass, tournament shape, '90s
262      **DISNEYLAND BLAST TO THE PAST**, imprint seal, plastic, tournament shape, 1988
   This yo-yo carries a copyright date of 1988. The reverse face has the same seal as Duncan Imperial #352. There is a wood version of this yo-yo. See #732. These were made for a Disneyland summer event in 1988.
263  07  **DISNEYLAND (Mickey's face)**, decal seal plastic, butterfly shape, '80s
264  07  **(Donald Duck)**, imprint seal, plastic, Festival, tournament shape, metal axle, '70s
   For discussion, see Pluto #299.
265  07  **(Donald Duck)**, paper sticker seal, wood, Hallmark, tournament shape, pegged string, '70s
   This was marketed as a beginner yo-yo by Hallmark. The blister card had a pre-printed price of $1.00. For an example of the blister card, see Plate 21, #265.
266  07  **(Donald Duck/Mickey)**, sculpted faces, plastic, Arco, metal axle, early '90s
   This was retailed on blister card, Stock No. 6065.
- **Donald Duck (on target)** - *See Spectra Star section #1432.1.*
267  07  **EURODISNEY**, plastic seal, brass, tournament shape, '90s
268  07  **(Goofy)**, imprint seal, plastic, Festival, tournament shape, '70s
   For discussion, see Pluto #299.
269  08  **Goofy, sculpted face**, plastic, Monogram Products, late '90s
   A series of sculpted Disney character yo-yos was released in 1997 by Monogram Products. This series included Goofy, Mickey #289 and Minnie #297.
270  07  **(Goofy)**, paper sticker seal, wood, Hallmark, tournament shape, three-piece, pegged string, '70s
271  07  **HAPPY BIRTHDAY**, paper insert seal, Festival, tournament shape, view lens, '70s
   The seal has a silhouette of Mickey Mouse standing next to a present. This yo-yo was retailed on blister card with a pre-printed price of $1.29. See Plate 21, #271.
272      **(Hunchback Of Notre Dame)**, imprint seal, plastic, Canada Games, tournament shape, '90s
   This is from the 1996 animated movie by Disney. See also Spectra Star section #1442.1.
273      **JIMMY DODD RETURNING MOUSEKATOP**, paper sticker seal, wood, Admiral, tournament shape, '50s
   The seal on this yo-yo shows famous lead Mouseketeer Jimmy Dodd wearing mouse ears. Along with #293, it was the first licensed Disney yo-yo made by the Admiral Toy Company in 1956-57. This model was a free wheeling yo-yo sold loose in a display box. An example of this yo-yo is shown next to this section's title.
- **Little Mermaid** - *See Spectra Star section #1446.*
280  06  **(Ludwig Von Drake)**, imprint seal, plastic, Humphrey shape, '90s
   This yo-yo was not retailed. It was given as a premium to Walt Disney World employees to help promote safety.
281  06  **(Marsupilami • Houbadoos)**, sculpted, plastic, tournament shape, '90s
   This was a Pizza Hut give away in the middle '90s. Two other Marsupilami toys were also released by Pizza Hut, a glow ball and jump rope.
282  07  **MICKEY & CO.**, imprint seal, glow plastic, Humphrey shape, early '90s
283  07  **MICKEY MOUSE AND MINNIE MOUSE**, paper insert seal, Duncan, tournament shape, view lens, early '60s
   This was a series of three Duncan Imperial Jr. Disney models released in the early '60s. The Imperial Jr. seal is on the reverse face. This yo-yo retailed on a blister card with the original pre-printed price of 59 cents, Stock No. 1451. See display card, Plate 13, #283. This display card came both as a slotted card and as a blister card. The blister card is more rare. See also #294 and #290.
284  07  **(Mickey Mouse)**, decal seal, plastic, puck shape, light-up, '90s
   This yo-yo retailed in a polybag with a saddle header card. Several different saddle header cards have been used with this yo-yo.

*Continued ...*

## YO-YO LISTINGS

● **DISNEY** *Continued ...*

285    *07*    **(Mickey Mouse), decal seal, plastic, bulge face shape, light-up, '80s**
This yo-yo was made by Billco Inc. and carries a copyright date of 1986. Billco is the same company that produced the Canada Games Light-Up Yo-Yo. See #1154.

286    *07*    **(Mickey Mouse), imprint seal, plastic, Festival, tournament shape, metal axle, '70s**
For discussion of the Mickey Pals series, see Pluto #299.

287    *07*    **(Mickey Mouse), paper sticker seal, plastic, tournament shape, metal axle, '80s**

288    *07*    **(Mickey Mouse), paper sticker seal, wood, tournament shape, three-piece, pegged string, '70s**
This yo-yo was produced by Festival for retail by Hallmark.

289    *08*    **(Mickey Mouse), sculpted face, plastic, Monogram Products, late '90s**
This yo-yo was first released in 1997.

290    *07*    **(Mickey Mouse • 3 faces), paper insert seal, plastic, tournament shape, metal axle, view lens, early '60s**
The Duncan Imperial Jr. seal is on the reverse face. This yo-yo was retailed on a display card with an original pre-printed price of 59 cents, Stock No. 1451.

291    *07*    **(Mickey Mouse • 5 Mickeys), paper insert seal, plastic, Festival, tournament shape, view lens, pinball, '70s**

292    *07*    **MICKEY MOUSE CLUB, imprint seal, plastic, Festival, tournament shape, late '70s**
This yo-yo was first released in 1977. The same design was used on a butterfly shaped yo-yo, but with a larger seal.

293           **MICKEY MOUSE CLUB, paper sticker seal, wood, Admiral, tournament shape, 1956**
This is a pegged string yo-yo with Mickey's face on the seal. Along with #273, these were the first Mickey Mouse Club yo-yos produced and very possibly the first licensed Disney yo-yos ever made. This Mickey Mouse Club Yo-Yo was sold loose out of a counter display box and was also included in a Mouseketeer outdoor play set, which included a paddleball, jump rope, top, jacks, and ball.

294           **MICKEY MOUSE CLUB (Mickey's face), paper insert seal, plastic, tournament shape, view lens, early '60s**
The Duncan Imperial, Jr. seal is on the reverse face. This retailed with a pre-printed price of 59 cents on display card Stock No. 1451. The display card can be seen on Plate 13, #283. There was a series of three Duncan Imperial Jr. Disney models in the early '60s. This is considered a more rare yo-yo than the other two Disney Imperial Jr. styles. See #283 and #290.

       •    **Mickey Mouse (on skateboard)** - *See Spectra Star section #1447.*

295    *07*    **(Mickey Mouse on unicycle), paper sticker seal, plastic, butterfly shape, '80s**
If the Disney seal is removed, there is a Prospin Stinger seal underneath.

296    *07*    **(Mickey Mouse with yo-yo), paper sticker seal, wood, tournament shape, three-piece, pegged string, '70s**
This yo-yo was advertised as a "Disney Yo-Yo for Beginners." It was retailed on a Hallmark blister card with a pre-printed price of $1.25.

297    *08*    **(Minnie Mouse), sculpted face, plastic, Monogram Products, late '90s**
This yo-yo was released in 1997.

298    *07*    **(Orange Bird), imprint seal, plastic, butterfly shape, '70s**
The Festival Company produced the Orange Bird Yo-Yo in the middle '70s. This bird was a Disney character created for the Orange Growers Association during the '70s when Disney had an Orange Growers pavilion at Walt Disney World, Florida. Yo-Yos were retailed on blister cards and came with four different seal designs: "Have A Nice Day, Come Fly With Me, I Love You, Be Happy and Smile." A tournament shaped model was also produced with the same seals.

299    *07*    **(Pluto), imprint seal, plastic, Festival, tournament shape, '70s**
Festival, a division of Union Wadding, had the Disney license for yo-yos in the '70s. Their most popular line was the Mickey Mouse and His Pals Series featuring Mickey, Donald, Goofy, and Pluto. Produced for several years, the seals remained the same, unlike the blister cards that have three variations. The original blister cards all had a price mark of $1.00 in the upper right hand corner of the card. The original 1972 cards had a face forward view of the character with the yo-yo centered in the mouth. The only exception was the Goofy card that was in partial profile. This was followed by a brief transition where all characters were in partial profile. These had the $1.00 pre-printed price as shown on Plate 21, #299. In 1973, the pre-printed price of a $1.00 was dropped. This card was used until the early '80s when all four characters were shown on the blister card. The earliest cards marked $1.00 are considered the rarest of the series and are the most difficult to find. Monogram Products owns the original Festival molds and still has the Disney license to reproduce these yo-yos.

300    *07*    **(Roger Rabbit), paper sticker seal, plastic, butterfly shape, late '80s**
This is a German-made yo-yo based on the 1988 Disney movie "Who Framed Roger Rabbit."

301    *07*    **(Rescuers), paper insert seal, plastic, Festival, tournament shape, view lens, pinball game, 1977**
This yo-yo is based on the animated movie released by Disney in 1977. The blister card called this a "Puzzler Yo-Yo."

       •    **Splash Mountain** - *See Hummingbird section #743.*

       •    **The Walt Disney Studios** - *See Hummingbird section #748.*

302           **(Toy Story), imprint seal, plastic, Canada Games, tournament shape, '90s**
This yo-yo is based on the 1995 Disney computer animated film. The yo-yo pictures Buzz Lightyear on the seal. See also Spectra Star #1479 and Humphrey #814.

## YO-YO LISTINGS

303    *07*    **WONDERFUL WORLD OF COLOR (multi-colored plastic embedments), Duncan, butterfly shape, early '60s**
This was a distinctive colorful yo-yo made by embedding objects in clear plastic. Duncan experimented with embedments on two yo-yo styles. These were the "World of Color" series and the "Mardi Gras" series. The Duncan Company produced both series in the early '60s.

The Wonderful World of Color yo-yos by Duncan are considered the most desirable of the Disney yo-yos. They all have a gold leaf hot stamped seal on a concave disk which inserts into the face of the yo-yo. Convex surface variations exist, but are considered extremely rare. On the reverse face, there is also a concave disk with a hot stamped seal reading "Duncan Yo-Yo Return Top."

The four variations retailed on the same display card with a pre-printed price of $1.00. Ludwig Von Drake was on the card, Stock No. 701. See the display card Plate 13, #303.

304    *07*    **WONDERFUL WORLD OF COLOR (multi-colored embedded stars), plastic, Duncan, flywheel shape, early '60s**
The flywheel shaped Wonderful World of Color yo-yos are considered more rare than the butterfly versions.

305          **WONDERFUL WORLD OF COLOR (silver and gold embedded stars), Duncan, flywheel shape, early '60s**
This silver and gold variation is more rare than the multi-colored star flywheel shaped version. See #304.

306    *07*    **WONDERFUL WORLD OF COLOR (white plastic and gold glitter embedments), Duncan, butterfly shape, '60s**
This retailed on the same card as #303.

## DUNCAN

Duncan yo-yos are the most famous line of yo-yos ever produced in the United States. Donald F. Duncan, Inc. was started in Chicago by Donald F. Duncan at the beginning of America's first yo-yo craze in 1929. Duncan, an entrepreneurial genius, realized the potential of the yo-yo and quickly jumped into the market by producing his first yo-yo line, the "O-Boy."

The person who introduced the yo-yo to Don Duncan may never be known. Duncan reported it was a young Filipino he first saw playing with the yo-yo. By some newspaper accounts in the 1930s, Tony Flor, a Duncan demonstrator, is given credit for being this Filipino. Other reports of Pedro Flores as the person Duncan first saw with the yo-yo are probably not true, but it is highly likely that it was a Flores yo-yo that was being played.

In 1930, Duncan bought the yo-yo trademark from Pedro Flores, the leading producer of yo-yos at that time. Reports of the purchase price range from $25,000 to over one quarter of a million dollars. Most feel the price was closer to the $25,000 range, but this was still a fortune during The Depression. When questioned about the buy out, Flores said he was more interested in teaching yo-yo skills than making yo-yos. Flores went on to run some of the earliest contest campaigns for Don Duncan in the 1930s. In these early contests, Duncan sold both the Flores and the Duncan yo-yo.

With his headquarters based at 1500 S. Western Ave., Chicago, Duncan heavily promoted his traveling Filipino yo-yo demonstrators by running contests and promotions around the country. The function of demonstrators was not to sell yo-yos, but to create consumer demand. By getting exposure for the yo-yo, demonstrators were highly successful in creating a nationwide craze. Not only did these yo-yo promotions and contests spread like wildfire throughout the U.S., but they were also taken over to Europe where they met with equal success. Throughout the '30s Duncan dominated the American yo-yo market.

Duncan's early success in promoting yo-yos was due, in large part, to his mastery of free publicity. He used the technique of combining contest campaigns with local newspaper sponsorships. The sponsoring newspapers benefited by requiring the entrants to sell subscriptions (usually three) for contest eligibility. They, in turn, provided free publicity and also prizes. Some newspapers even provided free yo-yos to contestants meeting eligibility requirements. Using this technique, Duncan convinced William Randolph Hearst, the biggest newspaper magnate of the early 20th century, to use yo-yo contests to stimulate circulation of his newspapers. Some of Duncan's biggest campaigns in the '30s were in conjunction with cities that had Hearst controlled newspapers.

Duncan also tapped into recognition surrounding celebrity exposure. Celebrities such as Douglas Fairbanks, Mary Pickford, and Baseball Hall-of-Famers Lou Gehrig and Hack Wilson were all photographed with yo-yos in hand. Paid promotions using popular icons such as "Our Gang," were used in the promotion of the "Gold Seal" and "O-Boy" yo-yos. If a town was without a visible celebrity, public officials did nicely for publicity shots. Mayors, police chiefs, city health commissioners, all were recruited to promote local yo-yo campaigns.

*Continued ...* ▶

### DUNCAN Continued...

*Duncan Filipino Demonstrators in 1931.*

The yo-yo craze briefly subsided during the years of WWII when material for producing yo-yos was hard to find. Although yo-yos continued to be produced, they were in limited quantities and many of the yo-yo demonstrators were off to war. Prior to WWII, Duncan purchased their yo-yos from Baurle Brothers, a wood turning company in Chicago. Following the war, in 1946, Duncan opened his own factory in Luck, Wisconsin and was back to manufacturing yo-yos. New competitors had entered the yo-yo field, but did not succeed at dethroning Duncan as the king of yo-yos.

The '50s again saw Duncan running nationwide campaigns. Now corporate sponsorships entered the picture. Companies like Pepsi, Coke, and Sealtest Ice Cream all sponsored Duncan contests. Incentives, such as entry coupons for free products, stimulated contest participation. Pressure from competing companies' promotions were answered with an increase in the array of prizes awarded. In some years, Duncan regional champions were given an all expense paid trip to Disneyland to attend the national championship tournament. The awards for the 1962 upstate New York championships were: first prize: round trip to Disneyland for two, second prize: 17 inch RCA TV set, third prize: Stereo set, fourth prize: transistor radio.

Unless the market area was very large, the demonstrators worked alone. Duncan, over the years, had varying numbers of full-time demonstrators in the field. In 1962, Duncan employed 27 full-time professionals. In some metropolitan areas, demonstrators would often recruit a "crew" of part-time demonstrators to run contests around the city. These part-time demonstrators were generally older teenagers and former local champions. Many of these part-time demonstrators went on to be full-time employees of Duncan or other yo-yo companies. Full-time traveling demonstrators were paid a salary, but also received an incentive bonus if sales figures exceeded a certain amount. A typical salary for a demonstrator in the '60s was 50 dollars a week plus 100 dollars for expenses. The bonus level was based on the cost of advertising, expenses planned for that particular campaign and the expected sales. Some of the demonstrators referred to this number as the "nut." If you exceeded the level it was called "cracking the nut." Typically, the demonstrators got a one-percent bonus of the previous month's promotional sales. The amount was then divided by the number of demonstrators in the field.

A typical regional campaign lasted between five and seven weeks. Demonstrators were sent to northern cities in September and October and southern cities in the spring and winter. During the summer months, July and August, promotions slowed down. Duncan discovered that these months, as well as before and after Christmas, were the worst times for promotions. During the first two weeks in a market area, the demonstrator would meet with the wholesalers and set up the campaign plan. They would identify the best retailers in the area and arrange contest times and promotions. TV ad time was then bought on local television. Ads included a ten second tag identifying the retailer sponsoring the contest. The demonstrator would then offer to perform on the local children's television show on which the ad time was purchased. This usually led to several extra free minutes of television airtime. The demonstrator would also contact the yo-yo retailer, explain that the commercial had a

*Continued...*

tag line with their store name, and ask if they would place a newspaper ad for the contest. This was usually agreed to, maximizing Duncan's marketing dollar.

Duncan had yearly meetings with their traveling demonstrators, which could last from a few days to a week. These meetings were used to disseminate new marketing strategies and long term goals. Meetings were also used to check for proficiency in the Duncan items that were being promoted. In the '60s, the demonstrators were expected to be proficient in not only yo-yos, but also Bang-a-Balls, and tops. Top contests, like yo-yo contests, were considered in the realm of the Duncan Demonstrator.

One of the more noble actions of the Duncan Company, now almost forgotten, was their commitment to entertaining sick children. From the '30s on, Duncan encouraged demonstrators to offer their services to crippled children's hospitals, Tuberculosis sanitariums, and pediatric wards. Duncan provided free yo-yos to ill children and, although sometimes publicity occurred, this was not the motivation. The fact that these demonstrations were financially supported by the company and enthusiastically done by demonstrators reflects strongly on the character of the company and the people that worked for it.

In 1957, Don Duncan retired from yo-yo making and handed the reigns of the company over to his two sons, Don Jr., and Jack. Throughout the late '50s and early '60s, Duncan continued to be the top producer of yo-yos. Much concern was given to other companies shadowing the promotions and skimming off sales from Duncan marketing during this period. The executive offices became very secretive about the areas where promotions were planned. Duncan frequently gave the demonstrators only three days prior notice regarding the location of the next promotion. In addition, cheaper non-promoted lines, were distributed in an attempt to foil other companies' low price items.

Although Duncan used television for marketing during the '50s, this was generally limited to live demonstrator performances in studios. Then, in 1959, Duncan began experimenting more aggressively with paid television advertising. It was an immediate success. Don Duncan, Jr. himself appeared on national television on August 12, 1961, on the popular television show "What's My Line?" The panelists failed to guess his occupation as America's top manufacturer of yo-yos. Alan Bonito, of the Bruns Agency of New York City, was in charge of purchasing airtime for the Duncan promotions. This nationwide television advertising campaign heavily increased the demand for yo-yos in the early '60s. An example of how successful these campaigns were was the promotion in Nashville, Tennessee. In a two month period, Duncan sold 350,000 yo-yos. The population of Nashville at that time was only 322,000. At peak production, the Luck plant was producing 60,000 yo-yos per day with 640 employees. In 1962 alone, Duncan sold more than 45 million yo-yos.

*Don Duncan and demonstrator Joe Dervera entertaining at a hospital for sick children. Clipping from the Toronto Evening Telegram, 1931.*

*Joe Radovan, an early Duncan demonstrator, with boxes of Gold Seal yo-yos just delivered to the local store, 1934.*

**DUNCAN** *Continued ...*

Unfortunately, the success of yo-yos in the early '60s may have led, in part, to Duncan's downfall. In order to fill the seemingly bottomless need for yo-yos, Duncan's plant worked 24 hours a day, paying heavy overtime costs plus premium prices for materials and supplies. This lower profit margin, combined with the expense of television promotions and expansion of product lines, also strained the company's finances. Perhaps the ongoing expense of the Duncan vs. Royal trademark trial was the coup de grâce which stripped Duncan of the Yo-Yo trademark protection. By the time the suit was settled, in 1965, Duncan had spent hundreds of thousands of dollars attempting to preserve its trademark rights.

On May 26, 1965, Flambeau Plastics Corporation filed suit against Donald F. Duncan for $38,827. Two days later, four other creditors filed a petition against Duncan for involuntary bankruptcy. It was granted. Donald F. Duncan Inc. was sold to Business Assets Corporation of Chicago. During the Duncan golden years (1929-1965), when scores of other yo-yo manufacturers had come and gone, Duncan was responsible for 85% of all yo-yos produced, a truly amazing marketing feat.

In 1966, Duncan underwent a liquidation auction of company assets which included trademark rights. Although the factory and lathes were purchased immediately, the Duncan name was not purchased until two years later. Duncan yo-yos were not produced until 1968 when the Flambeau Company, which was the plastic company that previously supplied Duncan plastic yo-yos, bought the rights to the Duncan yo-yo line name. Flambeau owned the molds and machinery already, all they needed was the Duncan name. The Flambeau Company, run by Bill Sauey, immediately began production of plastic yo-yos, but the era of Duncan wood yo-yos had ended. Don Duncan died in 1971, but Flambeau Products continues to produce the Duncan line of plastic yo-yos to this date.

*Store window display, circa 1962.*

*Duncan demonstrators Bob Babb and "Skeets" Beebe demonstrate on a local children's show in Binghamton, N.Y., in the early '60s.*

## YO-YO LISTINGS

## TIPS ON DATING

The words "Pat. Pend." were used on many yo-yos in the early '30s. These "Pat. Pend" yo-yos are some of the earliest Duncans. Duncan never owned a patent on the yo-yo, but he did receive a trademark on the name "yo-yo" in 1932. If a Duncan Yo-Yo has "Return Top" stamped on the seal, then the yo-yo was made between 1956 and 1965. Production of wooden Duncans stopped in 1965. There were no wooden Duncans produced until a new "Super" was released in 1996. Display cards prior to 1965 list the Duncan address as Luck, Wisconsin.

Plastic yo-yos with a hot stamped seal reading "Yo ® Yo" were produced prior to 1965 and are considered "Duncan Golden Years Yo-Yos." None of the Flambeau plastic yo-yos have the registration mark associated with the word "yo-yo." Flambeau began using a ® mark associated with the word "Duncan" in 1969. Prior to Flambeau, Duncan did not use a registration mark with its name.

Many early plastic Flambeau yo-yos from 1968 - 1976 were printed with a "hot stamp" process which indented the seal into the plastic. After 1976, this "hot stamp" process was discontinued, so the imprinted surface cannot be felt on recent models. Around 1979, Flambeau began using bar codes on blister cards. There was a short transition period when some cards could be found with both bar codes and pre-printed prices. By the early '80s, bar coded cards had completely replaced pre-printed price cards.

Prior to 1986, the Flambeau blister cards all had the address listed as Baraboo, Wisconsin. After 1986, the address on the cards was changed to Middlefield, Ohio. Skin pack blister cards from the Middlefield, Ohio address all have two letters following the item number.

*Duncan's first blister packed yo-yo was released in 1958.*

## PACKAGING

Like most collectibles, yo-yos with the original packaging have added value. From 1929 to 1958, all Duncan yo-yos were sold out of display boxes. In the '60s, Duncan yo-yos were retailed both out of display boxes and on cards. From the '70s on, Duncan Flambeau yo-yos were retailed on blister cards. Changes in retailing forced emphasis away from graphics on counter boxes to eye-catching graphics on the display cards. Although Flambeau did produce counter display boxes which held carded packaged yo-yos, the artwork never equaled that of the earlier Duncan display boxes. (See also the Counter Display Box section.)

The first Duncan yo-yos displayed on a blister card were the Duncan Jeweled and Duncan Litening yo-yos in 1958. Yo-Yos retailed on display cards were standard after the late 1950s. In the '60s, most of the carded yo-yos were retailed in polybags. The '70s saw Flambeau changing to blister cards universally. Interestingly, it is not uncommon to find a blister carded yo-yo with packaging artwork different from the yo-yo contained on the blister card. Errors in packaging occurred during peak demand periods when orders were backlogged and yo-yos were put in any packaging available just to get orders filled.

# YO-YO LISTINGS

 **YO-YO MAN LOGO (MR. YO-YO)**

Characters are often associated with advertising of a product. For instance, Tony the Tiger and the Pillsbury Doughboy instantly bring to mind the products they represent. In 1950, Duncan introduced the Mr. Yo-Yo character to help with the promotion of Duncan yo-yos. Much like the evolution of Mickey Mouse, the Mr. Yo-Yo character evolved over the years.

In 1929, when Duncan first started producing yo-yos, no character trademark was associated with the yo-yos. At the time, the marketing angle was the "gold seal" on Duncan yo-yos. Kids were told to buy the yo-yo with the "gold seal." Duncan contests were limited to individuals owning the Duncan Gold Seal Yo-Yo. The original Gold Seal Yo-Yo was the big "G" genuine gold decal seal. (See #491.) These were the standard tournament yo-yos used throughout the '30s.

Following WWII, Duncan had many competitors in the yo-yo industry, such as Cheerio, Royal, Goody, etc., so new marketing strategies were sought. In 1947, the original yo-yo character was introduced. A human body, with a yo-yo head and square nose, but no other facial features, became known as "Mr. Yo-Yo." Mr. Yo-Yo was featured on advertising, trick books, awards and sometimes on the yo-yos. (See Duncan 77 #486.) This character was the symbol of Duncan yo-yo promotion throughout the '50s.

In 1960, times had changed. Many new lines of yo-yos were being introduced and a more modern character was sought. Don Duncan, Jr., contracted cartoonist Morrie Brickman to produce a more "modern" Mr. Yo-Yo character. The new Mr. Yo-Yo now had a normal head and face, but was more diversified in his attire to reflect the yo-yo line being promoted. For example, on the Satellite yo-yo, which was considered a new "space age toy," Mr. Yo-Yo was dressed in a space helmet. With the Mardi Gras yo-yos, he was dressed in a Mardi Gras costume. Each yo-yo line in the early '60s had a uniquely clothed yo-yo man on the display card.

When Duncan went bankrupt in 1965 and was bought by Flambeau in 1968, Flambeau used a similar yo-yo man character but added nostrils to the nose. This "Nostril Mr. Yo-Yo" was seen on some display cards in the early '70s. Around 1975, the "Rosy Cheek" Mr. Yo-Yo first appeared, similar in design to his predecessor. The rosy cheeks became his identifying mark. A variation of this "Rosy Cheek" Mr. Yo-Yo character continues to be used today.

## THE EVOLUTION OF MR YO-YO

## YO-YO LISTINGS

## DUNCAN ADVERTISING

Early on Duncan realized the value of the yo-yo as an advertising medium. Although there is evidence of advertising on yo-yos in the pre-WWII era, serious use of yo-yos in advertising didn't begin until the '50s. The miniature Beginners yo-yos were the most common Duncan ad yo-yos in the '50s and early '60s. With the Flambeau takeover and the change to plastics, the Imperial became the standard Duncan ad yo-yo. Even so, over the years, almost every style of Duncan yo-yo has been used for advertising. Some Duncan produced ad yo-yos did not have a Duncan logo. This was very common if the advertiser desired to use both faces of the yo-yo for graphics. Most desirable of the Duncan ads are the few rare, tin whistling ads and the standard size, wood tournament models. In the '80s and '90s, ad yo-yos by Duncan Flambeau were typically made of opaque plastic rather than the more common translucent plastic used for yo-yos sold retail. This was done to increase the visibility of the ads. The listings below represent only a small sampling of Duncan advertising yo-yos.

| | | |
|---|---|---|
| 307 | | A.M. DAVISON'S, die stamped seal, wood, tournament shape, late '50s |
| 307.8 | 01 | (Big Yella yo-yo), imprint seal, plastic, Flambeau, tournament shape, 1978 |

This was a Kellogg's Sugar Corn Pops premium offer in 1978. This yo-yo has a Duncan Imperial seal on the reverse face. See #352.

| | | |
|---|---|---|
| 307.9 | 10 | BOSCO BEAR YO-YO, die stamped seal, wood, tournament shape, miniature, pegged string, early '60s |

This ad promotion for Bosco Bear chocolate syrup was one of the largest wood ad promotions ever produced by Duncan. An example of this yo-yo was exhibited during the Duncan vs. Royal yo-yo trademark trial. It was shown because it was a Duncan produced yo-yo, but did not have the Duncan name on the seal, only the word "yo-yo."

| | | |
|---|---|---|
| 307.10 | 10 | BURGER KING, imprint seal, plastic, Flambeau, tournament shape, 1979 |

This yo-yo has a Duncan Imperial seal on the reverse face. See #352. This was a nationwide campaign. If you bought an order of fries, you could also purchase the yo-yo for 45 cents.

| | | |
|---|---|---|
| 307.11 | | CALEDONIAN, imprint seal, plastic, Flambeau, tournament shape, '80s |
| 307.12 | 45 | (Campbell's kid play kit), 1963 |

This is a Campbell Soup promotion that features a yo-yo with the Campbell Soup Kid character on the face. The yo-yo is a plastic tournament shape with a metal string slot. The reverse face reads "Duncan Return Top." The kit also came with a wooden top, miniature paddleball and trick book. This kit could originally be obtained for 50 cents and three Campbell Soup labels. The offer expired May 31, 1963. Values are listed separately for the complete kit and the loose yo-yo.

| | | |
|---|---|---|
| 307.13 | | CHRYSLER CORPORATION, die stamped seal, wood, tournament shape, late '50s |
| • | | Circus World - *See Clown section #219.* |
| 307.14 | 10 | (Cocoa Puffs), imprint seal, glow plastic, Flambeau, butterfly shape, 1995 |

This yo-yo pictures Sonny the Coca Puffs bird, and was a premium for Cocoa Puffs cereal. Duncan also ran a co-promotion offering their first model of a glow butterfly yo-yo "free" with two Cocoa Puffs coupons. This promotion was featured on the Cocoa Puffs cereal boxes in 1995.

| | | |
|---|---|---|
| • | | Coke - *See Coca-Cola section #231, #235, #236, #242, and #243.* |
| 307.15 | 10 | COOLO-COOLO, gold leaf stamped seal, wood, tournament shape, miniature, '60s |

This has a little face Mr. Yo-Yo Beginners seal on the reverse face. This style of miniature ad yo-yos was common in the '50s and early '60s.

| | | |
|---|---|---|
| 307.16 | 10 | (Dig 'Em Frog), imprint seal, plastic, Flambeau, tournament shape |

This yo-yo has a Duncan Imperial seal on the reverse face. See #352. The yo-yo was a Sugar Smacks premium.

| | | |
|---|---|---|
| 307.17 | | DORAL • QUAZEPAM, imprint seal, plastic, Flambeau, tournament shape, '90s |

This yo-yo does not have a Duncan logo, but it came with a mini Duncan trick book. On the reverse face it reads, "easy to start, easy to stop." This is an example of a yo-yo using both faces for advertising with out the Duncan logo. Doral is a tranquilizer.

| | | |
|---|---|---|
| • | | Dr. Pepper - *See Soft Drink section* |
| 307.19 | | EVERSWEET ORANGE JUICE, gold leaf stamped seal, wood, tournament shape, pegged string, early '60s |

This is a different seal from 307.20. The little face Mr. Yo-Yo Beginners seal is seen on the reverse face #314.

| | | |
|---|---|---|
| 307.20 | 10 | EVERSWEET, die stamped seal, wood, tournament shape, three-piece pegged string, early '60s |

The little face Mr. Yo-Yo Beginners seal is seen on the reverse face. See #314. This was a premium given free with the purchase of Eversweet Orange Juice.

| | | |
|---|---|---|
| 307.21 | | FAO SCHWARZ, paper insert seal, plastic, Flambeau, slimline shape, view lens, '90s |

This model was sold by the famous toy store chain. Two different seals are available, one with four rocking horses and one with thirteen. The Duncan Eagle Professional seal is on the reverse face. See #448.3. This was retailed on a blister card, Stock No. 3270AA.

| | | |
|---|---|---|
| 307.22 | 10 | FEDERAL REALITY, paper insert seal, plastic, Flambeau, slimline shape, view lens, '80s |

This has the Duncan Eagle Professional seal on the reverse face. This is one of the many slimline ad yo-yos produced by Flambeau in the '80s and '90s. Ad yo-yos such as this have limited collectible value, unless the advertising is cross-collectible.

*Continued ...* ▶

### YO-YO LISTINGS

○ **DUNCAN ADVERTISING** *Continued ...*

307.23  *10*  **(Franco-American), paper insert seal, plastic, Flambeau, slimline shape, view lens, '80s**
This 1981 Franco-American promotion came with a trick book. See #1926.

307.24  *10*  **GM PARTS, imprint seal, plastic, Flambeau, tournament shape, '90s**
This yo-yo has a Duncan Imperial seal on the reverse face. See #352.

307.25  *10*  **HONEY & NUT, paper sticker seal, plastic, Flambeau, butterfly shape, '80s**
This was a promotion for Honey Nut Cereal. Duncan yo-yos with paper sticker advertising seals like this seal have only minimal collectible value, unless the advertising is highly cross-collectible.

307.26  **HANNAY REELS, imprint seal, plastic, Flambeau, butterfly shape**

307.27  *10*  **RETAIL CIGARS UNION LOCAL 324, die stamped seal, wood, tournament shape, early '60s**
The little face Mr. Yo-Yo Beginners seal is seen on the reverse face. See #314.

307.28  *10*  **JUMPING JACKS AT McALPHIN'S, die stamped seal, wood, tournament shape, early '60s**
The little face Mr. Yo-Yo Beginners seal is seen on the reverse face. See #314.

307.29  *10*  **KIST BEVERAGES, gold leaf stamped seal, wood, tournament shape, three-piece, '50s**
Duncan made this yo-yo for a Kist promotion, but the yo-yo does not carry a Duncan logo on the seal or reverse face.

307.30  *10*  **KITTY CLOVER POTATO CHIPS, gold leaf stamped seal, wood, tournament shape, miniature, pegged string, '60s**
This yo-yo is somewhat unusual as the Duncan name was inserted on the advertising face. The more typical pattern is to have the ad on one face and the Duncan seal on the reverse face.

307.31  **(Kool-Aid), imprint seal, plastic, Flambeau, tournament shape, '80s**

307.32  **KROGER DOG FOOD, imprint seal, plastic, Flambeau, tournament shape, '90s**

307.33  **LITTLE DARLIN'S ROCK AND ROLL PALACE, paper insert seal, plastic, Flambeau, slimline shape, view lens, '90s**

307.34  **MISTER PEANUT, paper insert seal, plastic, Flambeau, slimline shape, '80s**
There are at least three different Planters Peanuts yo-yos. See #52 and #872.

307.35  *10*  **MOBIL, hot stamped seal, plastic, tournament shape, metal string slot, early '60s**
The Duncan Little Ace seal is on the reverse face. See #418.1. The Mobil ad Little Ace is more common than the standard line Little Ace. See #418.

307.36  *10*  **MODERN WOODSMAN, gold leaf stamped seal, wood, butterfly shape, miniature, pegged string, '60s**
This yo-yo has a little "G" Beginners seal on the reverse face, see #310. The majority of miniature, butterfly, wood, advertising yo-yos were produced between 1960 and 1964. Most of these miniature butterfly ad yo-yos used the little "G" Beginners seal #310 on the reverse face.

307.37  *10*  **MONROE, hot stamped seal, plastic, Flambeau, butterfly shape, early '70s**
The psychedelic butterfly seal is found on the reverse face. See #330. This yo-yo was used as a promotion for Monroe Shock Absorbers.

307.38  *10*  **NATURE VALLEY GRANOLA BARS, imprint seal, plastic, Flambeau, tournament shape, '80s**
This yo-yo has a Duncan Imperial seal on the reverse face. See #352.

307.39  **NEW ERA POTATO CHIPS, gold leaf stamped seal, wood, tournament shape, miniature, early '60s**

307.40  **NEW HAVEN REGISTER OFFICIAL YO-YO, wood, '30s**
This is a rare advertising yo-yo from the '30s. One of these yo-yos was used as an exhibit during the Duncan vs. Royal Yo-Yo trademark trial.

307.41  **NICKELODEON, imprint seal, plastic, Flambeau, tournament shape, 1995**
This is a promotional yo-yo from Nickelodeon Studios, Orlando. The reverse face has a Duncan Neo seal. This yo-yo carries a copyright date of 1993.

307.42  *10*  **O'DUTCH POTATO CHIPS, gold leaf stamped seal, wood, tournament shape, miniature, three-piece, '60s**

307.43  **PEPSI, die stamped seal, wood, tournament shape, '50s**
The little "G" Beginners seal is on the reverse face. See #310.

307.44  **PRO, paper insert seal, Flambeau, plastic, slimline shape, view lens, '80s**
PRO is an acronym for Premium Representatives Organization. "Professional Duncan" yo-yo is on the reverse face.

307.45  **QUALITY CHECK ORANGE JUICE, die stamped seal, wood, tournament shape, three-piece, early '60s**
The little face Mr. Yo-Yo is seen on the reverse face. See #314.

307.46  *10*  **RICE KRISPIES, imprint seal, glow plastic, Flambeau, tournament shape, '80s**
A Duncan Imperial glow seal is on the reverse face. Unlike standard glow yo-yos (#385), this yo-yo has black ink. Although sometimes listed as "rare" by dealers, this is a very common yo-yo.

307.47  **RED DOT, gold leaf stamped seal, wood, tournament shape, miniature, pegged string, early '60s**
The little face Mr. Yo-Yo Beginners seal is seen on the reverse face. See #314.

307.48  *10*  **RED GOOSE SHOES, gold leaf stamped seal, wood, tournament shape, one-piece, early '50s**
This is an example of a standard size tournament shaped ad yo-yo produced in the early '50s. It has a Duncan Super seal #507 on the reverse face. Yo-Yos of this size and shape are more valued than the miniature versions. An example of this yo-yo is shown next to this section's title.

• **Sell Jen** - *See Duncan Tin section #485.*

307.49  **SEVEN UP, die stamped seal, wood, tournament shape, '50s**

307.50  *10*  **SCHLITZ, decal seal, plastic, Flambeau, butterfly shape, '70s**
The psychedelic butterfly seal is found on the reverse face. See #330.

307.51  *10*  **SCHWEIGERT, gold leaf stamped seal, wood, butterfly shape, miniature, early '60s**

*Continued ...* ▶

## YO-YO LISTINGS

| | | |
|---|---|---|
| 307.52 | *10* | **SO GOOD POTATO CHIP CO.**, die stamped seal, wood, tournament shape, miniature, early '60s |

The little "G" Beginners seal is on the reverse face. See #310.

307.53  *10*  **STOPPENBACH "MEATS OF EXCELLENCE,"** die stamped seal, wood, butterfly shape, miniature, early '60s
The little "G" Beginners seal is on the reverse face. See #310.

307.54  *10*  **(Sugar Crisp)**, paper seal, plastic, Flambeau, miniature, 1980
This snap together Post Sugar Crisp cereal premium came in a cellophane bag. A Duncan logo appears on the reverse face. The cereal box also had a mail-in offer for yo-yo. See #447.2.

307.55  **SZABO**, paper insert seal, plastic, Flambeau, slimline shape, view lens

307.56  **TEDDY SNOW CROP**, die stamped seal, wood, tournament shape, late '50s
This is a rare Duncan ad seal. An example of this yo-yo was exhibited during the Duncan vs. Royal Yo-Yo trademark trial.

307.57  **TURKISH TAFFY THE MAGIC CLOWN**, die stamped seal, wood, tournament shape, late '50s
This is a rare Duncan ad seal. Turkish Taffy also ran some yo-yo contest promotions with Duncan.

307.58  **THE ELECTRIC YO-YO**, imprint seal, plastic, Flambeau, tournament shape, '90s
TAITO and Duncan are imprinted on the reverse face.

307.59  *10*  **WHIRLPOOL**, die stamped seal, wood, tournament shape, '50s
Standard size tournament models like this are more valuable than miniature advertising yo-yos.

307.60  **WIENERS PRIDE**, gold leaf stamped seal, wood, miniature, '60s

307.61  *10*  **WORLD FAMOUS CLYDESDALES**, paper insert seal, plastic, Flambeau, slimline shape, view lens, '90s
For other Clydesdale yo-yos, see #804.

307.62  *10*  **YO-YO CHAMP FRANKIE SMITH**, foil sticker seal, plastic, Flambeau, butterfly shape
This has an unusual gold leaf hot stamped Duncan seal on the reverse face seen only on advertising yo-yos.

## DUNCAN BEGINNER

Duncan has produced Beginner yo-yos since the '30s. Beginner models were less expensive to purchase than tournament models and typically had pegged strings, which greatly limited the trick function of the yo-yo. Beginner yo-yos were wooden, three-piece models with pegged strings. The original Beginners models were standard in size, being larger than juniors, but smaller than tournament models. The classic color scheme for Beginners was one half red, the other half black, but other color variations exist. Some Duncan Beginner models came with a pegged red or blue string. (Pegged string yo-yos should be collected with the original string.)

The Beginners were originally retailed from counter display boxes for 10 cents each and carried a stocking number of 44. Later, when Duncan made some juniors larger and some Beginners smaller, miniature sized Beginners were retailed out of boxes marked with the stocking number of 33. (The stocking numbers 33 and 44 refer to the size of the yo-yo, not the model.)

In 1968, Flambeau took over Duncan and the era of making Duncan Beginners yo-yos solely out of wood ended. Flambeau completely dropped the wood line and began producing plastic beginner models exclusively.

This section includes only yo-yos that have the word "Beginner" or "Beginners" on the seal. Other yo-yos might be considered "beginner" models because they have pegged strings, but those are listed elsewhere.

308  *16*  **(Big "G") GENUINE**, gold leaf stamped seal, wood, tournament shape, pegged string, '30s and '40s
Big "G" (Genuine) refers to the large "G" in the word "Genuine" that extends to the base of the seal. This is the first Duncan yo-yo to use the word "Beginners" on the seal. Production of this began in the '30s and replaced the O-Boy #446.1 as Duncan's standard line beginner yo-yo. Although some of the O-Boy models were red and black fixed string yo-yos, they did not read "Beginners" on the seal.

309  *16*  **BEGINNER T.M.**, molded seal, plastic, Flambeau, puck shape, early '70s
This was retailed on a blister card with a pre-printed price of 59 cents, Stock No. 3263.

310  *16*  **(Little "G") GENUINE**, gold leaf stamped seal, wood, tournament shape, miniature, '50s and '60s
Little "G" (Genuine) refers to the small "G" in the word "Genuine." All these yo-yos were three-piece pegged string models. Although sold retail, this yo-yo was used most frequently for advertising.

*Continued ...* ▶

● **DUNCAN BEGINNER** *Continued ...*

311　　　**(Little "G") GENUINE, gold leaf stamped seal, wood, tournament shape, three-piece, '50s and '60s**
　　　　This larger version of the Little "G" was seldom used for advertising, unlike the miniature bearing this seal (#310). This was the standard retail version. This yo-yo replaced the big "G" Beginner #308 as Duncan's standard line Beginner.

312　　　**(Little "G") GENUINE, gold leaf stamped seal, wood, butterfly shape, miniature, three-piece, '60s**
　　　　Miniature butterfly Beginners yo-yos were usually associated with either advertising or the Happy Birthday Yo-Yo. This has the same seal as #310.

313　*16*　**(Sunburst), hot stamped seal, plastic, Flambeau, tournament shape, plastic axle, '70s**
　　　　Some collectors refer to this yo-yo as the "Sunburst Beginner" because of the seal design. However, the word "Sunburst" does not appear on the seal.

314　*16*　**(Yo-Yo Man face • Mr. Yo-Yo), gold leaf stamped seal, wood, tournament shape, early '60s**
　　　　This is the first appearance of the '60s style Mr. Yo-Yo on the seal. All these yo-yos were three-piece pegged string models. Some collectors refer to this as the "little face Mr. Yo-Yo Beginners." They were retailed on a display card with an original price of 29 cents, Card #44. At least two different 29 cent display cards, and one 39 cent, are known. Although the 39 cent card is more recent, it is considered more rare since fewer were made than the 29 cent cards. See both examples of the 29 cent display card on Plate 9 and Plate 13, #314.

315　*16*　**(Yo-Yo Man face • starred eye), gold leaf stamped seal, wood, tournament shape, early '60s**
　　　　The word "Beginners" forms Mr. Yo-Yo's mouth, two stars the eye's, and a circle for the nose. One of the rarest of the beginners models. Some collectors refer to this as the "Starred Eye Beginners." This yo-yo came on its own unique star eyed card. An example of this yo-yo is shown next to this section's title.

316　*16*　**(Yo-Yo Man standing • Mr. Yo-Yo), imprint seal, plastic, Flambeau, puck shape, metal axle, early '70s**
　　　　This was the first appearance of Mr. Yo-Yo on the seal of a Flambeau yo-yo. Flambeau had not yet changed the design to the rosy cheek Mr. Yo-Yo. The ® mark appears behind the word "Duncan" on Mr. Yo-Yo's shirt, which helps identify a loose yo-yo as a Flambeau. This was retailed on a blister card with an original pre-printed price of 59 cents, Stock No. 3263. See the display card Plate 13, #316.

317　*16*　**(Yo-Yo Man standing • Mr. Yo-Yo), hot stamped seal, Flambeau, puck shape, early '70s**

## DUNCAN BUTTERFLY

The Duncan Butterfly Yo-Yo began production in 1958. By reversing the cutting dies on the tournament style yo-yo, a butterfly shaped yo-yo could be produced. The butterfly shape allows for easier performance of many string tricks. Butterfly shaped yo-yos continue to be produced to this day.

Duncan did not invent the idea of the butterfly shaped yo-yo. Patents for butterfly shaped yo-yos existed as early as 1878. W. M. Katz received the first known patent for a butterfly shaped yo-yo. Credit is currently given to Wayne Lundberg for introducing the idea to Duncan. Don Duncan, Jr. is given credit for naming the Butterfly Yo-Yo. Duncan Flambeau still owns the trademark for the name "Butterfly."

In the famous Duncan vs. Royal Yo-Yo trial, not only was the word "yo-yo" an issue, but also the word "Butterfly." Although Duncan did loose trademark rights to the word "yo-yo," the company did retain the exclusive rights to the name "Butterfly." Before the trial, Royal did produce yo-yos with the Butterfly name. After the court case, it discontinued the use of the name "Butterfly." Companies producing butterfly shaped yo-yos use other descriptive names, for example: Hi-Ker uses "Flat Top," Festival "Dragonfly," and Royal's "Thunderbird."

After Flambeau took over Duncan in 1968, the original wooden butterfly line was dropped and replaced by plastic. The earliest Flambeau Butterfly was similar to its wooden predecessor by having a small centrally placed butterfly emblem on the seal. A few years later, the size of the butterfly emblem increased to cover the face of the yo-yo. Flambeau Butterflies were made out of opaque plastic until sometime in the '80s when translucent plastic was used. The opaque plastic butterfly yo-yos have faces a different color from the rim. The following are butterfly shaped yo-yos produced by Duncan.

　　　•　**Batman and Robin -** *See Duncan Super Heroes section #463.*

318　*11*　**BUTTERFLY, gold leaf stamped seal, wood, early '60s**
　　　　The letters of the word "Butterfly" form the butterfly emblem. This was a very limited production. It is considered one of the rarest of the Duncan Butterflies.

*Continued ...* ▶

## YO-YO LISTINGS

319    *12*    **BUTTERFLY (on triangle), imprint seal, plastic, Flambeau, late '90s**
Production of this model began in the late '90s. The yo-yo is retailed on blister card, Stock No. 3058NP. See Plate 12, #319.

320        **(Butterfly on triangle), imprint seal, plastic, Flambeau, late '90s**
This yo-yo is similar to #324, except the seal does not have the word "Butterfly." This yo-yo was reportedly made for foreign markets, but a few were accidentally placed on English cards and retailed in the states. The only other Butterfly made by Duncan that shows a butterfly without the lettered word "Butterfly" is the Gold Wings. See #326.

321    *12*    **(Delicate wing) BUTTERFLY, gold leaf imprint seal, plastic, Flambeau, '70s**
"Delicate Wing" refers to the thin veins in the butterfly wing. This yo-yo began production in 1977 and stopped in 1980. This yo-yo needs to be stored carefully, as the seal easily wears off the opaque plastic. It was retailed on a blister card with an original pre-printed price of $1.29, see Plate 12, #321A. More recent carded versions sold without a pre-printed price and had a bar code on the reverse (#321B).

322    *11*    **DUNCAN TOPS (with Mr. Yo-Yo logo), gold leaf stamped seal, wood, '60s**
This is a rare butterfly from the late '50s early '60s. Some versions of this model have been found with a Royal seal on the reverse face.

323    *11*    **EXPERT AWARD, silver leaf stamped seal, wood, '60s**
This yo-yo was released in the early '60s. It has an eagle on the seal rather than a butterfly. It had a much smaller production run than the standard wood Butterfly. In addition to being used as an award yo-yo, it was also sold retail. This is considered one of the most collectible of the butterfly shaped yo-yos. It was originally retailed with a pre-printed price of 69 cents, in a polybag with a saddle header card. See Plate 13, #323.

323.1    *11*    **EXPERT AWARD (fish scale finish), silver leaf stamped seal, wood, '60s**
This is the same seal as #323, but has fish scale finish. This is a paint finish made by spraying multi-colored nylon filaments into the paint. The fish scale finish is a very rare variation. Other Duncan yo-yos were made with a fish scale finish and are also considered rare. One can be seen in the National Yo-Yo Museum in Chico, California.

324    *12*    **(Fat wing) BUTTERFLY, imprint seal, plastic, Flambeau, '90s**
"Fat Wing" refers to the thick appearing wings on the butterfly logo that was used in the '80s and '90s. Production began in 1981, but stopped in the middle '90s. The yo-yo was retailed on a blister card with the original Stock No. 3058, then after 1986 with Stock No. 3058NP. Early models were opaque plastic. More recent models are translucent plastic. Three different cards are shown on Plate 12, #324A, #324B, and #324C.

325        **FLAT TOP RETURN TOP, gold leaf stamped seal, wood, late '50s**
This is one of the rarest of the Duncan butterfly shaped yo-yos. The seal is stamped "Flat Top" rather than "Butterfly." The yo-yo was manufactured for a foreign release in the late '50s or early '60s. Only one production run of the yo-yo was made. An example of this yo-yo can be seen in Chris Cook's book, "Collectible American Yo-Yo's."

326    *11*    **GOLD WINGS, molded seal, plastic, Flambeau, '70s**
Although this has a butterfly emblem hot stamped in the face, the word "Butterfly" is not used. Though it has not been documented, this butterfly style yo-yo may have been produced for foreign markets.

- **Happy Birthday** - *See Duncan Miscellaneous section #399.*

327    *11*    **(Large) BUTTERFLY, hot stamped seal, plastic, Flambeau, early '70s**
This yo-yo is somewhat of an enigma. It was reportedly not produced by Donald F. Duncan, Inc., but by Flambeau. If this is true, it is the only Flambeau yo-yo with the ® mark over the word "yo-yo" not behind the Duncan name. This model was retailed on at least two different blister cards. See an example Plate 12, #327. This yo-yo replaced design #332. The original price was $1.29.

328    *14*    **MARDI GRAS (quarter moon and glitter embedments), hot stamped seal, metal axle, early '60s**
This is one of the more rare Duncan plastic yo-yos. This model began production in the early '60s and stopped in 1965. This yo-yo is similar to the Disney Wonderful World of Color, but has different embedments. The quarter-moon embedments make this yo-yo unique. The seal is on a removable concave plastic insert that fits in the face of the yo-yo. This style of Mardi Gras is more difficult to find than the tournament shape. See #420.

329        **MARDI GRAS (plastic and glitter embedments), hot stamped seal, metal axle, early '60s**
This has the same design as #328, but different embedments. This variation is slightly less valuable than the quarter moon style, but is still more rare than the Mardi Gras #420.

- **Miniature Ad Butterfly** - *See Duncan Advertising section #307.36, #307.51, and #307.53.*

330    *11*    **(Psychedelic) BUTTERFLY, decal seal, plastic, Flambeau, early '70s**
The pop art style Butterfly, also referred to as the "pastel butterfly," was briefly produced in the early '70s. The last production of this yo-yo occurred around 1975. This is considered the most collectible of the Flambeau line of Duncan Butterfly yo-yos. It was retailed on a blister card with an original pre-printed price of $1.00, see Plate 12, #330. Later versions retailed for $1.29. The blister cards are marked Stock No. 3058.

331    *11*    **(Small) BUTTERFLY (glitter enamel), gold leaf stamped seal, wood, one-piece, '60s**
This model is the first standard line butterfly yo-yo made by Duncan. The "Small Butterfly" model was first produced in 1958 and production continued until 1965. These yo-yos were retailed both on display cards and loose out of counter boxes. The original price for this yo-yo was 69 cents. At least three different display cards were used with this yo-yo. Display cards were marked Stock No. 707, or Stock No. 707B. See Plate 12, #331. An example of this yo-yo is shown next to this section's title.

332    *11*    **(Small) BUTTERFLY, hot stamped seal, plastic, Flambeau, early '70s**
This is believed to be the first plastic butterfly produced by Flambeau. It was retailed on a blister card with an original pre-printed price of $1.00. For an example of the blister card, see Plate 12, #332.

*Continued ...* ▷

*The History and Values of Yo-Yos*

## YO-YO LISTINGS

**DUNCAN BUTTERFLY** *Continued ...*

333  *11*  **STRING PAK BUTTERFLY, embossed seal, plastic, Flambeau, '70s**
This yo-yo has a snap out face that opens to reveal a compartment to hold spare strings. This yo-yo was produced briefly in the middle '70s. The yo-yo was retailed on a blister card with a pre-printed price of $1.49.

- **Wonderful World of Color** - *See Disney section #303, #304, #305, and #306.*

## DUNCAN CATTLE BRAND

Duncan Flambeau produced the Cattle Brand series in 1977 and 1978 which featured famous cattle brands of the Old West. The mold was very similar to the one used to produce the Duncan Specials. The yo-yo has a hot stamped classic cattle brand on one face with the Duncan brand on the reverse. Cattle Brand yo-yos were simulated wood grain plastic. Two color variations of the wood grain were used, a dark walnut shade and a lighter oak style. Collectors collect both wood grain styles as well as all the different brands. These hard plastic yo-yos are fairly durable and can be found loose in fairly good condition. For a yo-yo to be considered in mint condition, it must have complete paint throughout the entire cattle brand. The paint on these yo-yos had a tendency to chip off or not stick in the brand. These yo-yos retailed on a display card with a pre-printed price of $1.29 and a bar code on the reverse of the card. (See Plate 13, #336.)

334  *16*  **(Bar S), hot stamped seal, tournament shape, late '70s**
This brand was used in southwest Texas. It is the most difficult of the cattle brand yo-yos to find.

335  *16*  **(Crossed W), hot stamped seal, tournament shape, late '70s**
This is one of the earliest Texas brands.

336  *16*  **DUNCAN (brand both faces), hot stamped seal, tournament shape, late '70s**
It is unclear whether this was a manufacturing error or intentional. This yo-yo is difficult to find on its original card. See Plate 13, #336 for an example of a Cattle Brand display card.

337  *16*  **MK, hot stamped seal, tournament shape, late '70s**
This is the brand of S.A. Maverick, founder of the republic of Texas.

338  *16*  **(Running W), hot stamped seal, tournament shape, late '70s**
This brand was used by the famous King Ranch.

339  *16*  **TC, hot stamped seal, tournament shape, late '70s**
This is the brand of Thomas O' Connor, a pioneer rancher.

340  *16*  **(Trunk handle), hot stamped seal, tournament shape, late '70s**
This brand is unusual in that it is from a ranch run by a woman, rare in 1838. It is slightly more difficult to find. An example of this yo-yo is shown next to this section's title.

341  *16*  **(T fork), hot stamped seal, tournament shape, late '70s**
This brand was used in the 1880s. It is slightly more difficult to find.

342  *16*  **XIT, hot stamped seal, tournament shape, late '70s**
This brand comes from what used to be the largest fenced cattle ranch.

## DUNCAN GOLD AWARD

This was a plastic line of tournament shaped yo-yos produced by Duncan Flambeau. Although Duncan frequently painted wooden yo-yos gold and called them Gold Award yo-yos, it was not until Flambeau took over that "Gold Award" appeared on the seal. The first Gold Award was one of the original five yo-yos produced in the first Duncan Flambeau line in 1969. The other four include the Imperial, Beginner, Trickster, and Special. Flambeau Gold Award yo-yos were used as contest prizes and were also sold retail. Flambeau discontinued its Gold Award line in 1995.

*Continued ...*

## YO-YO LISTINGS

343    14    **(Laurel wreath), imprint seal, tournament shape, glitter embedments, '80s and '90s**
This is the most recent version of the Gold Award. Multiple blister card variations exist. Earlier versions of the blister card are marked with a Stock No. 3266. More recent ones, after 1986, are marked Stock No. 3266NP. This is the last of the Gold Award line. It was discontinued in 1995.

344    14    **(Loving cup), hot stamped seal, plastic, tournament shape, early '70s**
This is the earliest and rarest of the plastic Gold Awards. In the early '70s, these butterscotch colored "gold" yo-yos were used as prizes in Duncan Flambeau contest kits. Some collectors call these "Butterscotch Loving Cups." An example of this yo-yo is shown next to this section's title.

344.1    14    **(Loving cup • glitter embedments), hot stamped seal, plastic, tournament shape, '70s**
At least three different versions of display cards for this yo-yo are known. Earlier cards had a pre-printed price of $1.00. More recent cards were marked $1.29. All blister cards were marked Stock No. 3266. Some collectors call these "Glitter Loving Cups." The value given in Appendix II of this book is for a yo-yo in mint condition on the $1.00 display card.

## DUNCAN IMPERIAL

Imperial Yo-Yos have been produced since 1954. The original Imperial has a chevron in front of the word "Imperial" and "Tenite" written beneath the chevron. Tenite refers to the type of translucent plastic used to make the yo-yo. Collectors call this style of yo-yo the "Chevron Imperial." In 1962, Duncan changed the chevron logo to the Fleur-de-lis emblem. Collectors call these " Fleur-de-lis Imperials."

All Imperials were made out of plastic and were tournament shaped. When introduced in the '50s, these plastic Imperials had a higher retail price than their wooden counterparts. Although this is not the first plastic yo-yo to be produced by Duncan, it is the most successful of the plastic yo-yo lines. The success of the plastic Imperial line was one of the reasons Flambeau elected to buy the Duncan trademark in 1968. Duncan had used Flambeau for the manufacturing of its plastic lines prior to its bankruptcy.

Some collectors refer to all yo-yos with a tournament shape as "Imperials," but Imperial should be used only to describe yo-yos bearing the word "Imperial" on the logo. Collectors should be aware that the Imperial Toy Company also produces yo-yos with the word "Imperial" on the seal. These are unrelated to Duncan Imperials.

345    11    **(Chevron • gold metallic finish), hot stamped seal, late '50s**
This is the rarest of the Duncan Imperial yo-yos. This model does not have the word "Tenite" on the logo. It was used as an award yo-yo, but was also sold retail. There may have been a metallic silver version as well, but this has not been confirmed.

346        **(Chevron • marbleized plastic), hot stamped seal, late '50s to early '60s**
The word "Tenite" does not appear on this logo.

347    11    **(Chevron) TENITE, hot stamped seal, '50s to early '60s**
This is the original standard line Duncan Imperial logo. Imperials were the first successful plastic yo-yo line produced by Duncan. This hot stamped seal was used as early as 1954 and continued until 1962. A variety of translucent colors were used. Under the chevron, the word "Tenite" appears which describes the type of plastic used in the manufacture of this model. An example of this yo-yo is shown next to this section's title.

348    11    **(Chevron) TENITE (glitter finish), hot stamped seal, '50s**
This is a rare glitter variation. Glitter is not embedded in the plastic, but on the surface as a finish. This model is difficult to find in good condition because the glitter finish tends to come off. The yo-yo has the same seal as #347.

349    11    **(Fleur-de-lis) YO ® YO, hot stamped seal, early '60s**
This Fleur-de-lis logo replaced the chevron logo in 1962. The registration mark ® appears below the word "yo-yo." This identifies this piece as an original "Golden Age" Duncan, not Duncan Flambeau. Duncan lost the trademark status of the word "yo-yo" before Flambeau purchased the company. None of the Flambeau yo-yos have a registration mark with the word "yo-yo." This model continued production until 1965. The yo-yo was retailed on at least two different display cards with an original price of $1.00, Stock No. 400. See Plate 13, #349.

350    11    **(Fleur-de-lis) YO ® YO (marbleized plastic), hot stamped seal, early '60s**
This marbleized version is the most difficult to find of the Fleur-de-lis Imperials. This yo-yo, like #349, carries a registration mark ® below the word "yo-yo" that identifies this as pre-Flambeau. See Plate 13, #350 for display card.

*Continued ...*

## YO-YO LISTINGS

● **DUNCAN IMPERIAL** *Continued ...*

351    *11*    **(Fleur-de-lis), hot stamped seal, Flambeau, '70s**
This was the only yo-yo model that continued to be produced following Flambeau's purchase of Duncan's trademark in 1968. This yo-yo was made from 1968 thorough 1976. It can be distinguished from the original "Golden Age Duncan" Fleur-de-lis by the lack of the ® mark below the word "yo-yo." It can also be distinguished from the recent "Fleur-de-lis" versions by the indentations left by the hot stamped seal. Several different blister cards exist, Stock No. 3269. Earlier cards are marked with a pre-printed price of $1.00. More recent cards show a price of $1.29.

352    *11*    **(Fleur-de-lis) MADE IN USA, imprint seal, Flambeau, '90s**
This is the Imperial currently produced. The imprint seal was introduced in 1976 and the word "yo-yo" was replaced by "Made in USA." The seal has been the same for nearly 20 years, except for the late '80s when some were made in Mexico. See #352.2. These do not carry the "Made in USA" mark. Originally the blister cards are marked with Stock No. 3269, then after 1986 with Stock No. 3269NP.

352.1      **(Fleur-de-lis), USA MADE • WORLD'S #1, imprint seal, Flambeau, late '90s**
This Fleur-de-Lis variation made its first appearance in 1997. Only a few of these were included in the shipments which featured #352. These are retailed on the same card as #352.

352.2    *11*    **(Fleur-de-lis), imprint seal, Duscan, '90s**
Duncan's sister company, Duscan, produced these yo-yos in Mexico. The seal is the same as #352, but missing "Made in USA" on the logo. In the late '80s, some were retailed in the United States. These have a hard bubble blister card in English. In 1994, these were also retailed in the United States on Spanish blister cards, apparently in violation of Duncan's agreement with its Mexican sister company.

•    **Glow Imperials** - *See Duncan Light and Glow section.*

352.3      **HYPER IMPERIAL, imprint seal, Flambeau, late '90s**
This is a Duncan series marketed in Japan by the Bandai Company, famed for its creation of the Power Rangers. Some of these yo-yos were retailed in the United States, but only in limited quantities. In Japan the hyper-series yo-yo seals are in English, but the packaging is in Japanese.

•    **Imperial (Franklin Mint) Reproduction** - *See Franklin Mint section.*

353      **RETURN TOP, hot stamped seal, early '60s**
This is a rare "Imperial" yo-yo. It is believed to have been produced during a brief transition period between the Chevron Imperial and the Fleur-de-lis Imperial yo-yos.

## DUNCAN IMPERIAL JUNIOR

This is a distinctive line of yo-yos with a dome shaped view lens on both faces. The yo-yo was listed as a new product in Duncan's 1962 catalog. It was marketed to younger players not ready for the full sized Imperial Yo-Yo. Artwork features the '60s style Mr. Yo-Yo in five different scenes wearing a three point crown. One scene is currently unidentified; the four known styles are listed below. The yo-yos have fixed strings, and although most had metal axles, there were some with wooden axles. Imperial Jr. yo-yos were retailed on display cards with a pre-printed price of 49 cents. Cards were marked Stock No. 450. Although a plastic model, the Imperial Jr. was not continued by Flambeau.

354    *17*    **(Mr. Yo-Yo with flowers), paper insert seal, plastic, metal string slot, tournament shape, view lens, early '60s**
355    *17*    **(Mr. Yo-Yo with bees), paper insert seal, plastic, metal string slot, tournament shape, view lens, early '60s**
356    *17*    **(Mr. Yo-Yo with birds), paper insert seal, metal string slot, tournament shape, view lens, early '60s**
357      **(Mr. Yo-Yo fishing), paper insert seal, plastic, metal string slot, tournament shape, view lens, early '60s**
An example of this yo-yo is shown next to this section's title.

# YO-YO LISTINGS

## DUNCAN JEWELS

Every child who played with a yo-yo in the '40s, '50s, and '60s remembers their jewel yo-yo. This was the yo-yo most prized by its owner. Production of Duncan Jewel yo-yos began in the '30s prior to WWII. Five rhinestones were used on both faces; after WWII, four rhinestones became standard. Some of the earlier models had slightly larger rhinestones (6mm) than the 1950s style (4mm). Standard line yo-yos were turned into jewel yo-yos before the name "Jeweled" appeared in 1953. The last wood Duncan Jewels were made in 1965.

When making jewel yo-yos, Duncan drilled the holes before the yo-yo was painted. The rhinestones were inserted by hand after the paint was applied. By closely inspecting around the edges of the holes, paint can usually be seen inside the drilled holes. Some exceptions to this rule may exist, especially with the oversized special award jewel yo-yos that were individually made. Other companies that produced jewel line yo-yos did not necessarily follow this rule, such as Goody, where the yo-yos were painted prior to the drilling of the jewel holes.

Flambeau reintroduced the Duncan Jewel line in the '70s as a plastic model. The multi-jewel style was dropped in place of a single 8mm-centered jewel. The large central rhinestone did have a tendency to detach. Since this was a small part hazard, this may have had some influence in the subsequent dropping of jewel type yo-yos. In 1990, Flambeau dropped the Jewel Yo-Yo in favor of the Neo line.

- **3 Jewel Luck-E JA-DO** - *See Duncan Miscellaneous Wood section #445.*

**358**    *15*    **(4 jewel • crossed flag), gold leaf stamped seal, wood, tournament shape, one-piece, early '60s**
The last of the wood jewel yo-yos produced by Duncan. Paint should be visible inside the drilled holes where the rhinestones are inserted.

**359**    *15*    **(4 jewel little "G") GENUINE, gold leaf stamped seal, wood, tournament shape, one-piece, early '50s**
This was a standard line tournament #499 made into a jewel model prior to the introduction of the Jeweled series.

**360**    *15*    **(4 jewel) JEWELED, gold leaf stamped seal, wood, tournament shape, one-piece, '50s and '60s**
This classic Duncan Jeweled Tournament yo-yo was produced from 1953 through the '60s. It is the best known and easiest to find of the wood Duncan Jewel models. Along with the Litening, this was one of the first two yo-yos retailed on a blister card by Duncan in 1958. An example of this yo-yo is shown next to this section's title.

**360.1**    *15*    **(4 jewel) JEWELED (fish scale finish), gold leaf stamped seal, wood, tournament, one-piece, '50s**
This is the same as the standard line 4 Jeweled #360, but has a unique fish scale finish. This is a rare paint finish made by spraying multi-colored nylon filaments into the paint. Fish scale finished yo-yos are very rare.

**360.2**    *15*    **(4 jewel) JEWELED (gold finish), die stamped seal, wood, tournament shape, one-piece, '50s and '60s**
The gold finished Jeweled is the most desirable of the standard Jeweled line, #360. Paint color does make a difference with this gold painted Jeweled. This was an award yo-yo given out to contest winners. Although contest promotions advertised that the gold Jeweled was not available in stores, some were retailed.

**360.3**      **(4 jewel) JEWELED (pearlescence paint), gold leaf stamped seal, wood, tournament shape, one-piece, '50s**
This is the same yo-yo as #360, but with pearlescence paint.

**361**    *15*    **(4 jewel) JEWELED (Mr. Yo-Yo figure), gold leaf stamped seal, wood, tournament shape, '50s**
This rare yo-yo was briefly produced in the '50s. There are only two Duncan woods that read "Jeweled" on the seal, this one and #360. This version is much more rare than #360.

**362**    *15*    **(4 jewel) SUPER, gold leaf stamped seal, wood, tournament shape, one-piece, early '50s**
This was a standard yo-yo line made into jewel yo-yos prior to the production of the Jeweled line. Prior to 1956, many were made with metallic pearlescence paint.

**363**    *15*    **(5 jewel big "G") GENUINE, gold leaf stamped seal, wood, tournament shape, one-piece, '30s**
Big "G" (Genuine) refers to the large "G" in the word "Genuine" that extends to the base of the seal. This is a pre-WWII jewel yo-yo. The first Duncan jewel model was the five jewel. Rhinestones for this model were larger than those for the four jewel model. The last of the five jewel Duncans were made in the late '40s. The change was a cost cutting measure. It took less time to drill four holes and was less expensive to use smaller and fewer rhinestones.

**364**      **(9 jewel big "G") GENUINE, decal seal, wood, tournament shape, 1932**
This rare Gold Seal Yo-Yo may have been used just for publicity shots. One is shown in a contest publicity shot from 1932 with Filipino demonstrator Frank Funtanila. It is unclear whether this model was ever retailed.

**365**    *15*    **(9 jewel • big "G") GENUINE, die stamped seal, wood, tournament shape, jumbo, '30s**

*Continued ...* ▶

## YO-YO LISTINGS

● **DUNCAN JEWELS** *Continued ...*

Big "G" (Genuine) refers to the large "G" in the word "Genuine" that extends to the base of the seal. This is a rare special presentation yo-yo given to dignitaries attending yo-yo promotions. This yo-yo is nearly 5" in diameter, compared to the 4-1/4" diameter of standard jumbos. Unlike most other Duncan jumbos, this has an airbrushed stripe. Oversized yo-yos photographed well. Both Duncan and elected officials enjoyed the benefits of free publicity that came with presentation ceremonies. It was always easier to get the support for a city wide promotion from the local government if the mayor or other officials got a little publicity along the way. These yo-yos were not retailed.

366    *15*    **(Cat eye • jewel), imprint seal, plastic, Flambeau, tournament shape, metal axle, '80s**
Flambeau changed the Jewel line in the '80s by adding animals to the seal. The Jewel colors remained the same as #369: ruby, emerald, sapphire, topaz and amethyst. Yo-Yos had two variations. Some had the logo and jewel on one face and others had the logo and jewels on both faces. The cat model seems to be slightly more desired than the snake (#367) or eagle (#368). The black cat version, on Plate 15 #366B, may be the most difficult of this series to acquire. This yo-yo retailed on a blister card, Stock No. 3264NP.

367    *15*    **(Snake • jewel), imprint seal, plastic, Flambeau, tournament shape, metal axle, '80s**
For discussion, see #366.

368    *15*    **(Eagle eye • jewel), imprint seal, plastic, Flambeau, tournament shape, metal axle, '80s**
For discussion, see #366.

369    *15*    **TM JEWEL, imprint seal, plastic, Flambeau, tournament shape, metal axle, '70s and '80s**
Flambeau re-introduced the Jewel line in the '70s with this plastic model. This model has the trademark, TM, after the word "Jewel." The style was changed from four small (4mm) jewels, as found in wooden jewel yo-yos, to one large jewel (8mm) centered on the face of the yo-yo. With this model, the jewel colors do make a difference. The ruby color is the most desirable, followed by emerald, sapphire, topaz, and finally amethyst. These yo-yos were all retailed on blister cards. See Plate 9, #369.

 **DUNCAN JUMBO**

Duncan made oversized wooden yo-yos from the early '30s to the '60s. The most common size for jumbo yo-yos was 4-1/2" in diameter. These yo-yos were not retailed, with the exception of the "Executive Yo-Yo." Yo-Yos were made for awards or for publicity photographs. Duncan learned early on that publicity was everything when it came to promoting contests. Standard sized yo-yos showed up poorly in publicity photographs, so oversized presentation yo-yos were used. These yo-yos were also frequently given as gifts to local officials. Later award yo-yos, from the '50s and '60s, did not have a decorative stripe, but early jumbos from the '30s had a paint stripe across the face.

- **9 Jewel Big "G" (Genuine)** - *See Duncan Jewels section.*
- **Champion (eagle seal)** - *See Duncan Miscellaneous Wood section.*

370    *14*    **EAGLE 999, foil sticker seal, wood, tournament shape, late '50s**
This yo-yo is more rare than the Jumbo Award with the 77 seal. See #372. This, like #372, was an award yo-yo and not retailed. It is believed that this Eagle 999 seal was only used on jumbo sized yo-yos.

371        **EXECUTIVE, brass seal, wood, tournament shape, '50s**
This was retailed in the same executive gift box as #374.

372        **(Jumbo award) 77, decal seal, wood, tournament shape, '50s**
These oversized yo-yos were used as awards in contests. They were not sold retail and, therefore, are very desirable to collectors. Most of them were produced in the '50s. Some may be found with scenes or names carved by demonstrators. Jumbo Award yo-yos with airbrushed stripes are considered rare variations.

372.1        **(Jumbo award) 77 (gold finish), decal seal, wood, tournament shape, '50s**
This is the same yo-yo as #373, but with a metallic gold paint finish. Gold painted Jumbo Award yo-yos are considered more collectible than other finishes. An example of this yo-yo is shown next to this section's title.

373    *14*    **(Jumbo award) 77 (metallic pearlescence), decal seal, wood, tournament shape, '50s**
This yo-yo is the same as the Jumbo Award #372, but has an impressive metallic paint finish. These are considered more desirable than the standard flat paint models.

374    *46*    **VIP, metal seal, wood, tournament shape, '50s**
This yo-yo was retailed in the Duncan executive gift box. It does not have Duncan on the seal, rather "VIP" in silver letters. This yo-yo is more rare than other versions of the executive. This yo-yo could be special ordered with initials in silver similar to the "VIP."

*Continued ...* ▶

# YO-YO LISTINGS

## DUNCAN JUNIOR

Like Duncan Beginner yo-yos, Juniors were designed for younger, inexperienced players. These yo-yos all had fixed strings allowing for easy return. The original Juniors were small versions of the Beginners models. They were retailed with a stock number of 33. (See also the Imperial Juniors #354, #355, #356, and #357.)

| | | |
|---|---|---|
| 375 | 16 | **O-BOY JR., die stamped seal, wood, tournament shape, '30s** |
| | | This is believed to be the first Duncan Junior Yo-Yo and the only wood Duncan with Junior abbreviated "Jr." This model was produced from the late '20s to early '30s and is the rarest of the Junior line. The original price for the O-Boy Jr. was 5 cents. |
| 376 | 16 | **(Big "G") GENUINE, gold leaf stamped seal, wood, tournament shape, miniature, pegged string, '30s - '50s** |
| | | Big "G" (Genuine) refers to the large "G" in the word "Genuine" that extends to the base of the seal. Big "G" Juniors replaced the O-Boy Jr. #375. Production began in the '30s and continued into the early '50s. These were typically red on one half, black on the other. The same seal was used on a standard sized model as well. |
| 377 | 16 | **(Italic) JUNIOR, gold leaf stamped seal, wood, tournament shape, miniature, pegged string, '60s** |
| | | The word "Junior" is done in italic print. This yo-yo was also used in a Crest toothpaste promotion in the '60s. The yo-yo was given out on a blister card which said "Free Genuine Duncan Yo-Yo when you buy Crest extra large size." The card value is listed for the Crest blister card. Collectors call this yo-yo the "Italic Junior." An example of this yo-yo is shown next to this section's title. |
| 379 | 16 | **(Midget) JUNIOR RETURN TOP, embossed seal, plastic, midget size, '50s** |
| | | This yo-yo is a two-piece snap together Post Cereal premium. The yo-yo came unassembled in a small paper packet with assembly instructions printed on the outside. Although sometimes reported as a Cracker Jack prize, this is not documented and is unlikely. This is one of only two midget sized yo-yos produced with a Duncan logo. The other is a wood version. See #440. |
| 380 | 16 | **(Yo-Yo Man • Mr. Yo-Yo), imprint seal, plastic, Flambeau, puck shape, fixed string, '70s** |
| | | This yo-yo has Patent No. D231707 embossed on both faces. It is the only Junior produced by Flambeau and the last of the Junior series. The seal is identical to #316, except the word "Beginner" was replaced with the word "Junior." |

## DUNCAN LIGHT-UP AND GLOW

All the yo-yos in this section are Duncan Flambeau creations with the exception of the 1950 Big "G" Electric Lighted. This model was produced before Flambeau bought the Duncan trademark in 1968. Flambeau began production of its own Light-Up Yo-Yo, the Satellite, in 1972.

Prior to Flambeau, the Duncan Company never produced a glow yo-yo. Glow plastic concentrate became available in the late '30s, but it was not until the early '50s that glow plastic toys were produced. Glow plastic toys were a fad in the late '60s and Flambeau jumped on the bandwagon producing a Glow Imperial Yo-Yo. The Glow Imperial has been a consistent performer in sales and continues to be produced nearly 30 years later. Over the years, changes in the Glow's seal have occurred, making this a collectible series. All Glow Imperials are tournament shaped and retail on a variety of blister cards. Early blister cards were marked with Stock No. 3257, after 1986, cards have Stock No. 3257NP.

The original Glow had florescent orange paint on the seal and a "Sunrise Design." This design was abruptly discontinued and changed to the more familiar 8 Ray "Distant Star" design. Florescent orange paint was the standard color until 1972 when it was changed to red. The same 8 Ray Distant Star logo was still used after the color change. In 1976, the hot stamped seal process changed to an imprint seal which did not leave impressions in the plastic. The print color remained red, but the "Distant Star" design changed to a 10 Ray Star. This 10 Ray "Distant Star" logo has remained unchanged since 1976. In the middle '90s, the paint color changed from a flat red to a metallic red.

Duncan also produced a Hyper Glow Yo-Yo for the Japanese market in 1997. Only a few Hyper Glows have been retailed in the United States.

*Continued ...*

*The History and Values of Yo-Yos*

● DUNCAN LIGHT-UP AND GLOW *Continued ...*

381   *11*   **(Big "G") GENUINE ELECTRIC LIGHTED, molded seal, plastic, tournament shape, light-up, 1950**
Big "G" refers to the large "G" in the word "Genuine." Originally advertised by Duncan as the "Electric Yo-Yo," it was retailed out of its own multi-colored counter display box which held one dozen yo-yos. The display box carries a copyright date of 1950. The "Electric Yo-Yo" was the first battery operated Duncan Light-Up produced. Production started in 1950. This is believed to be the only year the yo-yo was retailed. It is one of the earliest plastic yo-yos produced by Duncan and one of the earliest Light-Up yo-yos produced by any company. This is considered the most collectible of all Light-Up yo-yos. Collectors call the yo-yo the "Duncan Electric."

382   *11*   **DUNCAN (satellite), embossed seal, plastic, Flambeau, butterfly shape, light-up, '70s - '90s**
This AA battery operated Duncan Light-up yo-yo has been produced from the same mold since 1972. The word "Satellite" is on the blister card, not the yo-yo. This yo-yo should not be confused with Duncan's wood satellite series from the '60s. A loose yo-yo has little collectible value because there is no way to tell an early model from one recently produced. This yo-yo should be collected in the original packaging, since this is the only way to date earlier yo-yos. The yo-yos with packaging marked $2.00, and the one in the clear art view box marked $3.00, are the most desirable of the packaged Satellites. Early to more recent cards are marked 3268, 3268AA, 3268NP. See Plate 9, #382. In 1994, when stores began retailing Hong Kong produced Light-Up yo-yos at a price lower than the cost of manufacturing the Satellite, Flambeau dropped the line. Collectors call the yo-yo the "Duncan Satellite Light-Up."

383   *11*   **GLOW (sunrise design), orange hot stamped seal, glow plastic, Flambeau, tournament shape, late '60s**
The first Duncan Flambeau Glow design was introduced around 1969. This seal was quickly discontinued and replaced by the 8 Ray Distant Star pattern. This yo-yo is the most difficult to find of the Glow Imperials. For an example of the blister card, see Plate 13, #383.

384   *11*   **GLOW (8 ray distant star), orange hot stamped seal, glow plastic, Flambeau, tournament shape, early '70s**
The original Distant Star pattern had orange paint which was used until 1972. This yo-yo is easier to find than the "sunrise design" #383.

385   *11*   **GLOW (8 ray distant star), red hot stamped seal, glow plastic, Flambeau, tournament shape, '70s**
Orange paint was switched to red paint in 1972. The example shown is an original factory prototype used in the testing of the new red paint. Hot stamped seals continued until 1976. At least two different blister cards are known to exist with the original price of $1.29, Stock No. 3057.

386   *11*   **GLOW (10 ray star), imprint seal, glow plastic, Flambeau, tournament shape, '70s - '90s**
This style of Glow yo-yo has been produced since 1976. The 10 Ray models were never hot stamped, so impressions will not be felt. This model replaced #385. These yo-yos were retailed on blister cards with the original Stock No. 3057, then after 1986 with Stock No. 3057NP.

387       **GLOW (10 ray distant star), metallic red imprint seal, glow plastic, Flambeau, tournament shape, mid '90s**
In the mid-'90s, the paint for this model was changed from flat red to metallic red. More than one blister card exists, Stock No. 3057NP.

388       **HYPER GLOW (10 ray distant star), imprint seal, glow plastic, Flambeau, tournament shape, late '90s**
This model was first produced in 1997 for the Japanese market, however, a small quantity were released in the United States on English cards. An example of this yo-yo is shown next to this section's title.

•   **Rice Krispies** - *See Duncan Advertising section #307.46.*

## YO-YO LISTINGS

### DUNCAN MISCELLANEOUS ITEMS

For other Duncan related items, see the following sections: Awards, Boxes, Patches, Pins, Posters, String Packs, Trick Books, and Videos.

| | | |
|---|---|---|
| 392 | | **(Bumper sticker) NO STRINGS ATTACHED • YO-YO PLAYERS INTERNATIONAL, 1980** |
| | | This bumper sticker was sold through Duncan's Yo-Yo Players International Newsletter in 1980. |
| 393 | *46* | **(Cap) DUNCAN, 1980** |
| | | This Cap was sold through Duncan's Yo-Yo Players International Newsletter in 1980. |
| 394 | | **COLLECTORS CHEST, early '60s** |
| | | This set included four yo-yos: the Satellite, Mardi Gras, Disney Wonderful World of Color, and Chevron Imperial. In addition to the yo-yos, the set included one string pack and a 25 cent Duncan trick book. The original retail price of the collectors set was $4.00. |
| 395 | *47* | **(Contest announcement fliers), '50s** |
| | | These 8.5 x 11 sheets were posted or mailed before contests and announced times, locations, and awards. The fliers shown are from the middle '50s. |
| 396 | *52* | **(Contest announcement fliers), '30s** |
| | | These one sheet contest fliers from the '30s are rare. |
| 397 | *47* | **(Contest ads and promotions • newspaper)** |
| | | Original clippings from newspapers announcing contests and promotions have varying values depending upon their age, size, condition, and graphics. |
| 398 | | **(Duncan contest kit), early '60s** |
| | | This contest kit contained the following items: two sets of 1st through 3rd award patches, trick sheets, two contest posters, the rule sheet, three yo-yos, and four string packs. These kits were made for marketing contests with recreational and educational programs. |
| 398.1 | | **(Duncan contest kit), '70s** |
| | | This kit contained one eagle champion patch, four first, second, and third place patches, four butterscotch gold award yo-yos, a poster, trick sheets, and extra strings. This contest kit was supplied to recreational directors for free in the early '70s. This kit was shown in the Duncan Flambeau film, "How to Run a Yo-Yo Contest," featuring Duncan yo-yo demonstrator, Barney Akers. |
| 398.2 | | **(Duncan contest kit) YO-YOLYMPICS, 1979** |
| | | This was the competition kit used by Duncan Yo-Yolympics contest directors. The kit contains four 1979 champion yo-yos that were awarded to contest winners, two posters, trick sheets, spare strings, and a sheet of competitor stickers. |
| 399 | *14* | **HAPPY BIRTHDAY (party favors), '60s** |
| | | The Happy Birthday yo-yo was first introduced by Duncan in 1963. These were retailed as a blister card party pack of six, wood, miniature, butterfly shaped yo-yos. The Happy Birthday yo-yo has a gold leaf stamped seal with a beginners logo on the reverse face. The yo-yo has a pegged string and was made in the early '60s. The display card has an original pre-printed price of $1.00 and is marked Stock No. 1708. |
| 400 | *44* | **IMPERIAL (yo-yo display stand with six imperial chevron tenite yo-yos), '50s** |
| | | This is a rare store display. In the '50s, the "new plastic" yo-yos had a higher price than the wooden tournament models. The value listed is for a display complete with all yo-yos. |
| 401 | *9* | **(Key chain), late '90s** |
| | | This key chain has a miniature plastic Duncan yo-yo with a Fleur-de-lis imperial seal. The seal reads, "World's #1." The key chain is retailed on a blister card and in a polybag with a saddle header card. |
| 402 | | **(Personalized kit)** |
| | | This kit came with a wooden butterfly yo-yo, trick book, and sheet of stick-on gold foil letters for the buyer to personalize their yo-yo. The original price for this kit was $1.00. |
| 403 | | **(Post cards)** |

*During the 1950s, these Duncan Yo-Yo Club ID cards were distributed in the Chicago area.*

Continued ...

## YO-YO LISTINGS

● **DUNCAN MISCELLANEOUS ITEMS** *Continued ...*

404	(Transfer • iron-on) Duncan, late '70s
Several different Duncan T-shirt iron-on transfers exist. They all read "Yo-Yo Champion" with the Duncan logo centered. The number of stars, one star, three stars or five stars, identifies these transfers. Value is given for unused transfers.

405	(Trick award cards), 1997
This is a series of ten trick cards given as awards to players upon successful demonstration of the tricks. The card front is in English and features Alex Garcia, a Team High Performance player. On the other side, the tricks are described in Japanese.

There are 10 trick levels, level one being the most difficult and level ten the easiest. These cards were very popular in Japan. Value is given for a complete set of ten cards. An example of these trick cards is shown next to this section's title. For other trick trading cards, see #1738.

406	*45*	(Trick kit), early '60s
This kit contains a Little Ace Yo-Yo, a wooden Duncan Tornado Top, a Duncan Hand Ball, and a 1962 Trick Book. For other play kits, see Campbell Kid Play-Kit in Duncan Advertising section #307.12.

407	*17*	VOTE FOR DUNCAN YO-YO, paper sticker seal, early '60s
This sticker was released by Duncan in the early '60s. It is unlikely that it was intended for placement on yo-yos, but this did occur. A yo-yo with this sticker seal is displayed in the National Yo-Yo Museum, but it is attached to a plastic non-Duncan yo-yo of recent make. There is no evidence that a yo-yo with this seal was ever retailed. The value listed in this book is for an unused sticker.

408	(Window decal) YO-YO PLAYERS INTERNATIONAL 1980
This decal was sold through Duncan's Yo-Yo Players International Newsletter in 1980.

409	(Yo-Yo pencil), '60s
This yo-yo is similar to the Hasbro Yo-Yo Pencil. A plastic yo-yo fits over the eraser end of a pencil. This yo-yo is midget sized without any markings or seals and can be separated from its pencil holder. The pencil is stamped with "Duncan Yo-Yo Pencil" in gold leaf.

## DUNCAN (MISCELLANEOUS PLASTIC)

The plastic yo-yos listed in this category do not fit in other Duncan sections.

410	40'S PHYSICAL FITNESS DEVICE, paper insert seal, Flambeau, slimline shape, view lens, '80s
This is a gift yo-yo that has a copyright of 1987. There were several different slogans in this series. The yo-yo was retailed in a polybag with a small trick book.

411	*9*	ALIEN, paper insert seal, glow plastic, Flambeau, slimline shape, view lens, late '90s
This is the second slimline style plastic yo-yo made out of glow plastic by Flambeau. It was introduced in 1998.

412	*17*	COLORAMA, paper insert seal, tournament shape, view lens, early '60s
This plastic, dome shaped view lens yo-yo was briefly produced in the early '60s. The yo-yo has the word "Colorama" molded into the plastic view lens and a metal string slot axle. The style is the same as the Imperial Jr's, also produced during the early '60s. The paper insert has a multi-colored pattern. This is considered one of the more valuable of the plastic Duncan yo-yos.

•	Dinosaurs - *See Foreign Section #596.*

413	DUNCAN YO-YO RETURN TOP, hot stamp seal, coaster shape, multi-colored star embedments, '60s
This is the same style yo-yo as the Disney Wonderful World of Color #304. The seal is a gold leaf hot stamp on a removable concave plastic insert. The reverse face has a blank concave plastic insert. This was retailed in a polybag with a Duncan Imperial display card. The original price was $1.00. The card is marked Stock No. 1400.

414	*17*	(Eagle) GENUINE DUNCAN'S CHAMPION YO-YO, paper insert seal, slimline shape, view lens, '50s
In the early '50s, Duncan purchased the plastic mold used for the Roy Rogers and Trigger Yo-Yo #184. These yo-yos were planned for distribution in foreign markets through Duncan's D.R.I. (Duncan, Russell, Ives) division. This foreign division dissolved in 1958, after the death of Tom Ives. The mold was acquired by Jack Russell and used in many of his early promotions. This model may have also been released in the United States. The paper insert seal has an illustration of a standing eagle and the Duncan Company referred to this yo-yo's mold as the "Eagle Mold." The edge of the paper insert seal can be seen on the Eagle models. This distinguishes it from the more recent Flambeau plastic slimline models. The Flambeau slimline models have insert seals that fill the entire view lens.

*Continued ...* ▶

# YO-YO LISTINGS

| | | |
|---|---|---|
| 415 | 14 | **(Galaxy), diffraction insert seal, Flambeau, slimline shape, view lens, late '90s** |

This model was introduced in 1994 and has the hairy Mr. Yo-Yo on the seal. It does not read "Duncan" or "Galaxy" on the yo-yo, only on the blister card, Stock No. 3272GH. See Plate 9, #415.

| | | |
|---|---|---|
| 415.1 | | **Glo-Yo, hot stamped seal, tournament shape, late '60s** |

Although this yo-yo was named Glo-Yo, it was not made of glow plastic. This Glo-Yo was made out of a standard opaque plastic. It is believed to have been produced for only one year, 1969, and retailed on a display card with the pre-printed price of $1.00.

| | | |
|---|---|---|
| 416 | 17 | **HOOT MON', hot stamped seal, tournament shape, metal string slot axle, '60s** |

This is the Duncan Canadian version of the Little Ace. It was retailed on a display card with an original price of 79 cents, Stock No. 1098. It does have a ® mark after the word "Duncan." During the Duncan "Golden Age," pre-Flambeau, only foreign yo-yos made by Duncan had a ® mark with the word "Duncan."

| | | |
|---|---|---|
| 417 | 24 | **HYPERPRO 1997, paper insert seal, slimline shape, view lens, 1997** |

Collectors call this yo-yo the "Flaming Skull." It was made for distribution in Japan, but a small number were retailed in the United States.

| | | |
|---|---|---|
| 418 | 17 | **LITTLE ACE, hot stamped seal, tournament shape, metal string slot axle, '60s** |

This yo-yo was produced from 1962 through 1965. Two different seals exist, this style and one used with a Mobile ad promotion. See #418.1. This style was included in the Duncan Play-Kit which came with a top, paddleball, and trick book. See Plate 45, #406.

| | | |
|---|---|---|
| 418.1 | 17 | **LITTLE ACE, hot stamped seal, tournament shape, metal string slot axle, '60s** |

This logo was used on the reverse face of the Mobil ad yo-yo. See #307.35.

| | | |
|---|---|---|
| 419 | 14 | **LIL' CHAMP, embossed seal, Flambeau, coaster shape, '80s** |

This yo-yo has an embossed rosy cheek Mr. Yo-Yo on the reverse face and was dropped from Flambeau's product line in 1990.

• **Mardi Gras (butterfly)** - *See Duncan Butterfly section #328.*

| | | |
|---|---|---|
| 420 | 14 | **MARDI GRAS, hot stamped seal, tournament shape, metal axle, early '60s** |

Mardi Gras yo-yos were produced from 1961 through 1964. To attract the attention of girls, Duncan created the Mardi Gras. The ad literature said, "The Mardi Gras was especially designed to capture the hearts of young women all across America." These multi-colored plastic yo-yos were made by fusing various colors of tenite plastic and glitter. The yo-yos were described as 3-D confetti colored plastic yo-yos. They are considered one of the more collectible of plastic Duncans. Twelve different patterns were produced with two glitter types: small glitter style #420 and long glitter style #421. The long glitter style is less common. The 9 different colored patterns of the small glitter styles are: red, white, and blue; yellow and dark blue; black and white; red and white; black, red and white; green, red and white; white in clear translucent; white in green translucent; and white in hot pink translucent. Display cards came with a pre-printed price of $1.00 on card, Stock No. 700. See Plate 9, #420.

| | | |
|---|---|---|
| 421 | 14 | **MARDI GRAS (long glitter style), hot stamped seal, tournament shape, metal axle, early '60s** |

This yo-yo is less common than the small glitter style #420. Collectors sometimes call this the "Long Glitter Mardi Gras."

| | | |
|---|---|---|
| 422 | 11 | **MEL•YO•DEE, embossed seal, Flambeau, butterfly shape, early '90s** |

Production of the Mel•Yo•Dee Yo-Yo began in 1986. The Mel•Yo•Dee Yo-Yo plays the Duncan jingle when spun. Like the Duncan Flambeau "Satellite" Light-Up, it requires a AA battery on each side. The Mel•Yo•Dee was retailed on a blister card, Stock No. 3500AA. When Hong Kong began retailing sound producing yo-yos at a price lower than the cost of manufacturing the Mel•Yo•Dee, Flambeau dropped line. See Plate 9, #422.

| | | |
|---|---|---|
| 423 | 9 | **MIDNIGHT SPECIAL, imprint seal, Flambeau, tournament shape, '90s** |

Production of the Midnight Special began in 1989. All Midnight Specials are made of black opaque plastic. They have been retailed on at least two different blister cards, Stock No. 3059NP.

| | | |
|---|---|---|
| 424 | | **NO MO YO-YO, paper insert seal, tournament shape, view lens, early '60s** |

This yo-yo was reportedly made right before the closing of the Duncan factory in 1965. It has a Pony Boy body with a clear view lens. The paper insert seal shows a picture of a sad-eyed dog with "No-Mo Yo-Yo" imprinted on it. This was a very limited production and it is unclear as to how many of them exist. This yo-yo was apparently not sold retail. It was only given to Duncan employees.

| | | |
|---|---|---|
| 425 | 14 | **NEO, imprint seal, Flambeau, tournament shape, metal axle, '90s** |

The Neo line was released in 1989. Five color variations of this yo-yo exist: green, yellow, pink, orange, and watermelon. It was developed as the replacement yo-yo for the Flambeau Jewel line. Neos were originally intended for a three season release, but they continue to be produced. These yo-yos are retailed on blister cards, Stock No. 3436PK. The yellow Neo color was discontinued by Flambeau in 1995.

| | | |
|---|---|---|
| 426 | | **NEW YORK, imprint seal, Flambeau, tournament shape, '90s** |

This was made as a souvenir yo-yo.

| | | |
|---|---|---|
| 427 | 13 | **PAC-MAN, flasher seal, Flambeau, slimline shape, view lens, early '80s** |

This yo-yo was based on the popular '80s video game. It was briefly produced in 1982 while the Pac-Man animated TV series was on the air. This is a flasher style yo-yo similar to the Super Hero line. It was retailed on a blister card, Stock No. 3272.

| | | |
|---|---|---|
| 428 | 14 | **PONY BOY (reflective disk faces), embossed seal, tournament shape, fixed string, '50s** |

Production of the Pony Boy started in 1954 and briefly ran through the middle '50s. The yo-yo is made out of styrene and has a plastic axle. There is a small BB inside the yo-yo that produces sounds when the yo-yo is played. It was retailed loose out of a counter display box which held 36 yo-yos. The original price of this yo-yo was ten cents.

| | | |
|---|---|---|
| 429 | 16 | **(Smile face), imprint seal, Flambeau, tournament shape, '90s** |

*Continued ...* ▶

## YO-YO LISTINGS

● **DUNCAN (MISCELLANEOUS PLASTIC)** *Continued ...*

**429.1**  *16*  **TRICKSTER**, hot stamped seal, Flambeau, tournament shape, late '60s
In 1969 when Flambeau began production of Duncan yo-yos, this model was one of the original five yo-yo lines released. It was also the first to be dropped from production. It retailed on a card with a pre-printed price of 59 cents and a Stock No. 3264.

**430**  *13*  **TRON**, flasher seal, glow plastic, Flambeau, slimline shape, view lens, early '80s
This is a licensed yo-yo from the 1982 sci-fi computer generated movie TRON. The yo-yo was briefly produced in 1982. This is a flasher style yo-yo similar to the Super Hero line. This was retailed on a blister card, Stock No. 3274. It is believed to be the first slimline style yo-yo made from glow plastic.

**431**  *17*  **VELVET 1972** (nylon flocked finish), hot stamped seal, Flambeau, tournament shape, metal axle, '70s
Two styles of the Velvet exist, one dated 1972 and the other non-dated. The non-dated style is considered to be slightly more difficult to find than the dated style. Color definitely makes a difference in desirability, as the red and blue velvet yo-yos have become more sought after than the other colors.

**432**  **VELVET 1972**, red or blue
The red and blue variations are considered more desirable than the other colors.

**433**  **VELVET** (no date • nylon flocked finish), hot stamped seal, Flambeau, tournament shape, mid '70s
Although produced after the dated 1972 version, this yo-yo is more difficult to find. The Velvet was retailed on a blister card with a pre-printed price of $1.49, Stock No. 3260. See Plate 12, #433.

**434**  *17*  **VELVET** (no date), red or blue
The red and blue variations are considered more desirable than the other colors.

**435**  **WORLD CLASS**, embossed seal, plastic, Flambeau, butterfly shape, '80s and '90s
This was a heavy plastic butterfly style yo-yo with metal inertia rings embedded near the rim to increase spin time. The yo-yo also has a slightly concave Teflon coated axle to reduce friction and keep the string centered. The credit for the development of the "World Class" is given to Charles Lanius. This yo-yo was released in response to Duracraft's long spinning Pro Yo. The original production was in 1980. This yo-yo was re-released in 1990 with new-old stock. Mint in package value is for the yo-yo in the original release packaging. An example of this yo-yo is shown next to this section's title.

## DUNCAN (MISCELLANEOUS WOOD)

The wood yo-yos listed in this category do not fit in any of the other Duncan sections.

**436**  *43*  **(Autographed) DON DUNCAN JR.**, ink seal, tournament shape, miniature, 1992
This limited edition yo-yo is autographed in black ink by Don Duncan Jr. These yo-yos are from the original unfinished yo-yo blanks produced by the Luck plant in the early '50s. The yo-yo comes with a certificate of authenticity signed by Don Duncan, Jr. The yo-yos are signed and numbered up to 950.

**436.1**  *20*  **(Big "G") GENUINE DUNCAN YO-YO**, die stamped seal, tournament shape, '30s
Big "G" refers to the large "G" in the word "Genuine." This is very similar to the #493 seal, but is missing the word "Tournament." This yo-yo was selected by the Franklin Mint as one of their three reproductions released in 1994. For details on identifying this original yo-yo from the reproduction, see #618.

**436.2**  *14*  **(Big "G") GENUINE YO-YO**, gold leaf stamped seal, tournament shape, one-piece, late '20s
This is believed to be one of Duncan's earliest yo-yos. Like the O-Boy #389.2, this seal does not bear the Duncan name. It is also possible that this could have been a Flores yo-yo.

•  **Big "G" (Franklin Mint)** - *See Franklin Mint section #618.*

**437**  *14*  **CHIEF**, foil sticker seal, tournament shape, three-piece, 1959
This was a briefly produced beginners style yo-yo released by Duncan in 1959. The seal featured an Indian's head with full headdress. This model, and the Rainbow, are the only Duncan models known to have multi-colored foil seals. This is one of the most desirable of Duncan's beginner style yo-yos. Like most foil sealed yo-yos, this seal has a tendency to detach.

**438**  *14*  **CHAMPION (eagle seal)**, foil sticker seal, tournament shape, one-piece, '50s
This is a rare award yo-yo. The foil seal features an embossed eagle and shield. This award yo-yo was used by Duncan promoter Randy Brown in Chicago area promotions. Both silver and gold sticker seals exist. The demonstrator usually applied the seals in the field. This seal is most commonly found on standard size, pearlescence, award yo-yos, but were occasionally applied on jumbo award yo-yos.

**439**  **DON DUNCAN (signature)**, gold leaf stamped seal, tournament shape, one-piece, late '50s
This yo-yo was produced in very limited quantities in the late '50s. The Duncan Company gave out this model in promotional packs at toy trade show conventions. The yo-yo was not known to have been retailed, but was used occasionally as an award yo-yo. The signature on the yo-yo is that of Don Duncan, Jr., made from a die stamp. An example of this yo-yo is shown next to this section's title.

*Continued ...* ▶

## YO-YO LISTINGS

440    *14*    **(Little "G") GENUINE YO-YO DUNCAN, gold leaf stamped seal, wood, tournament shape, midget, '50s**
This is the smallest wood yo-yo manufactured with a Duncan logo. It was included in gift packages such as the "Perk Up Get Well" gift package and "My Merry Toy Closet" which included several other miniature toy items. It is doubtful that this yo-yo was ever retailed individually. Production was from the late '50s into early '60s.

441    *16*    **DUNCAN TOPS, silver leaf stamped seal, tournament shape, three-piece, pegged string, '30s**
The yo-yo shown is a typical red and black beginners model. For some reason, this yo-yo is difficult to find with a seal that has been fully imprinted in the face. The upper right corner often has a weak strike. This seal has been found on both standard and miniature sized models.

441.1    **DUNCAN TOPS, foil sticker seal, tournament shape, miniature, three-piece, '30s**
This yo-yo is the same design as #441, except this model has a foil seal.

442    *43*    **GENUINE DUNCAN FAMILY COLLECTION, paint seal, tournament shape, three-piece, '90s**
This yo-yo is sold at the Chico Yo-Yo Museum. It is part of a set that includes a pin back button and a miniature felt pennant.

443    *14*    **GENUINE YO-YO RETURN TOP DUNCAN, gold leaf stamped seal, tournament shape, pegged string, '60s**
This seal was sometimes used on the reverse face of advertising yo-yos in the early '60s. The seal was not as common as the other "standard line" seals of the '60s.

444    *14*    **"LITENING," paper sticker seal, tournament shape, one-piece, '50s**
The Litening was produced from 1956 to late 1959. A two-coat painting process gave a crackle (lightning) appearance to the surface. The outer enamel layer was a rapid drying paint which formed cracks in the surface allowing the base color to show through. The outer paint coat has a tendency to chip easily. Along with the Duncan Jeweled, this was the first yo-yo to be packaged on a blister card. This yo-yo on the original blister card is a very rare find. The Franklin Mint made a reproduction of this yo-yo in 1994. See #616.

- **Litening (Franklin Mint)** - *See Franklin Mint section #616.*

445    *14*    **LUCK-E JA-DO CONTEST TOP, die stamped seal, tournament shape, one-piece, '50s**
The four leaf clover designed JA-DO was produced for a short time beginning in 1951. "JA" is from the first two letters in Jack Duncan and "DO" is from the first two letters in Don Duncan, Jr. This was an inexpensive, non-promoted yo-yo released by Duncan to undercut the low priced yo-yos of competitors. There was a jeweled model of the JA-DO with 3 jewels in a row on the reverse face. JA-DOs ordinarily are one solid color, but some half-and-half JA-DOs have been found. Some JA-DOs may be found with a Pepsi decal over the seal. These were used for a Pepsi promotion in the '50s.

- **Super Practice Return Top** - *See Duncan Tournament section #510.*

## DUNCAN O-BOY

The O-Boys were some of the first Duncan yo-yos produced. The O-Boy line started in the late '20s and continued to be produced into the '30s. Yo-Yos with the words "Pat. Pending" are considered to be some of the earliest models. It is believed that #446.2 is the first yo-yo ever produced by Duncan.

446          **O-BOY DUNCAN "PAT PENDING," die stamped seal, wood, tournament shape, early '30s**
This yo-yo is believed to be the first O-Boy with the Duncan name on the seal. This yo-yo was replaced by the #446.1 seal which did not have "Pat. Pending" lettered on it. This model is more difficult to find than #446.1.

446.1    *16*    **O-BOY DUNCAN (without pat pending), gold leaf stamped seal, wood, tournament shape, early '30s**
This is the classic Duncan Beginners style yo-yo produced throughout the '30s. These models were half black and half red.

446.2    *14*    **O-BOY, gold leaf stamped seal, wood, tournament shape, one-piece, 1929**
This model is thought to be the first yo-yo Duncan ever produced. It is perhaps the rarest of Duncan's standard line. Duncan's name does not appear on the seal and the shape of the yo-yo resembles the early Flores style. This yo-yo was illustrated in one of Duncan's earliest trade ads where it was described as "The Sensational New Ever Spinning Toy." An example of this yo-yo is shown next to this section's title.

- **O-Boy Jr.** - *See Duncan Junior section #375.*
- **O-Boy Whistling** - *See Duncan Tins section #482 and #483.*

446.3          **THE O-BOY, die stamped seal, wood, tournament shape, 1929**
Similar to #446.2, this is one of Duncan's earliest produced yo-yos. It also has the early Flores style tournament shape. The yo-yo was only briefly produced and is considered rare.

## YO-YO LISTINGS

### DUNCAN "OLYMPICS" YO-YOLYMPICS

The Yo-Yolympics promotion was one of the last nationwide contest promotions sponsored by Duncan Flambeau. This campaign was run in '79 and '80. Like all previous Duncan contests, the ages were limited to 16 and under, with two categories, 12 and under, and 12 and over. Playground contest winners went to city finals. The four city finalists went to a state competition. Nine regional contests followed, with winners going to the Nationals hosted by Marriott's Great America in Illinois. The 1st prize was a $1,000 scholarship, 2nd prize $500, and 3rd prize $250.

A video on how to run a Yo-Yolympics was produced at the Six Flags Over Georgia Amusement Park in Atlanta. Duncan promotion director Doug Beringer and demonstrator Lance Lynch were featured in the video. Unfortunately, the marketing company that Duncan hired to produce the video did not understand the particulars of marketing yo-yos. The promotion was unsuccessful in recruiting contestants for yo-yo competitions and was considered somewhat of a disaster.

Collectors should be aware that yo-yo competitor stickers, provided to Yo-Yolympics participants, sometimes end up on blank yo-yos.

|       |    |                                                                                           |
|-------|----|-------------------------------------------------------------------------------------------|
|       | •  | **Competition Kit Yo-Yolympics** - *See Duncan Miscellaneous Items #398.2.*               |
| 447   | 17 | **(Competitor sticker seal)**                                                             |

Competitor stickers were paper stickers given to all entrants in the competition. A sheet of thirty stickers was included in each contest kit. See #398.2. Many of these seals wound up on non-Duncan yo-yos. These yo-yos have little collectible value other than for the seals themselves. Values are listed for complete sheets and individual unused stickers.

| 447.1 | 17 | **1979 CHAMPION, paper insert seal, plastic, Flambeau, slimline shape, view lens** |

This was an award yo-yo given as a prize to contest winners. Four of these were provided in every contest kit. See #398.2.

| 447.2 | 17 | **1980 SUGAR CRISP, paper insert seal, plastic, Flambeau, slimline shape, view lens** |

This yo-yo was given out at Yo-Yolympics contests. It could also be received through the mail by sending one dollar and two proof of purchase seals to Post Sugar Crisp cereal. The cereal box featured an advertisement for the yo-yo. See #307.54.

| 447.3 | 17 | **DUNCAN YO-YOLYMPICS, paper insert seal, plastic, Flambeau, slimline shape, view lens** |

This yo-yo was retailed with two other yo-yos as the "Yo-Yolympics" championship set. It came with a trick book, a Light-Up (#382), and a Butterfly (#321).

| 447.4 | 17 | **SUGAR CRISP • YO-YOLYMPICS, paper insert seal, plastic, Flambeau, slimline shape, view lens** |

An example of this yo-yo is shown next to this section's title.

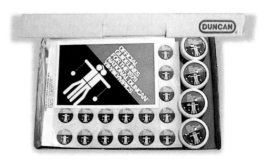

*This official Duncan contest kit, #398.2, was used nationwide for promoting the 1979 National Duncan Yo-Yolympics.*

# YO-YO LISTINGS

## DUNCAN PROFESSIONAL

The original Duncan Professional was released in 1971. This model was a Duncan Flambeau response to the All-American Toy Company's release of a gold and silver metallic plated plastic yo-yo in 1971. The All-American Yo-Yo did not survive the Duncan Flambeau's competition.

The original Professional was a tournament shaped model, but around 1974, Duncan Flambeau changed the Professional to the slimline shape which was cheaper to produce. The slimline shaped yo-yo went on to be used in the Superhero, Advertising, Galaxy, and most recently the Alien yo-yo lines.

448    14    **(Diamond) PROFESSIONAL (metallic gold finish), embossed seal, tournament shape, metal axle, 1971**
This was a specially molded plastic yo-yo with a metallic finish. The yo-yo was used as both an award yo-yo and sold retail. This was Duncan's response to the All-American Yo-Yo that also began production in 1971. The original Duncan Professional retailed in an individual display box. The display box had a pre-printed price of $2.00.

448.1    14    **(Diamond) PROFESSIONAL, foil insert seal, plastic, slimline shape, view lens, '70s and '80s**
This is the first time the slimline shape was used as its own separate line. Production began around 1974 and continued through early '80s. On the original display box the price was pre-printed as $2.00. See Plate 31, # 448.1. An example of this yo-yo is shown next to this section's title.

448.2         **(Diamond) PROFESSIONAL (metallic silver finish), embossed seal, tournament shape, 1971**
Like the metallic gold version #448 this was used as an award yo-yo and sold retail in a display box.

448.3    14    **(Eagle) PROFESSIONAL, paper insert seal, plastic, slimline shape, view lens, '90s**
The Professional slimline series changed from the #448.1 style to this seal in mid '80s. This seal continues to be produced by Flambeau. Unlike previous Professionals, this was retailed on a blister card with Stock No. 3270AA. Flambeau dropped the white color Professional in 1995.

448.4    10    **PROFESSIONAL DUNCAN, paper insert seal, slimline shape, view lens, '80s**
This yo-yo was not retailed, but was used as a promotion for the slimline style of advertising yo-yo. Sometimes this seal was used on the reverse face of an ad yo-yo. This yo-yo would be given out as a promotional item at trade show conventions.

448.5    10    **PROFESSIONAL DUNCAN ASI 54560, paper insert seal, slimline, view lens, '90s**
Like #448.4, this yo-yo was used as a promotional sample for the slimline style of advertising yo-yos.

## DUNCAN RAINBOW

In 1956, Duncan produced the Rainbow series. Rainbows were typically one-piece wooden yo-yos. Some two-tone, three-piece models do exist. The two-tone beginners models are less valued than the one-piece models. Most Rainbows have a half and half enamel finish instead of the standard airbrushed paint stripe. All Duncan Rainbows say "return top" on the seal.

The original Duncan Rainbo, from the '30s, was made of tin and did not have a "W" at the end of rainbow. (See #484.) Many other companies have produced yo-yos with the word "Rainbow" on the seal.

    •    **Rainbo** - *See Duncan Tin section #484.*

449    20    **RAINBOW, gold foil sticker seal, tournament shape, 1956 - late '50s**
Rainbows were one of the few yo-yos on which Duncan used multi-colored foil, sticker seals. The only other Duncan multi-colored, foil sticker seal is the Duncan Chief #437. These foil seal yo-yos are difficult to find because the seal detaches easily. Most Rainbows have half-and-half paint instead of paint stripes. Some models were retailed in a polybag with a header card and a pre-printed price of 49 cents.

*Continued ...* ▶

## YO-YO LISTINGS

● DUNCAN RAINBOW Continued ...

| | | |
|---|---|---|
| 449.1 | *20* | **RAINBOW, gold leaf stamped seal, tournament shape, 1956 - late '50s** |
| | | This was also used as a Post Cereal mail-in premium. The yo-yo came with a small Post Cereal Duncan trick book which is very rare. See #1927. |
| 449.2 | *20* | **RAINBOW, silver foil sticker seal, tournament shape, 1956 - late '50s** |
| | | This yo-yo has the same seal as #449, but in silver. Some were retailed in a polybag with a header card and a pre-printed price of 49 cents. An example of this yo-yo is shown next to this section's title. |

# DUNCAN SATELLITE

Duncan Satellite yo-yos were introduced in 1960 at the New York Toy Fair. The series was introduced as the M-1 Satellite Yo-Yo, named after the stock ordering number for the yo-yo. Several versions of Satellite yo-yos were produced from 1960 through 1965. All Satellite yo-yos are made of wood and have a glitter paint finish. The end of the wood Satellite series came with the closing of the Duncan plant in 1965. The uniqueness of the "flying saucer shape," and it being one of the "space race toys" of the '60s, make the Satellite a desirable cross-collectible yo-yo. Satellites were retailed loose, on display cards, and in polybags with saddle header cards.

In 1972, Duncan Flambeau named their plastic Light-Up yo-yos "Satellites." Since these are not Satellite "flying saucer" shape, collectors do not consider them true Satellites. The Duncan Flambeau version of the yo-yo does not carry the name "Satellite" on the yo-yo's seal, only on the blister card. Collectors call these Satellites "Light-Ups."

| | | |
|---|---|---|
| 450 | *17* | **DUNCAN SATELLITE, gold leaf stamped seal, one-piece, '60s** |
| | | This Satellite is distinctive for having a flat center on the rounded face. Collectors sometimes call this the "Flat Face Satellite." |
| 450.1 | *17* | **DUNCAN YO-YO RETURN TOP, gold leaf stamped seal, one-piece, '60s** |
| | | This has a rounded face and the same seal as the "Shrieking Sonic Satellite," but does not whistle. It is harder to find than the Shrieking Sonic Satellite #450.2. |
| 450.2 | *17* | **DUNCAN YO-YO RETURN TOP (shrieking sonic satellite), gold leaf stamped seal, one-piece, '60s** |
| | | This yo-yo produces a high-pitched whistling sound when it is spinning. The sound, according to advertising literature, is "a high tone, simulating radio waves to satellites whirling around the Earth." Holes in the rim generate air currents that pass through a whistle mounted in the center of the yo-yo face. The yo-yo was introduced in 1961, but was listed as a new product in Duncan's 1962 catalog. This model continued to be produced through 1965. It was retailed on a display card reading "Shrieking Sonic Satellite," with a pre-printed price of $1.00. The display card is marked Stock No. 500. See the display card Plate 13, #450.2. |
| 450.3 | *17* | **(Saturn and star) Duncan (centered), gold leaf stamped seal, one-piece, '60s** |
| | | This is the rarest of the standard line Duncan Satellites. This yo-yo is easy to identify because it is the only Satellite with the Duncan name centered. |
| 451 | *17* | **(Saturn and stars) SATELLITE, gold or silver die stamped seal, one-piece, '60s** |
| | | This model is the original M1 Satellite. Two paint patterns were used for this yo-yo, a solid color with glitter paint (Plate 17, #451A) and a two color pattern with the rim a different color than the face (Plate 17, #451B). The aluminum flecked finish was described in ad literature as "to suggest the galaxy of outer space." Check the seal carefully for missing parts because the round face frequently did not get a full strike. Care must be taken in storage as the seal has a tendency to wear off the round face. Display cards carry a pre-printed price of 69 cents. See display card on Plate 13, #451. |
| 452 | *17* | **SEATTLITE SPACE NEEDLE, gold leaf stamped seal, sculpted shape, '60s** |
| | | Although not a true Satellite yo-yo, many collectors consider this yo-yo part of the Duncan Satellite series. This yo-yo is the most cross-collectible of the Satellite yo-yos. Compared to the other Satellites, this model has a unique flying saucer shape. It is sculpted to resemble the top of the Space Needle in Seattle. This yo-yo was produced as a souvenir for the 1962 World's Fair in Seattle. It was later sold at the Space Needle. The Seattlite Space Needle was retailed on a display card in a polybag with a pre-printed price of a $1.00. See the display card Plate 13, #452. An example of this yo-yo is shown next to this section's title. |

# YO-YO LISTINGS

## DUNCAN SPECIAL

The first Duncan Special was the Duncan Special 44. It was a wooden, pegged string yo-yo. The years of its release and production remain unclear. When Flambeau again began production of Duncan yo-yos in 1969, the "Special" was one of the original five yo-yos released. Since the Flambeau version was plastic, it was completely re-designed and looked nothing like the original Special 44. It no longer carried the number 44 and now had a slip string. The Duncan Flambeau Specials were marketed as advanced beginner models and were all retailed on blister cards. Specials were dropped from the Duncan Flambeau product line in 1990.

453　　*16*　　**(4 point center star), hot stamped seal, plastic, Flambeau, tournament shape, metal axle, '69 - '70s**
　　　　　　In 1969, when Flambeau began production of Duncan yo-yos, this model was one of the original five yo-yo lines released. It was retailed on blister cards with a pre-printed price of 79 cents. At least two different cards are known, both are labeled Stock No. 7265. The cards with the "nostril - Mr. Yo-Yo" are the rarest. Some collectors call this the "4 Point Star Special." Production of this yo-yo stopped in the middle '70s.

454　　*16*　　**(4 star), imprint seal, plastic, Flambeau, tournament shape, metal axle, '80s**
　　　　　　Production of these yo-yos began in the middle '70s. They were retailed on blister cards. Early cards have the Stock No. 3262 (Plate 9, #454); more recent cards are marked Stock No. 3262NP.

455　　*16*　　**(5 star with 8 point center star), imprint seal, plastic, Flambeau, tournament shape, metal axle, '80s**
　　　　　　This model is sometimes called the "8 Point Star Special."

456　　*16*　　**SPECIAL 44, gold leaf stamped seal, wood, tournament shape, three-piece, pegged string**
　　　　　　This is Duncan's original Special. It's the only wood version and the only Special seal made prior to Flambeau. The years of production of this model are still unclear. An example of this yo-yo is shown next to this section's title.

## DUNCAN SPORTSLINE

In 1964, the year before Duncan was forced into bankruptcy, the AMF Bo-Yo was introduced. The Bo-Yo was a promotion for Amflite's bowling supply line. The yo-yo was promoted at bowling centers and at the 1964 New York World's Fair. The AMF Bo-Yo was the impetus for the development of the Duncan Sportsline series.

In 1965, Duncan released the five model Sportsline series. These sports ball shaped yo-yos were all plastic and didn't play as well as the other Duncan yo-yos. Even though a television advertising promotion of the Sportsline series took place in 1965, Duncan was forced into bankruptcy before a major campaign could be completed. Sportsline yo-yos retailed on blister cards with an original price of $1.00.

457　　*11*　　**(Baseball), hot stamped seal, ball shape, '60s**
　　　　　　This model retailed on a blister card, Stock No. 1090. See Plate 9, #457.

458　　*11*　　**(Basketball), hot stamped seal, ball shape, '60s**
　　　　　　This yo-yo is the most difficult to find of the Duncan Sportsline series. It retailed on a blister card, Stock No. 1050.

459　　*12*　　**(Bowling ball) 1 AMF (in triangle), molded seal, ball shape, '60s**
　　　　　　This is the rarest of the two Duncan bowling ball yo-yos. It was the prototype for the Bo-Yo. The yo-yo does not have Duncan or Bo-Yo hot stamped on it. The yo-yo was retailed on a Duncan card, but not the Sportsline blister card. See Plate 12, #459. The card carries a copyright date of 1964 which precedes the Sportsline series copyright date by one year. The yo-yo came with a single sheet of yo-yo tricks and bowling tips.

460　　*11*　　**BOWLING BALL • AMFLITE, molded seal, ball shape, '60s**
　　　　　　The words "Duncan Bo-Yo" are molded into the reverse face of this yo-yo. This molded seal does not have any paint in the recesses. This model was retailed on a Sportsline blister card with Stock No. 1070. This variation is more common than #459. See the blister card on Plate 12, #460. An example of this yo-yo is shown next to this section's title.

*Continued ...*

## YO-YO LISTINGS

● **DUNCAN SPORTSLINE** *Continued ...*

461    *11*    **(Eight ball), imprint seal, ball shape, '60s**
                     Like the Basketball, this yo-yo is more difficult to find than the others in the Duncan Sportsline series.

462    *11*    **(Golf ball), hot stamped seal, ball shape, '60s**
                     This model retailed on a blister card, Stock No. 1060. See Plate 9, #462.

## DUNCAN SUPER HERO

The Super Hero line of Duncan yo-yos is the most cross-collectible of the Duncan Flambeau lines. The series originally started in 1978 with the production of Batman and Superman. The original yo-yos were sold in a clear hard plastic showcase. A bat logo sticker sealed the Batman case; a Superman logo sticker sealed the Superman case. Yo-Yos in original showcases with intact seals are very difficult to find. The original Batman was a butterfly shaped yo-yo and the only Butterfly produced for the Super Hero series. All other Super Hero yo-yos were of the slimline style. In 1979, after the successful introduction of the Batman and Superman yo-yos, other Super Heroes became available such as Wonder Woman, The Hulk, and Spider-man.

In 1980, Duncan introduced its "Magic Motion," a flasher type of yo-yo. The same characters were used as in the original Super Hero line. Even though the flasher series was produced more recently, these are more valued by collectors than the non-flasher styles.

After the first year (1978), the packaging was changed from the hard plastic showcases to hanging plastic and cardboard display boxes. Non-flasher series boxes were silver. See Plate 31, #471. The Flasher boxes were yellow and orange and marked "Magic Motion." (See Plate 31, #464.) For other Super Hero yo-yos, see the Character section.

463    *31*    **(Batman and Robin), paper sticker seal, plastic, butterfly shape, 1978**
                     When this yo-yo was licensed, "The New Adventures of Batman" was a popular animated TV series. Of the Duncan Super Heroes series, this is the most difficult to find. It retailed in a hard plastic showcase box with a bat sticker. An example of this yo-yo is shown next to this section's title.

464    *18*    **BATMAN, flasher seal, slimline shape, view lens, early '80s**
                     This model has a #465 seal on the reverse face. See the Magic Motion display box, Plate 31, #464.

465    *18*    **(Batman), paper insert seal, slimline shape, view lens, late '70s**
                     There are two different non-flasher versions of the Batman yo-yo.

466    *18*    **(Hulk), flasher seal, slimline shape, view lens, early '80s**
                     This model has a #467 seal on the reverse face.

467    *18*    **HULK, paper insert seal, slimline shape, view lens, late '70s**
                     There are two different non-flasher versions of the Hulk yo-yo.

468    *18*    **SPIDER-MAN, flasher seal, slimline shape, view lens, early '80s**
                     This model has a #469 seal on the reverse face.

469    *18*    **SPIDER-MAN, paper insert seal, slimline shape, view lens, late '70s**
                     There are two different non-flasher versions of the Spider-Man yo-yo.

470    *18*    **(Superman), flasher seal, slimline shape, view lens, early '80s**
                     This model has a #471 seal on the reverse face.

471    *18*    **(Superman), paper insert seal, slimline shape, view lens, late '70s**
                     Two different non-flasher graphics exist for the Superman yo-yos. They were used on yo-yos in several combinations and appear on the reverse face of the flasher models. The first Superman, in 1978, came in a hard plastic showcase box with a Superman logo sticker seal sealing the showcase. In 1979, this was changed to the silver hanging display box. See this display box Plate 31, #471.

472    *18*    **WONDER WOMAN, flasher seal, slimline shape, view lens, early '80s**
                     This model has a #473 seal on reverse face.

473    *18*    **WONDER WOMAN, paper insert seal, slimline shape, view lens, late '70s**
                     There are two different non-flasher versions of this graphic. One was always used on the reverse of the flasher model #472.

## YO-YO LISTINGS

## DUNCAN TINS

Duncan tin yo-yos began production around 1932 with the Duncan O-Boy Whistling Yo-Yo. The O-Boys were immediately followed by larger lithographed tin whistler yo-yos known to collectors as the "First Series." The earliest versions of the tin Duncans have "Pat. Pend." imprinted somewhere on the seal. Later versions do not. All Duncan tins are tournament shaped and, with the exception of the Rainbo, are whistling yo-yos. They have two pairs of air holes on the front rim and one pair on the reverse rim. Air passing through the holes creates a whistling sound as the yo-yo spins.

Duncan never manufactured any tin yo-yos themselves, but farmed them out to the Cayo Manufacturing Company of Benton Harbor, Michigan. Cayo also produced their own line of tin yo-yos in the '30s and '40s. Cayo yo-yos are the same shape and size of some Duncan tin models. Production of tin yo-yos stopped from 1941 through 1945 due to the war effort. Cayo continued to produce tin whistling yo-yos for Duncan after the war, but only in solid color designs. Production of Duncan tins completely stopped by 1950.

Duncan also retailed yo-yos for Louis Marx in the '30s. In the early '30s, Marx approached Don Duncan, Sr. with a large quantity of tin yo-yos that were not selling. He asked Duncan if he would work for him. Duncan declined, but agreed to sell the yo-yos on commission. These are believed to be the yo-yos called the "Second Series." It is also believed that Duncan sold these yo-yos for Marx in England and France under the Lumar name.

The most prized tin yo-yos are the multi-colored lithographed yo-yos, called "lithos" by collectors. Later, one color tin Duncan whistlers were also produced. These are not as desirable as the "lithos" and collectors refer to the one color tins as "solids." It is rare to find a tin yo-yo in mint condition due to its tendency to rust.

- 474   *19*   **(Big "G") GENUINE WHISTLING, ink stamped seal, solid color, tournament shape, '40s**
  Big "G" (Genuine) refers to the large "G" in the word "Genuine" that extends to the base of the seal. Considered the "common" Duncan whistler, this is less sought after than the lithographed styles. Produced into the late '40s, several different solid colors exist, green and orange being the most common. Collectors refer to these yo-yos as "solids."
- 475   *19*   **(Big "G") GENUINE WHISTLING, ink stamped seal, solid color, tournament shape, '40s**
  Big "G" (Genuine) refers to the large "G" in the word "Genuine" that extends to the base of the seal. This has a similar seal to #474, but with concentric circles around the seal. Sometimes this yo-yo is referred to as the "concentric circle solid." This variation is far less common than #474.

### Duncan Whistling "First Series"

Whistlers, known as the "First Series," are multi-colored lithographed tin yo-yos from the early '30s that have a tournament shape. Like the other whistlers, they have air holes in the rims which give off a whistling sound when the yo-yo is played. The names of the yo-yos describe the lithograph patterns. These yo-yos are very popular among collectors and very difficult to find in near mint condition due to the yo-yo's tendency to rust. "First Series" yo-yos were 2-1/8" in diameter, larger than the "Second Series" models. "First Series" yo-yos all have the words "Pat. Pend." on the seal. These are considered some of the earliest of the Duncans. They were retailed out of a counter display box.

- 476   *19*   **(Hypno stripe), lithograph seal, tournament shape, early '30s**
  In 1932, this yo-yo was given to any child who could sell a three month subscription to the Daily and Sunday Tribune. This yo-yo has a big "G" Genuine Whistling seal. There are two seal variations for this yo-yo, the standard big "G" Genuine and the red seal. See #476.1. The red seal differs from the standard seal in that it is slightly smaller and the seal is printed on a background of red. Due to the lithograph pattern, collectors call this yo-yo the "Hypno Stripe Whistler."
- 476.1  *19*  **(Hypno stripe), red center • lithograph seal, tournament shape, early '30s**
  This yo-yo varies from the Hypno Stripe Whistler, #476, by having a red big "G" Genuine seal. Due to the lithograph pattern, collectors call this yo-yo the "Hypno Stripe Red Seal Whistler."
- 477   *19*   **(Hypno swirl), lithograph seal, tournament shape, early '30s**
  This yo-yo has a big "G" Genuine Whistling seal. Due to the lithograph pattern, collectors call this yo-yo the "Hypno Swirl Whistler."

*Continued ...* ▶

*The History and Values of Yo-Yos*

## YO-YO LISTINGS

**DUNCAN WHISTLING "FIRST SERIES"** *Continued ...*

478   *19*   **(Pinwheel), lithograph seal, tournament shape, early '30s**
This yo-yo has a big "G" Genuine Whistling seal. Due to the lithograph pattern, collectors call this yo-yo the "Pinwheel Whistler." Two versions of the pinwheel seal are known.

478.1   *19*   **(Ink blot), lithograph seal, tournament shape, early '30s**
This yo-yo has a big "G" Genuine Whistling seal. Due to the lithograph pattern, collectors call this yo-yo the "Ink Blot Whistler."

### Duncan Litho Whistler "Second Series"

These were the "Second Series" of tin whistlers released by Duncan. The "Second Series" are lithographed tin yo-yos that have a tournament shape. Like the other whistlers, they have air holes in the rims which give off a whistling sound when the yo-yos are played. The sizes were 2" in diameter compared to the "First Series" which were 2-1/8". These yo-yos are more common than the "First Series," but still highly collectible. The names describe the lithograph patterns. The "Second Series" does not have the words "Pat. Pend." on the seal like the yo-yos in the "First Series." These were retailed out of a counter display box which was redesigned from the "First Series." The original price was 15 cents each. They were packed 12 to a box. The stocking number for this yo-yo was 88.

479   *19*   **(Starburst), lithograph seal, tournament shape, mid-'30s**
This yo-yo has a big "G" Genuine whistling seal. Due to the lithograph pattern, collectors call this yo-yo the "Starburst Whistler."

480   *19*   **(Dart board), lithograph seal, tournament shape, mid-'30s**
This yo-yo has a big "G" Genuine Whistling seal. Due to the lithograph pattern, collectors call this yo-yo the "Dart Board Whistler." An exact copy of this pattern was also used for one of the Lumar yo-yos. See the Foreign section for a discussion of Lumar yo-yos.

481   *19*   **(Propeller), lithograph seal, tournament shape, mid-'30s**
This yo-yo has a big "G" Genuine Whistling seal. Due to the lithograph pattern, collectors call this yo-yo the "Propeller Whistler."

### Duncan O-Boy, Whistling Series

O-Boy whistlers are lithographed tin yo-yos that have a tournament shape. Like the other whistlers, they have air holes in the rims which give off a whistling sound when the yo-yo is played. These yo-yos were first released in 1932. They were the first of the Duncan "musical" yo-yos. At 1-3/4", these are smaller in size than later whistlers. Both O-Boy whistlers have the words "Pat. Pend." on the seal.

482   *19*   **(Quarter seal), lithograph seal, tournament shape, early '30s**
This has a quarter-size center logo.

483   *19*   **(Dime seal), lithograph seal, tournament shape, early '30s**
This has a dime-size center logo.

484   *19*   **RAINBO, lithograph seal, tournament shape, 1934**
The 1934 Duncan Tin Rainbo is considered one of the most collectible of all Duncan yo-yos. It has an internal metallic spinning disk that can be seen through a cut out window in the face of the yo-yo. Early Duncan ads used to refer to this yo-yo as a "color blending yo-yo in all the colors of the rainbow." The original price of a Rainbo was 15 cents and it was retailed out of a rare two color counter display box marked "Duncan Rainbow." There were 12 yo-yos per box. The yo-yo had a stocking number of 22 and this number is found on the display box. This yo-yo is very difficult to find in mint condition. Collectors call this the "Tin Rainbo" or the "Color Changer." An example of this yo-yo is shown next to this section's title.

485   *19*   **SELL JEN, ink stamped seal, tournament shape, '40s**
Duncan tin yo-yos were occasionally used as advertising yo-yos. This yo-yo has a standard whistler seal on the reverse face. See #474. Duncan tin ad yo-yos are rare finds. Only "solids" have been found with advertising. This is one of the earliest examples of the phrase "Why Be A Yo-Yo" being used on a promotional yo-yo.

## YO-YO LISTINGS

## DUNCAN TOURNAMENT

All Duncan yo-yos marked "Tournament Yo-Yo" are classic tournament shaped. When adults reminisce about their childhood yo-yoing days, this is the yo-yo most commonly described. Duncan produced millions of Tournament yo-yos. Over the decades, several different seals bore the word "Tournament" and most carried the familiar decorative airbrushed paint stripe on both faces.

The first Duncan Tournament yo-yo was the big "G" Gold Seal Tournament, which dominated the yo-yo market in the '30s. This was the yo-yo of choice for all contest play. Original Duncan Gold Seals should not be confused with the yellow decal Tournaments. The original Gold Seal is in fact a gold color, not yellow, and should be the only yo-yo referred to as a "Gold Seal." Yellow decal Tournaments have their own classification and can be separated into three styles: the 77's, the big "G," and the little "G." These are all called "yellow seals" by collectors.

Tournament models with decals are difficult to find in mint condition. Over time minute cracks form in these seals and chipping is very common. This was particularly true of the Day Glow 77. The seals adhere even less to the Day Glo paint than they do to the standard enamel based paint. Tournament models with decal seals occasionally had the seals applied by demonstrators in the field. The demonstrators would have a supply of blank yo-yos and strips of decals for application. This practice sometimes lead to the application of seals over the top of yo-yos that already had seals in place. Collectors should also be wary; there is at least one counterfeit Tournament 77 yo-yo in circulation which is being sold as an original.

486  *20*  **77 (broken Y pedestal), decal seal, wood, one-piece, '50s**
There are two versions of the Broken Y 77. It is called the "Broken Y" because there is a space halfway down on both Ys in the word "Yo-Yo." The two versions of this style of yo-yo are the "pedestal style," where Mr. Yo-Yo is standing on a small pedestal, and the "non-pedestal style" where the pedestal is absent. The pedestal style is more difficult to acquire.

It is often confusing to new collectors when older collectors use the number 77 to describe a yo-yo's shape or size. In the past, collectors referred to any tournament shaped yo-yo as a 77. In the 1950s, the stock numbers, 77, 44, and 33, often indicated Duncan sizes. The 77 was a 2-1/4" diameter yo-yo, the 44 was 2-1/8" and the 33 was 1-7/8". The 77 was typically a one-piece tournament shaped yo-yo while the 44s and 33s were three-piece beginners models. In the original sales literature, many yo-yos that were tournament shaped were identified by the stock number 77, so it became common to refer to all tournament shaped yo-yos as "77s." Collectors now only refer to "77s" as models that bear the number 77 at the base of their seal.

487  *20*  **77 (broken Y non-pedestal), decal seal, wood, one-piece, '50s**
See discussion under #486.

488  *20*  **77 (solid Y), decal seal, wood, one-piece, '50s**
In the Solid Y version of the 77, Mr. Yo-Yo does not stand on a pedestal. The Duncan Mr. Yo-Yo character first appeared in the late '40s. This logo was used throughout the '50s on Duncan awards, promotional literature, and some yo-yos.

489  *20*  **(77 fake), decal seal, wood, one-piece, '90s**
This yo-yo is a counterfeit made with the sole purpose of defrauding yo-yo collectors and they call it the "JT 77 Fake." The yo-yo blanks were original Royal yo-yo blanks without seals. An exact replica of the 77 seal was made in the '90s and placed on these yo-yos which were then marketed as originals. With careful observation, these can be easily identified from the originals.

There are several clues that can be used to identifying a JT 77 fake from an original. First, many of the JT 77s are in near mint condition. It is rare to have an original 77 in mint condition. Most have small chips or very fine age cracks in the decal. If you find a 77 in near mint condition, examine the seal carefully for the presence of age cracks. Second, the face of the fake yo-yo is wider than that of the original 77. Third, the registration mark ® and the letters in the word "Genuine" are larger on the JT 77 fake compared to the original 77. Finally, the color of the number 77 at the base of the seal should be the same color as the word "Tournament." In the JT 77 the word "Tournament" is in red and the number 77 is in black.

490  *20*  **77 (Day Glo), decal seal, wood, one-piece, '50s**
This yo-yo has an unusual specialty paint that was tried in the early '50s. The Day Glo's had a 77 seal, but were painted with fluorescent colors. Five colors were available: neon red, signal green, arc yellow, fire orange, and Saturn yellow. These yo-yos do not have the classic decorative stripe across the faces. The consistency of this particular paint caused the decal to easily detach. These yo-yos are very difficult to find with a seal completely intact. Collectors call these yo-yos "Day-Glo's" or "Day Glo 77s."

*Continued ...*

## YO-YO LISTINGS

● **DUNCAN TOURNAMENT** *Continued ...*

491    *20*    **(Big "G") GENUINE (gold seal), wood, one-piece, '30s**
Big "G" (Genuine) refers to the large "G" in the word "Genuine" that extends to the base of the seal. This is the archetypal Duncan tournament shaped yo-yo referred to as the original "gold seal." The decal seal has a gold background and comes in two styles. The first gold seal had letters with two colors, blue and black or red and black. These yo-yos are considered the earliest "gold seals." The seal was later changed to all black letters on a background of gold. The two color letter seals are valued more than the all black letter seal.

In the '30s, you could not enter Duncan contests unless you used an "original Duncan Gold Seal Yo-Yo." Production of this yo-yo started in 1929 and it was the most popular yo-yo throughout the '30s. The original price was 20 cents and they were retailed out of counter display boxes. Collectors refer to these yo-yos as "gold seals." An example of this yo-yo is shown next to this section's title.

492        **(Big "G") GENUINE (gold seal • black letters), wood, one-piece, '30s**
For discussion, see #491.

493    *20*    **(Big "G") GENUINE, gold leaf stamped seal, wood, one-piece, '30s**
Big "G" (Genuine) refers to the large "G" in the word "Genuine" that extends to the base of the seal. Although this was produced in the same time frame as the Gold Seal #491, it is not as highly valued.

494    *20*    **(Big "G") GENUINE, yellow decal seal, wood, one-piece, '30s and '40s**
Big "G" (Genuine) refers to the large "G" in the word "Genuine" that extends to the base of the seal. This seal has the same design as Gold Seal decal #491, but it is yellow instead of gold. This yo-yo should not be referred to as a "Gold Seal." Collectors call this yo-yo the "Big 'G' Yellow Seal."

494.1        **(Big "G") GENUINE DELUXE, hot stamped seal, plastic, '50s**
Big "G" (Genuine) refers to the large "G" in the word "Genuine" that extends to the base of the seal. This is a rare plastic pre-Flambeau variation.

495    *20*    **(Crossed flags), gold leaf stamped seal, wood, one-piece, '60s**
This model is the last of Duncan's wood Tournament line. It was produced from 1960 to early 1965. Although the production of this model stopped in 1965, so many were produced and warehoused that they continued to be retailed until the early '70s. This is the most easily found of the wood tournament line. Collectors call these yo-yos "Crossed Flags Tournaments." At least four different display cards were used with this yo-yo. Display cards carry the stock number of 77 or 107. Original pre-printed prices on display cards are 49 cents. More recent cards read 59 cents. For the original display card, see Plate #13. There is also a jewel version with this seal. See #358.

496        **(Crossed flags), gold leaf stamped seal, wood, jumbo, one-piece, '60s**
This yo-yo has the same seal as #495. Although not as large as the 4.5" diameter Duncan Jumbo Award yo-yos, this yo-yo is still a jumbo size. Few yo-yos of this size, bearing the Crossed Flag, exist. One can be seen in the National Yo-Yo Museum in Chico. It is unknown whether this yo-yo was ever retailed.

497        **(Crossed flags), hot stamped seal, plastic, early '60s**
This is a rare plastic pre-Flambeau yo-yo. The sales of this model, like those of the Little "G" Genuine #500, were apparently very poor. It is unclear as to why Duncan even produced this yo-yo.

498        **CHEERIO DUNCAN TOURNAMENT, die stamped seal, wood, '50s**
Duncan produced this Cheerio Yo-Yo after it bought out the Cheerio Company in 1954. This rare yo-yo has an airbrushed stripe which is unusual for Cheerio yo-yos. It is believed to be the only yo-yo with both "Duncan" and "Cheerio" on the seal.

•     **Imperial Tournament -** *See Duncan Imperial section.*

499    *20*    **(Little "G") GENUINE, gold leaf stamped seal, wood, one-piece, '50s**
Little "G" refers to the "G" in the word "Genuine." The "G" is the smallest letter in the word "Genuine." This model was the classic standard line tournament produced during the late '40s and '50s. Collectors refer to this yo-yo as the "Classic Little 'G'."

500    *20*    **(Little "G") GENUINE, hot stamped seal, plastic, metal axle, early mid '50s**
This was one of Duncan's first attempts at producing a plastic standard line yo-yo. It used the "Classic Little 'G'" seal. See #499. Sales had limited success, so the production run time was short. This is one of the most difficult of the Duncan standard line plastic yo-yos to find.

501    *20*    **(Little "G") GENUINE, yellow decal seal, wood, one-piece, '50s**
This has the same logo as the "Classic Little 'G'," but on a yellow decal seal. See #499. This yo-yo is more difficult to find than the die stamped seal version. This seal was also used during the '40s and '50s, the same period the 77 decal seal was in use, see #486. This decal seal, much like the 77, was prone to age cracks and is difficult to find in near mint condition. Collectors call this the "Little 'G' Yellow Seal."

502    *20*    **(Little "G") GENUINE (swirl paint finish), gold leaf stamped seal, wood, one-piece, '50s**
This is an unusual multi-colored swirl paint variation with the "Classic Little 'G'" seal #499. This line was briefly produced in the late '50s. This is the most difficult variation of the "Classic Little 'G'" seal to find. Collectors refer to these as "Swirls" or "Duncan Swirls." Fli-Back also produced a similar yo-yo known as a "Candy Swirl." See #573.

503    *20*    **(Little "G") GENUINE (suede finish), gold leaf stamped seal, wood, one-piece, mid '50s**
A special "flocked" paint coating gave this yo-yo a suede-like appearance. This is another rare variation of a yo-yo with the "Classic Little 'G'" seal #499. Duncan experimented with several unique paint styles in the middle to late '50s such as Day-Glo, Pearlescence, Swirl and Crackle. These paint variations had brief production runs and are highly coveted by collectors.

504    *20*    **(Loving cup), imprint seal, plastic, Flambeau, '70s**
This is the most recently produced Flambeau Tournament model. These yo-yos are called "Loving Cup Tournaments."

*Continued ...* ▶

## YO-YO LISTINGS

**505** *20* **(Loving cup • glitter embedments), hot stamped seal, plastic, Flambeau, '70s**
Some refer to this style of yo-yo as a "metal flake" or "glitter" yo-yo.

**506** *20* **PEARLESCENCE 888 (pearlescence paint), gold leaf stamped seal, wood, one-piece, late '50s**
Due to the expense of the special paint requirements, this unique yo-yo was only produced for a brief time. The paint had a pearl-like luster, from which the yo-yo derived its name. There are several different color variations. White is considered the most desirable. This is one of the few Duncan Tournaments which does not have the decorative paint stripe across the face. The yo-yo's stocking number, 888, is imprinted at the bottom of the seal. Both silver and gold leaf stamped seals can be found. Some collectors call these "Pearlescence" or "888s."

**506.1** **PEARLESCENCE 888 (fish scale finish), gold leaf stamped seal, wood, one-piece, late '50s**
This model has the same seal as #506, but with a fish scale finish. This rare paint finish is made by spraying multi-colored nylon filaments into the paint. This yo-yo is a rare pearlescence variation. An example of this yo-yo can be seen in the National Yo-Yo Museum in Chico, CA.

**507** *20* **SUPER, gold leaf stamped seal, wood, one-piece, early '50s**
This yo-yo was the original "Super" produced by Duncan. It was produced until 1956 when it was replaced with the Super Return Top. See #508.

**507.1** *20* **SUPER (splatter paint finish), gold leaf stamped seal, wood, one-piece, '50s**
This is the rarest variation of the Duncan Super #507. A contrasting color is sprayed onto the base color in a random fashion giving it a splatter appearance. There is no airbrushed stripe. Collectors call this yo-yo the "Super Splatter."

**508** *20* **SUPER RETURN TOPS, gold leaf stamped seal, wood, one-piece, '50s and '60s**
This model was produced from 1956 to the early '60s. It has "Return Top" imprinted on the seal, unlike the original Super. "Return Top" was added to many Duncan yo-yos in the late '50s and early '60s. It was Duncan's contention that "yo-yo" was the trademarked name and "Return Top" was the generic name for the toy.

**509** *20* **SUPER RETURN TOPS (re-release), gold leaf stamped seal, wood, Flambeau, three-piece, '90s**
This is the first wood Duncan yo-yo to be produced since 1965, and the only wood yo-yo that Flambeau has issued. The yo-yo is being produced by the What's Next Company and was released in 1996. The style and shape of the yo-yo is almost identical to the original Super, except that behind the word "Duncan" there is a registration mark ® that is not on the original. The originals have a registration mark ® below the word "Yo-Yo." This makes the re-release easy to differentiate from the original. The yo-yo is retailed in a plastic case and includes a re-release of the original Duncan trick book with a re-designed cover.

**510** *20* **SUPER PRACTICE RETURN TOPS, gold leaf stamped seal, wood, one-piece, '60s**
This is the rarest variation of the "Duncan Super Return Top." It has the same logo as #508, except "practice" replaces the word "Tournament" on the seal. This yo-yo was retailed in a polybag with a saddle header card and a pre-printed price of 49 cents.

**511** *20* **TOURNAMENT, hot stamped seal, plastic, Flambeau, early '70s**
The seal reads "Duncan Yo-Yo Tournament." The letters in the title are arranged in a circle. This was Flambeau's first tournament yo-yo. It retailed on a blister card with a pre-printed price of $1.00, Stock No. 3262.

**512** *20* **TOURNAMENT (big "T"), gold leaf stamped seal, wood, '60s**
The seal has the word "Tournament" centered and the letter "T" larger than the letters following.

**512.1** *20* **TOURNAMENT (big "T"), hot stamped seal, plastic, Flambeau, '70s**
The seal has the word "Tournament" centered and the letter "T" larger than the letters following. This was retailed on blister card with a pre-printed price of $1.49, Stock No. 3260.

**512.2** *20* **TOURNAMENT (big "T" • glitter embedments), hot stamped seal, plastic, Flambeau, '70s**
The seal has the word "Tournament" centered and the letter "T" larger than the letters following. The seal on this model has a logo like #512 and embedments like #505. This yo-yo was later replaced by #505.

# YO-YO LISTINGS

## DUNCAN WHEELS

The original design idea for a Wheel yo-yo was created by Gregory Hart (Englehart) in 1970, with his Hot Spin Wheel Yo-Yo. This non-Duncan series was first produced in the early '70s. The first model was the Road Runner #1605.1. In 1974, Greg Hart was introduced to Bill Sauey, the owner of Duncan Flambeau. The Duncan company was interested in producing a Wheel line of yo-yos. Hart worked with Harvey Lowe, a long time Duncan demonstrator, to design a Duncan Wheel yo-yo with better spin and control.

Duncan Flambeau first released "Duncan Wheels" in 1979. The original Duncan Wheels were opaque black plastic with a shiny chrome finish. (Care needs to be taken in storage of any chrome finish wheels, as the chrome imprinted letters are very prone to scratches.) These yo-yos were made of dense ABS plastic with most of the weight near the rim. This design gives it longer spinning action than the standard butterfly. Four "Mag Wheel" versions were made and these have continued to be used even with recent design changes. Duncan Flambeau produced a commercial promoting the Wheels yo-yo in the early '80s. In the video, the yo-yos had a shiny chrome finish, but the lettering on the faces was raised, white and said, "Duncan Wheels." It is not clear whether this yo-yo was retailed or used only in the promotional video. The most collectible of the Duncan Wheels is the "Dukes of Hazzard" series which is cross-collectible. A few advertising wheels were also produced, but these are uncommon.

| | | |
|---|---|---|
| 513 | *17* | **AUTOFINDERS (chrome finish), imprint seal, butterfly shape, '80s** |
| | | This is a rare example of a wheel ad yo-yo. |
| 513.1 | | **FIRE WHEELS (chrome finish), imprint seal, butterfly shape, 1995** |
| | | This is a German-made Duncan. It is a red plastic yo-yo with a chrome finish. These are no longer produced. |
| 514 | *17* | **GENERAL LEE (Dukes of Hazzard • chrome finish), butterfly shape, '80s** |
| | | This yo-yo is based on the "Dukes of Hazzard" TV series (1979-1985). This is the most common of the Dukes of Hazzard yo-yo series. There are four different hub styles. Luke Duke (Tom Wopat) and Bo Duke (John Schneider) are pictured on the blister card, Stock No. 3282DY. |
| 514.1 | *17* | **HAZZARD POLICE (Dukes of Hazzard • chrome finish), imprint seal, butterfly shape, '80s** |
| | | This is another Wheel version from the Dukes of Hazzard show. The Hazzard Police is the most difficult to find of the series. This yo-yo also had four different hub styles. |
| 514.2 | *17* | **DAISY'S CJ JEEP (Dukes of Hazzard • chrome finish), imprint seal, butterfly shape, '80s** |
| | | This is another yo-yo which also has four different hub styles from the Dukes of Hazzard series. This model is more difficult to find than the General Lee, but easier to find than the Hazzard Police yo-yo. The carded yo-yo is shown on Plate 13, #514.2. |
| 514.3 | | **HYPERWHEELS, imprint seal, butterfly shape, 1997** |
| | | This Wheel series was originally made for the Japanese market only, however, some were put on English blister cards and retailed in the states by mistake. An example of this yo-yo is shown next to this section's title. |
| 515 | *17* | **LONG SPIN WHEELS, imprint seal, butterfly shape, '90s** |
| | | After this original chrome style was discontinued, four hub styles in different colors were released in 1988. The colors were light gray, red and white. The Wheels with white hubs are reportedly discontinued. More than one version of the blister card exists. See Plate 9, #515. |
| 515.1 | | **LONG SPIN WHEELS, imprint seal, butterfly shape, metal rim inserts, '90s** |
| | | These are rare versions of #515. Metal rings are inserted on the face, near the rim. The seal is lettered on the metal inserts. |
| 516 | *17* | **LONG SPIN WHEELS (chrome finish), imprint seal, plastic, butterfly shape, '80s** |
| | | This was the original Duncan Wheels model produced in the '70s and '80s. There were four hub styles produced. During this period, Wheels were more popular than the standard butterflies because they were heavier and had a longer sleep time. The original wheels were retailed in a cardboard hanging display box. |

## YO-YO LISTINGS

### FESTIVAL

Wilfred Schlee Jr. founded the Festival Yo-Yo Company in 1965. Schlee left the Donald F. Duncan Company just months before it was forced into bankruptcy. The Union Wadding Company acquired Festival in 1966 and Schlee, Jr. ran the Festival division until 1981. Festival did not run yo-yo contests, but it did conduct promotions. Festival's only demonstrator was Bob Rule, a former Duncan demonstrator. Retailers of Festival yo-yos, such as Hallmark, would request a promotion and Festival would contact Rule.

The "Be-A-Sport" series was the first line of yo-yos released by the Festival Yo-Yo Company. These yo-yos were molded with the Festival logo and the words "Festival Professional Model." The "Be-A-Sport" series came in showcase display boxes and retailed on blister cards. The yo-yos packaged in the showcase display boxes are considered more desirable than the ones on blister cards. Original showcase boxes from the '60s were black and fuchsia and had a pre-printed price of $1.00. Showcase boxes in the early '70s were green and blue; in the late '70s the boxes were yellow and blue.

Although better known for their plastic yo-yos, Festival also produced a line of wooden yo-yos known as Zappers. Most of Festival's wood yo-yos were produced in Sweden and have "Sweden" die stamped on the reverse face. Three styles of Zappers were made, the "Big Zapper" (2-5/16" diameter), "Little Zapper" (2-1/8" diameter), and the "Zapper" (2" diameter).

From 1970 to 1982, The Union Wadding Company owned the rights to market Disney characters and produced a wide selection of Disney yo-yos. In the early '80s they also acquired licensing rights to Hanna Barbera characters. See also the Disney and Hanna Barbera sections.

In 1980, Union Wadding decided that yo-yos no longer fit in with their other merchandise lines and they elected to drop Festival. In 1981, they sold their Festival line to Monogram Products Inc. which continues to produce Festival and other licensed yo-yos such as Disney and Peanuts.

Collectors may be interested to know that Monogram Products still produces plastic yo-yos from the original Festival molds. The original Festival yo-yos from the '70s can be difficult to distinguish from recently produced models. Originals have paint in the molded recesses of the Festival seal. Paint was applied by hand over the surface of the yo-yo. The yo-yo was then placed in a tumbler to wear off the paint, except in the recesses. The recently produced Monogram versions have no paint in the Festival seal, but may have paint in other recesses of the yo-yo.

Monogram Products could theoretically reproduce any of the original Festival line, but company representatives say this is unlikely. There are a few items still being produced and some that could possibly be re-released in the future. It is highly unlikely that any of the wood or tin lines will be re-released. To guarantee authenticity of original models, it would be safest to collect plastic Festivals in original packaging.

| | | |
|---|---|---|
| 517 | 08 | **ALL AMERICAN YO-YO**, imprint seal, plastic, tournament shape, early '70s |
| | | There were two variations of the plastic Festival All American Yo-Yo. The one shown was produced until 1973 and was made of opaque plastic. The blister card carried a pre-printed price of $1.00. In 1973, the seal design changed to the All Star Champion #519. In 1977, they made an All American out of translucent plastic with a different seal. |
| 518 | | **ALL AMERICAN YO-YO**, paper sticker seal, wood, tournament shape, late '70s |
| | | This was a natural wood yo-yo made in the late '70s and early '80s. |
| 519 | 08 | **ALL STAR CHAMPION YO-YO (with Festival)**, imprint seal, plastic, tournament shape, '70s |
| | | This yo-yo was first introduced in 1973. Two similar styles of this seal exist, this one and one missing the word "Festival." Early blister cards have a pre-printed price in the upper right hand corner. See Plate 21, #519. The later cards have no price pre-printed. |
| 520 | 08 | **ALL STAR CHAMPION YO-YO (without Festival)**, imprint seal, plastic, tournament shape, '70s |
| | | Two styles of this yo-yo exist, this one and one marked "Festival" on the seal. See #519. |
| 521 | 08 | **(Baseball)**, colorless molded seal, plastic, Monogram, ball shape, metal axle, '90s |
| | | Monogram is still using the same Baseball yo-yo mold that Union Wadding used. Whereas some of the original Festival seals were colorless, all the Monogram versions are this way. Monogram Baseballs use red paint for both the stitches and the molded lettering reading "Official League." |
| 522 | 21 | **(Baseball)**, red or black molded seal, plastic, ball shape, '70s |
| | | Any Baseball yo-yo that has color on the "Festival" seal was produced by Festival. The Monogram versions are produced from the same mold, but have a colorless "Festival" seal. The presence of black or red printing in the Festival seal identifies this as an original "Festival." For this yo-yo on a Hallmark blister card, see Plate 21, #522. |

*Continued ...*

## YO-YO LISTINGS

**FESTIVAL** *Continued ...*

523    *08*    **(Baseball • red stitches), black molded seal, plastic, ball shape, '70s**
Two-color baseball yo-yos are considered the most desirable of the Festival baseballs. Monogram has not used two colors of paint on their baseball yo-yos.

524    *08*    **(Baseball • gold metallic finish), molded seal, plastic, ball shape, '70s**
This model is the rarest of the Festival baseball series by Union Wadding. Monogram has not made gold finish baseballs.

525    *08*    **(Basketball), black molded seal, plastic, ball shape, '70s**
This Basketball yo-yo was given out as part of a Yo-Yo Day promotion at a professional basketball game in the '70s. It was also available retail. For a short time in 1975, this yo-yo came on a Harlem Globetrotters blister card. See #526. This Globetrotters card carries a higher value than the standard Festival card. The basketball yo-yo has been re-released by Monogram. The re-released version does not have paint in the Festival seal.

525.1    **(Basketball), molded colorless seal, plastic, Monogram, ball shaped, '90s**
This is the basketball re-release made by Monogram. Unlike the original Festival basketballs made by Union Wadding, this has no paint in the Festival seal. The plastic color is also different from the original Festival basketball.

526    *21*    **(Basketball on Globetrotters card), '70s**
This is the same yo-yo as #525, but on the Globetrotters blister card.

527    *21*    **BIG ZAPPER, gold leaf stamped seal, wood, tournament shape, one-piece, '70s**
The Zapper Yo-Yo line was first introduced in 1967. The Big Zapper was the largest of the series being 2-5/16" in diameter. Three seal variations for the Big Zapper yo-yos exist. Original cards had a pre-printed price of 59 cents. These were later changed to 69 cents, then to $1.00.

528    **BIG ZAPPER, foil sticker seal, wood, tournament shape, '70s**

529    *08*    **BIG ZAPPER, paper sticker seal, wood, tournament shape, one-piece, late '70s**
Festival first started using paper sticker seals on this yo-yo in 1977.

530    **BOB RULE, paper sticker seal, plastic, tournament shape, '70s**
This yo-yo was both given away and retailed. Bob Rule was a former Duncan demonstrator who became Festival's only promotional demonstrator. An example of this yo-yo is shown next to this section's title.

531    *08*    **(Bowling ball), molded seal, plastic, ball shape, metal axle, '70s**
This yo-yo could be re-released by Monogram, but that is unlikely. Originals have white paint on the Festival seal and finger holes. For an example of a display box, see Plate 31, #531.

•    **Disney Character Yo-Yos -** *See Disney section.*

532    *08*    **(Disney Pals autographed baseball), molded seal, plastic, ball shape, '70s**
Produced in the middle '70s, this yo-yo displays the autographs of Mickey Mouse, Donald Duck, Pluto, and Goofy. It is a standard Festival Baseball #522 and is the hardest to find of the Festival Disney yo-yos.

533    *08*    **(Disney Pals autographed football), molded seal, plastic, sculpted, '70s**
This yo-yo displays the autographs of Mickey Mouse, Donald Duck, Pluto and Goofy. It is a standard Festival Football #536. It is hard to find in its original packaging, however, it is more common than the autographed Baseball version. See #532.

534    *21*    **DRAGONFLY, paper sticker seal, plastic, butterfly shape, metal axle, '70s**
The Dragonfly was Festival's version of a butterfly shaped yo-yo and could possibly be re-released by Monogram. The Dragonfly's original card carries a pre-printed price of $1.00. Later cards do not show a pre-printed price. If this Dragonfly yo-yo is re-released, it will be on a different, less valuable, blister card. Therefore, this model should be collected on its original blister card.

535    *08*    **(Eight ball), molded seal, plastic, ball shape, metal axle, '70s**
The original versions of this yo-yo have a white "8" as part of the molded seal.

•    **Flintstones -** *See Hanna Barbera section.*

536    *31*    **(Football), molded paint seal, plastic, sculpted football shape, metal axle, '70s**
Like the others in the "Be A Sport" series, this yo-yo was made in the '60s and '70s. This version has paint in the molded Festival seal. The paint is either white or black. These yo-yos retailed both in showcase boxes and on blister cards. Those in the showcase boxes carry a higher value. See Plate 31, #536. This mold is still being used to produce Festival Footballs for Monogram. See #537.

537    **(Football), molded seal without paint, plastic, Monogram, sculpted football shape, '90s**
This yo-yo is still produced by Monogram. It does not have paint in the Festival seal, unlike most originals. See #536. The plastic color used for the Monogram Football is different from that used for the original Festival Football.

538    *21*    **(Football on Joe Namath card), '70s**
This is a standard Festival Football #536, on a Joe Namath blister card. The blister card has a pre-printed price of $1.29. This card was briefly released in the middle '70s.

539    **(Football • gold metallic finish), molded seal, plastic, sculpted football shape, '70s**
Gold and silver Footballs on Joe Namath blister cards, or in showcase boxes, are considered the most collectible of the Festival line. Gold and silver versions of this yo-yo were packaged on both Joe Namath blister cards and in showcase boxes reading "Joe Namath Signature Yo-Yo." Although both the blister card and showcase box are considered rare, the showcase box is more desirable. Football yo-yos with this metallic paint finish are unlikely to be re-released by Monogram.

540    **(Football • silver metallic finish), molded seal, plastic, sculpted football shape, '70s**
This yo-yo is the same as #539, but with a silver finish.

*Continued ...*

## YO-YO LISTINGS

541 *08* **(Game yo-yo • baseball), paper insert seal, plastic, tournament shape, view lens, '70s**
Union Wadding used this mold to make several different Festival "pinball" game yo-yos. The standard line was called "Game Yo-Yo" and there were four in the series. The mold was also used to make Disney, Hanna-Barbara, and Peanuts versions which are all considered more collectible than the standard line game yo-yos.

542 **(Game yo-yo • golf), paper insert seal, plastic, tournament shape, view lens, '70s**
Like #541, this has a pinball game in the view lens.

543 *08* **(Game yo-yo) PINBALL, paper insert seal, plastic, tournament shape, view lens, '70s**
Like #541, this has a pinball game in the view lens.

544 *08* **(Game yo-yo) TOE-TAC-TIC, paper insert seal, plastic, tournament shape, view lens, '70s**
Like #541, this has a pinball game in the view lens.

545 *08* **(Golf ball), hot stamped seal, plastic, sculpted golf ball shape, metal axle, '70s**
Many early Festival Golf Ball yo-yos have a colorless seal. Unless they are in their original packaging, these Golf Ball yo-yos cannot be distinguished from the re-release versions. Monogram may re-release this yo-yo.

546 **(Hockey puck • any team logo), paper sticker seal, plastic, puck shape, metal axle, '70s**
The Hockey Puck yo-yos were introduced in 1972. They were retailed on blister cards and out of a counter display box. Seventeen different logo seals were produced for this series, 16 team logos and the NHL logo. The most popular among collectors is the NHL logo variation.

The Hockey Puck yo-yo was used to promote professional hockey games. The yo-yo was given out to kids who attended the game on "Hockey Puck Yo-Yo Night." Other products, like Coca-Cola, have occasionally used this yo-yo for promotions. This yo-yo could possibly be re-released by Monogram, but it is doubtful that the original team seals would be reproduced.

547 *08* **(Hockey puck • NHL logo), paper sticker seal, plastic, puck shape, metal axle, '70s**
For discussion, see #546.

548 *21* **(Joe Namath photo), photo sticker seal, plastic, tournament shape, '70s**
This yo-yo was produced in 1975 and came on a football shaped blister card with a price of $1.00. The reverse of the blister card carried a discount ad for the Joe Namath Football Camp. The discount ad expired on July 1, 1976. For other Joe Namath yo-yos, see #538 and #539.

549 *08* **LITTLE ZAPPER, foil sticker seal, wood, tournament shape, three-piece, pegged string, '70s**
This yo-yo was considered Festival's Beginner model. Smaller than the Big Zapper, this yo-yo is slightly larger than the Zapper. There are two seal variations. This is the considered more collectible than the paper seal version. See #550.

550 **LITTLE ZAPPER, paper sticker seal, wood, tournament shape, three-piece, '70s**
This paper seal was first used by Festival in 1977.

551 *21* **OLYMPIC, decal seal, plastic, tournament shape, '70s**

• **Rescuers** - *See Disney section #301.*

552 *21* **SCREAMER, paper sticker seal, tin, tournament shape, whistler holes, '60s and '70s**
The "Screamer" is a whistling yo-yo first released in 1967. Air holes in the rim cause a whistling sound when the yo-yo spins. The blister cards carried a pre-printed retail price of 59 cents. The design of the Screamer seal changed in 1972. See #553. For the blister card, see Plate 21, #552.

553 *19* **SCREAMER, paper sticker seal, tin, tournament shape, whistler holes, '70s**
This is the second version of the Festival Screamer. This version is more difficult to find than #552, even though it was more recently produced. Production began in 1972 and ended in 1976.

554 *08* **(Tennis ball), molded seal, plastic, ball shape, '70s and '90s**
The Tennis Ball yo-yo was not one of the original "Be-A-Sport" series, but was added in 1974. Of all the yo-yos in the "Be A Sport" series, this is the most difficult to find in its original showcase box. Monogram still produces this model. In order to guarantee that a Tennis Ball yo-yo is an original, it must be collected in its original showcase box. Two color variations were made, fluorescent green and white.

555 *21* **ZAPPER, foil sticker seal, wood, tournament shape, '60s and '70s**
The production of the Zapper began in 1967. The Zapper is 2" in diameter, smaller than both the Big Zapper and Little Zapper. The Zapper did not have a pegged string and retailed at a higher price than a Little Zapper. Zappers are more difficult to find than Big and Little Zappers because they were not in production for many years.

556 **ZAPPER, gold leaf stamped seal, wood, tournament shape, one-piece, '70s**

## YO-YO LISTINGS

## FLI-BACK

James Emery Gibson founded the Fli-Back Company in the early 1930s. The company originally did not make yo-yos, but produced two styles of paddleballs, the Fli-Back and the Sock-It. In 1946, the company diversified their product line by manufacturing other toys. This was when their first yo-yos were introduced. One of those first yo-yos was the Whirl King Top. (See the Whirl King section.)

In 1968, James Gibson died and his heirs sold the company to the Ohio Art Company. Ohio Art continued to produce Fli-Back yo-yos throughout the 1970s. Fli-Back produced one and three-piece wooden yo-yos as well as plastic models. The three-piece yo-yos frequently had pegged strings and two different colored halves. The most common color combination was red and blue. Fli-Back yo-yos had three principal seal designs: numbered seals, doughnut seals and eagle seals. The most common Fli-Back logo is the double circle or "doughnut seal" logo. Fli-Back also produced lines of yo-yo's that did not bear the Fli-Back name: Style 55 Champion Return Tops, Orbit, Orbit Away, Sock-It-Space Whirler and Whirl King.

| # | $ | Description |
|---|---|---|
| 557 | | 45, gold leaf stamped seal, wood, tournament shape, '60s |
| 558 | | 45 (with stars), gold leaf stamped seal, wood, tournament shape, '60s |
| | | Circling stars surround the number 45 on this model. The word "Fli-Back" is not lettered on the yo-yo seal. |
| 559 | | 55, gold leaf stamped seal, wood, tournament shape, three-piece, pegged string, '60s |
| | | This yo-yo is considered the most common of the numbered Fli-Back yo-yos. |
| 560 | | 60, gold leaf stamped seal, wood, tournament shape, three-piece, '60s |
| 561 | 22 | 65, gold leaf stamped seal, wood, tournament shape, three-piece, '60s |
| | | Produced in the early '60s, this model is more rare than the 55, but more common than the other numbered Fli-Backs. |
| 562 | | 823, die stamped seal, wood, tournament shape, metallic glitter finish, '60s |
| | | This is a rare model produced in the early '60s. |
| 563 | 32 | (Astro spin), foil sticker seal, plastic, tournament shape, early '60s |
| | | This blister carded yo-yo was released in 1962. The card features a graphic of an astronaut. The card carries the Stock No. #2099-B and has a pre-printed price of 79 cents. The yo-yo is plastic, with a hollow aluminum axle. The idea behind this design was to create a hollow axle that would vent out heat through a hole in the yo-yo's face, thereby cooling the axle. This yo-yo has a hole in each face, but the sticker seal only covers the hole on the one side. The seal is a foil sticker "doughnut seal" similar to #567. |
| 564 | 22 | (Double circle), gold leaf stamped seal, wood, tournament shape, '60s |
| | | Collectors refer to this type of logo as the "doughnut seal." This Fli-Back "doughnut" was the standard line model for many years. It is very common. |
| 564.1 | 22 | (Double circle), paint seal, wood, tournament shape, three-piece, pegged string, late '70s |
| | | This painted seal variation is believed to have replaced the earlier die stamped seal #564. |
| 565 | 22 | (Double circle • metallic glitter finish), die stamped seal, wood, tournament shape, three-piece, '60s |
| | | This yo-yo came in three color variations: green, blue and red. |
| 566 | 22 | (Double circle • skinny), die stamped seal, tournament shape, three-piece, pegged string, '70s |
| | | This double circle variation has both circles outside of the lettering. It is sometimes called the "skinny doughnut seal." |
| 567 | 22 | (Double circle), foil sticker seal, wood, tournament shape, three-piece, '70s |
| | | Collectors call this yo-yo the "foil doughnut seal." |
| 568 | 22 | (Double circle • double arch), foil sticker seal, wood, tournament shape, one-piece, '60s |
| | | This is an unusual variation of the "doughnut seal." This seal has a double arch through the doughnut seal. |
| 569 | 22 | (Double circle), embossed seal, plastic, tournament shape, wood axle, pegged string, late '70s |
| | | This is a translucent brittle plastic yo-yo and is one of the few low quality plastic yo-yos to have a wooden axle. |
| 570 | 22 | (Double circle • glitter embedments), imprint seal, plastic, tournament shape, '70s |
| 571 | 22 | (Eagle), gold leaf stamped seal, tournament shape, three-piece, pegged string, '60s |
| | | Collectors refer to Fli-Backs with eagles on the seals as "Fli-Back Eagles." This model is known as the "Circled Eagle." |
| 572 | 22 | (Eagle), foil sticker seal, wood, tournament shape, one-piece, '50s |
| | | These models were the best playing yo-yos produced by Fli-Back. The foil sticker seals were either red or blue. Collectors refer to these as "Fli-Back Eagles" or "Foil Eagles." These were retailed loose and sold in polybags with saddle header cards for 29 cents. |
| 573 | 22 | (Eagle • candy swirl finish), foil sticker seal, wood, tournament shape, one-piece, '50s |
| | | Because Duncan also produced a multi-colored paint swirl patterned yo-yo in the late '50s, #502, it is unknown which company released this style first. This is considered the most collectible of the Fli-Back line and it is very difficult to find with the foil sticker seal attached. Collectors call these "Candy Swirls" or "Candy Swirl Eagles." An example of this yo-yo is shown next to this section's title. |

Continued ...

## YO-YO LISTINGS

574  22  **(Eagle), gold leaf stamped seal, wood, tournament shape, one-piece, early '70s**
This Fli-Back Eagle does not have a circle around the logo. The seal reads "Championship Tournament." This version is less common than the "Circled Eagle" style. See #571.

575  22  **ORBIT- A-WAY, gold leaf stamped seal, wood, butterfly shape, '60s**
This is a wood butterfly with a metallic glitter paint finish. It was first produced by Fli-Back in 1962.

576  **ORBIT-A-WAY, plastic, tournament shape, aluminum axle, early '70s**

- **Whirl-King** - *See Whirl King section #1608, #1609.*
- **Style 55 Champion** - *See Style 55 Champion section.*

## FLORES

The Flores Yo-Yo was the first "yo-yo" manufactured in the United States. Its originator was Pedro Flores. Prior to Flores, the yo-yo was called a bandalore in the United States. One astute observer noted in the late 1920s, "We've all done the yo-yo before, but we never had a name for it."

Pedro Flores was a native of Vintar, Ilocos Norte, Philippines. He immigrated to the United States in 1915. He attended the High School of Commerce in San Francisco (1919 - 1920) and then took up the study of law at the University of California in Berkeley and the Hastings College of Law in San Francisco. Flores dropped out of school for reasons unknown and moved to Santa Barbara, California. He worked at odd jobs for years. When he started his yo-yo business, he was working as a bell-boy in Santa Barbara.

Flores developed his vision for the yo-yo when he read about a man that had made a million dollars selling a ball attached to a rubber band. He remembered the game "yo-yo," which was played for hundreds of years in the Philippines. Flores thought the yo-yo had good market potential in the United States. He was quoted as saying, "I do not expect to make a million dollars, I just want to be working for myself. I have been working for other people for practically all my life and I don't like it."

In early 1928, Flores came to Los Angeles and asked some wealthy Filipino countrymen for assistance in manufacturing yo-yos. His friends thought he was crazy, so he returned to Santa Barbara with only his dream. Being a true entrepreneur, at the age of 29, on June 9th 1928, Flores applied for and received a certificate to conduct business as the Yo-Yo Manufacturing Company in Santa Barbara. On June 23, 1928, he made one dozen yo-yos by hand and began selling them to neighborhood children. By November of 1928, his company had produced 2000 yo-yos and he was able to attract two American financiers, James Lewis and Daniel Stone of Los Angeles. Now, with the ability to produce machine-made yo-yos, four months later, over 100,000 yo-yos were produced. By November of 1929, newspaper accounts reported Flores' three factories were making 300,000 yo-yos daily and employing 600 workers. (These numbers are now debated.) These companies were the Flores and Stone Company, Los Angeles; The Flores Yo-Yo Corporation, Hollywood; and the Yo-Yo Manufacturing Company, Santa Barbara.

*Continued ...*

*Pedro Flores demonstrating the yo-yo, circa 1930.*

## YO-YO LISTINGS

● **FLORES** *Continued ...*

Thanks to Flores and his yo-yo spinning contest, yo-yoing became immensely popular in the United States in late 1928 and 1929. The yo-yo was promoted as the "Flores Yo-Yo, The Wonder Toy." The marketing slogan was, "If it isn't Flores, it isn't a yo-yo." Duncan used the same slogan after purchasing the yo-yo trademark from Flores.

Early contests did result in increasing the popularity of the yo-yo, but they were clearly different from modern contests. Endurance was most important during the initial Flores contests. Any individual who could keep his or her yo-yo spinning up and down for the longest time without missing was declared the winner. Many contests resulted in ties when stubborn competitors refused to quit after hours of continuous yo-yoing. When this happened, champions were determined by drawing straws. In addition to the endurance contest, participants competed to see who could throw their yo-yo the farthest with a complete return to the hand, and who could perform the most perfect spins in five minutes. Prizes were also awarded for fancy hand-made yo-yos and for the largest yo-yo. During these early contests, it was common to see working yo-yos made out of bicycle wheels and wooden barrel tops. Contests could be staged anywhere, but on November 22, 1929, the Gates Theater in Portsmouth, Virginia, became the first theater to hold a contest. For the rest of the '20s and '30s, theaters became the most popular sites for contests.

Some Flores yo-yos did have a slip string and could potentially do spinners, but the importance of this feature was not recognized right away. The yo-yo instructions said that the string should be wound tightly to allow for better return. Many early Flores yo-yo strings were made out of silk which resulted in less sleep action than later cotton strings. Prices for Flores yo-yos in 1929 ranged from $.15 to $1.50 each depending on the design and decoration. Flores employed Dorothy Carter as his chief designer during this production period.

By the end of 1929, a true yo-yo craze was spreading across the country. More yo-yo manufacturers began entering the arena including Don Duncan, Jr., Louis Marx and others. Although Duncan is now the name most frequently associated with the popularity of the yo-yo, Pedro Flores fueled the original yo-yo fire.

Although Pedro Flores was commonly described as the inventor of the yo-yo, he never personally claimed to have invented it. Flores always said the yo-yo was a centuries old Filipino game. He was also frequently described as the patent holder of the yo-yo, but yo-yos, "Bandalores," had already been patented in the United States dating back to 1866. Even though "Patent Applied For" and "Patent Pending" are often imprinted on Flores yo-yos, this was only a technique used to dissuade other toy companies from producing yo-yos. Pedro Flores never held any legal patent for the standard yo-yo. Flores did apply for and receive a trademark for the "Flores Yo-Yo" which was registered on July 22, 1930. Shortly after this Flores sold his interest in the yo-yo company to Donald F. Duncan.

It is unclear exactly what date the Duncan Yo-Yo Company acquired the Flores Yo-Yo name. It is currently believed that the transfer took place in 1930. It is documented that Duncan did have the Flores trademark legally assigned in 1932. For a brief period of time, in the early 1930s, the Duncan Corporation sold both Duncan Yo-Yos and Flores Yo-Yos. In early Duncan contests in 1931, either a "Genuine Flores Yo-Yo" or a "Genuine Duncan Gold Seal Yo-Yo" could be used in competition.

Pedro Flores reportedly sold his interest in his yo-yo manufacturing companies for more than one quarter of a million dollars. Other sources quote thirty thousand dollars. Whatever the amount, this was a fortune during The Depression. Flores was quoted as saying, "I am more interested in teaching children to use the yo-yos than I am in manufacturing of yo-yos." Flores followed through by becoming one of the key promoters during Duncan's early yo-yo campaigns. In 1931 and '32, Flores was instrumental in setting up many Duncan promotions. The Duncan contests had vastly changed from the initial contests run by Flores just two years previously. The contests now required correct performance of a series of tricks. Ties were broken by the number of "Loop the Loops" completed.

Flores remained involved with yo-yos his entire life. After leaving Duncan in the '30s, he reportedly set up the Bandalore Company which briefly made a Flores shaped yo-yo. After WWII, Flores helped Joe Radovan establish the Chico Yo-Yo Company and helped demonstrate Radovan's Royal yo-yos. In 1954, he started the Flores Corp. of America which briefly produced another line of Flores yo-yos. Linda Sengpiel, the first female professional demonstrator, worked with Flores for several years in the '50s and was his only demonstrator during this period. During the '50s, most of Flores' promotions and retail sales took place at Woolworth stores. Although the general public may not recognize his name, it was Pedro Flores who started the first yo-yo craze in America.

577     23     **(Big "F") FLORES, ink stamped seal, wood, tournament shape, one-piece, 1929 - '30**
This is the original Flores yo-yo from 1929 and 1930. This was America's first "yo-yo." The seal design has a large "F" in the name Flores. Flores yo-yos have many shapes and seals. Collectors refer to a Flores style yo-yo, like this model, as one with a narrow width and narrow string slot. For a profile view of a Flores yo-yo, see the Yo-Yo Shapes Plate. An example of this yo-yo is shown next to this section's title.

578         **(Big "F") FLORES, decal seal, wood, ball shape, miniature, 1929 - '30**
This yo-yo is not completely round, but it does resemble a ball shape more than a tournament shape. This is an early style Flores beginner model yo-yo from 1929 and 1930.

*Continued ...*

## YO-YO LISTINGS

579      **(Big "F") FLORES, decal seal, wood, tournament shape, 1929 - '30**
This yo-yo has the same logo design as the ink stamped seal. See #577. Because the seal on this model tends to chip easily, it is difficult to find with a complete seal.

580    *23*    **FLORES (Flores centered), die stamped seal, wood, jumbo, '50s**
Rumor has it that this die still exists, so it is possible that some of these yo-yos were produced more recently. Flores originally used these for promotions and to give as gifts to friends.

581      **FLORES ORIGINAL TOURNAMENT YO-YO TOP, die stamped seal, wood, jumbo, '50s**
This is another jumbo variation from the '50s and can be seen on display at the National Yo-Yo Museum in Chico, CA. This yo-yo has the same seal as #582.

582    *23*    **FLORES ORIGINAL TOURNAMENT YO-YO TOP, gold leaf stamped seal, wood, tournament shape, '50s**
Twenty five years after the release of the original Flores Yo-Yo, this model was produced by Flores second company "Flores Corp. of America." This yo-yo does not have a narrow width and string slot like the original. See #577.

583    *23*    **FLORES, decal seal, wood, tournament shape, 1929**
This has a different seal design than the Big "F" (Flores). The "F" in Flores is the same size as the other letters.

# FOREIGN

This book is geared toward yo-yos that were retailed in the United States. Undoubtedly collectors and dealers will come across foreign yo-yos and some examples are listed in this section.

Duncan has a sister company in Mexico, (Duscan, S. S. de C.V.) which produces Duncan yo-yos from components interchangeable with Flambeau. The yo-yos are identical in appearance, except the Mexican company sometimes used unusual colors of plastic, like brown. These yo-yos are not marketed in the United States, but apparently in 1994 some did trickle across the border. Unlike American made Duncans which retail only on "skin pack" blister cards, the Mexican made yo-yos are retailed on hard plastic blister cards. The Mexican Duncans also have "Hecho en Mexico" printed on the cards. The trick descriptions on the back of the cards are written in Spanish.

*A French street demonstrator in 1932, Paris, France.*

Duncan also had an arrangement with a German company that produced Duncan yo-yos overseas. Unlike the Mexican company, the tools used in producing these yo-yos were not interchangeable with the current Flambeau tools. This German company produced their own line of Duncan yo-yos with seals different from those on the American Duncans. The three most common model designs were the Butterfly, Wheels and Special. These yo-yos do say "Duncan" on the seal which can cause some confusion for collectors. This company went out of business in 1995, and the tools were returned to Flambeau. Flambeau is not currently using these molds.

Premier yo-yos are the more common foreign yo-yos filtering into the United States. Premier is the major yo-yo company in Mexico, with their headquarters located in Mexico City. Founded in the middle 1980s, Premier still promotes with traveling demonstrators. U.S. collectors currently have only minimal interest in the Premier line.

There are a few foreign companies with products of great interest to American collectors. The most notable is Cheerio which had its origins in Canada. All Cheerio yo-yos are considered highly collectible, whether marketed in Canada or the U.S. Cheerio has its own section in this book. There is also some interest in more recent Canadian yo-yo makers like National, Parker, and Canada Games. These companies also have their own section. Lumar, an English distributed yo-yo, originally produced by American toy giant Louis Marx, is often sought after by collectors. Louis Marx is best known for production of tin toys. He reportedly sold millions of yo-yos in England. Some lithographed patterns are the same as early Duncan whistling tin yo-yos. (See #480.)

*Continued ...* ▶

## YO-YO LISTINGS

● FOREIGN *Continued ...*

591      **66 WHISTLING MODEL, lithograph seal, tin, tournament shape, whistler holes, '30s**
This British model was manufactured by the Yo-Yo Distributing Company, LTD. An example of this yo-yo is shown next to this section's title.

592      **(Antigua), paint seal, wood, tournament shape, '80s**
This is a hand-painted yo-yo. Because hand-painted yo-yos are not dated, their value is based purely on aesthetics.

•      **Canadian Yo-Yos** - *See National, Parker and Canada Games section.*

593      **BRIO, gold leaf stamped seal, wood, tournament shape, '90s**
This yo-yo was produced by the German toy company famous for making wooden train sets.

594    24    **DMI YO-YO, imprint seal, plastic, bulge face shape, '80s**
This yo-yo was produced in Great Britain.

595      **DUMCAN PROFESSIONAL SUPER, plastic sticker seal, plastic, tournament shape, '80s**
This is a Duncan knock off yo-yo made in Mexico. They have changed the letter "N" to "M" in the word "Duncan."

596    24    **DUNCAN DINOSAURS, paper sticker seal, plastic, butterfly shape, '90s**
This was Duncan line of German yo-yos completely different from the American line. This Dinosaur series was not available in the U.S. and they are no longer produced. They were retailed in miniature cardboard "hat boxes" that had dinosaur graphics.

597    24    **DUNCAN (ying yang), imprint seal, plastic, butterfly shape, '90s**
This yo-yo is from the German line. It was not released in the United States.

598    24    **ELC 7738, paper sticker seal, plastic, tournament shape, '90s**
The reverse face has an embossed seal that reads "HP PLAST Spinner Made in Denmark."

599    24    **(El Salvador), painted, wood, tournament shape, '80s**
This is a hand-painted yo-yo.

600    24    **FREILAUF YO-YO, imprint seal, plastic, butterfly shape, '80s**
This is a German produced yo-yo.

601    24    **(French), silver medallion seal, plastic, tournament shape, '50s**
This yo-yo has beads in the attached medallion that produce a sound as the yo-yo is played. The yo-yo is made out of Bakelite plastic.

602    24    **(Guatemala), painted seal, wood, tournament shape, '80s**
This is a hand-painted yo-yo. Because hand-painted yo-yos are not dated, their value is based purely on aesthetics.

603    24    **HEROES, paper sticker seal, wood, tournament shape, '80s**
Heroes is a German toy company.

604    24    **KALMAR TRISSAN, silver leaf stamped seal, wood, tournament shape, '90s**
This yo-yo is made by Elfuerson's in Sweden. Kalmar is the name of the city where this yo-yo is produced.

605      **LUMAR (big "G") GENUINE, embossed seal, plastic, tournament shape, '70s**
Big "G" (Genuine) refers to the large "G" in the word "Genuine" that extends to the base of the seal. The seal, which is the same on both faces, reads "Championship Yo-Yo 99 Made in Great Britain." This yo-yo has a metal string slot axle.

606    24    **LUMAR (big "G") GENUINE, gold leaf hot stamped seal, plastic, tournament shape, '60s**
Big "G" (Genuine) refers to the large "G" in the word "Genuine" that extends to the base of the seal. The seal on this model is similar to #605, except that it is hot stamped. The reverse seal is embossed and reads "Lumar Yo-Yo Made in Great Britain."

607    24    **LUMAR BEGINNERS 33, lithograph seal, tin, tournament shape**
More than one version of the "Lumar Beginners" tin yo-yo exist.

608      **LUMAR BEGINNERS 34, lithograph seal, tin, tournament shape**
Many styles of Lumar Tins exist, some dating back to the early '30s.

609    24    **(Mexican yo-yos), paint seals, wood, varying sizes**
Colorful souvenir yo-yos have been made in Mexico for decades. They do not have logos or seals so there is no way to date them. They have little collectible value, except for the aesthetic value of the individual yo-yo. These are still being produced and sold retail in Mexico and the U.S. Mexican yo-yos of this style typically retail in the two to five dollar price range.

610    24    **PREMIER GALAXY, imprint seal, plastic, tournament shape, '80s**

611    24    **PREMIER MARIPOSA MONARCH, imprint seal, plastic, butterfly shape, '80s**
Mariposa means "butterfly" in Spanish.

612    24    **TARONGA ZOO, paper insert seal, plastic, slimline shape, view lens, '80s**
This yo-yo is from Sydney, Australia.

# YO-YO LISTINGS

## FORESTER

Forester produced a small line of wood, non-promoted yo-yos. The Forester line is believed to be from the early '60s. Forester caused some confusion with its Worldsfair yo-yo. (See #615.) It was not made for a World's Fair, but is identified as such by some dealers.

| | | |
|---|---|---|
| 613 | *39* | **FORESTER TOYS MFG. CO. INC**, gold leaf stamped seal, wood, tournament shape, '60s |
| 614 | *39* | **FORESTER WORLDSFAIR SPINNER**, die stamped seal, wood, tournament shape, '60s |
| | | An example of this yo-yo is shown next to this section's title. |
| 615 | *39* | **WORLDSFAIR**, die stamped seal, wood, tournament shape, '60s |
| | | This yo-yo was not associated with any World's Fair. |

## FRANKLIN MINT

In 1994, the Franklin Mint produced a collectors' series of vintage Duncan yo-yos. The original plan was to reproduce twelve models of classic Duncans. All twelve yo-yos were advertised in a few test markets. Because of poor initial sales, only three of the twelve were actually produced, in limited quantities. Although visually more appealing than the old classics, these reproductions were not exact duplicates of the originals and can be easily identified. Due to their limited production numbers, these yo-yos increased in value immediately after they were released.

| | | |
|---|---|---|
| 616 | *23* | **LITENING**, engraved metal seal, wood, tournament shape, three-piece, metal axle, 1994 |
| | | This yo-yo comes with a gold finished metal display stand and yellow string. It is easily distinguished from the original Litening #444, which had a paper sticker seal. |
| 617 | *23* | **IMPERIAL (chevron) TENITE**, imprint seal, plastic, tournament shape, 1994 |
| | | Unlike the original #347, this has a small rhinestone centered on the face. It does not have a hot stamped seal or a ® trademark associated with the word "yo-yo." An example of this yo-yo is shown next to this section's title. |
| 618 | *23* | **(Big "G") GENUINE**, gold inlay seal, wood, tournament shape, three-piece, 1994 |
| | | Big "G" (Genuine) refers to the large "G" in the word "Genuine" that extends to the base of the seal. This yo-yo has the same logo as #436.1. Of the three Franklin Mint reproductions, this model is the most difficult to tell from the original. Although it has a seal identical to the original, the seal is inlaid with a much wider font width than the original die stamped seal. The seal is also larger than the original. |

## YO-YO LISTINGS

# GOODY

The New York based Goody Manufacturing Company, like Royal, was pre and post-war competitor to Duncan. Goody, unlike Royal, did not use the word "yo-yo" on their toys, but called them "Filipino Twirlers." However, they did use the term "yo-yo" in some advertising literature. In 1958, Duncan wrote Goody a letter of protest which was largely ignored. Goody was a promoted line and did employ professional demonstrators. Goody, like many small yo-yo companies, shadowed the much larger Duncan and Cheerio campaigns.

Goody did produce several different jewel models. Unlike Duncan, who drilled jewel holes before painting the yo-yo, Goody drilled the jewel holes after the paint was applied. Rhinestones were glued into the drilled out holes. Although little is known about the history of the Goody Company, their yo-yos remain favorites among collectors.

| | | |
|---|---|---|
| 619 | 25 | **ATOMIC, paint seal, wood, tournament shape, '50s and '60s** |
| | | This is considered one of the rarest of the Goody yo-yos. |
| 620 | 25 | **(Little "G"), paint seal, wood, tournament shape, three-piece, '50s and '60s** |
| | | In this version, the "G" in the word "Goody" is the smallest letter. |
| 621 | 25 | **CHAMPION (three jewels), paint seal, wood, tournament shape, one-piece, '50s and '60s** |
| | | The three jewels may be different colors. Goody, unlike Duncan and Royal, tended to use rhinestones of varying colors in their jewel models. |
| 622 | | **COMET, paint seal, wood, tournament shape, '50s and '60s** |
| | | This yo-yo differs from #623 in that it has no jewels. It is common to find Goody yo-yos with jewels and without jewels bearing the same seal. |
| 623 | 25 | **COMET, two jewels, paint seal, wood, tournament shape, '50s and '60s** |
| | | This model was made both with and without jewels. |
| 624 | 25 | **(Big "G"), paint seal, wood, tournament shape, three-piece, nylon pegged string, '50s and '60s** |
| | | Big "G" refers to the large "G" on the seal that is shared by both the words "Genuine" and "Goody." |
| 625 | 25 | **(Goody Star before "G" • double circle), paint seal, tournament shape, three-piece, '50s and '60s** |
| | | This seal variation has a star before the letter "G" in the word "Goody." |
| 626 | 25 | **(Goody), paint seal, wood, tournament shape, miniature, three-piece, pegged nylon string, '50s and '60s** |
| | | All the letters in this Goody are the same size. This beginners model sometimes has paint stamps of animals on the reverse face of the yo-yo. It is unclear as to how many different animal stamps were used. At the present time, the butterfly, horse, monkey, elephant, lion, and Scottie dog are documented. These yo-yos carry a higher value than the un-stamped versions. Add twenty five percent to the value for animal stamped versions of this yo-yo. |
| 627 | | **JOY-O-TOP, paint seal, wood, tournament shape, '50s and '60s** |
| 627.1 | 25 | **JOY-O-TOY, paint seal, wood, tournament shape, '50s and '60s** |
| 628 | 25 | **MASTER (one jewel), paint seal, wood, tournament shape, one-piece, '50s and '60s** |
| 629 | 25 | **MASTER (3 star), paint seal, wood, tournament shape, three-piece, brass axle, '50s and '60s** |
| | | This yo-yo has an unusual hollow brass axle that extends through both yo-yo faces. Hollow axles were thought to cool faster than solid metal axles. The exposed axle in the string slot has a special coating to decrease string friction. The reverse face of the yo-yo has three stars on it and reads "Master." |
| 630 | 25 | **RAINBOW (seven jewel), paint seal, wood, tournament shape, '50s and '60s** |
| | | Although not the rarest of the Goody line, this yo-yo is the most famous and most desired by collectors. The seven jewels are typically different colors. An example of this yo-yo is shown next to this section's title. |
| 631 | 25 | **TROPHY (4 jewel), paint seal, silver, wood, tournament shape, '50s and '60s** |
| | | This silver colored yo-yo was presumably an award yo-yo for second place finishers in Goody contests. |
| 632 | | **TROPHY (4 jewel), paint seal, gold, wood, tournament shape, one-piece, '50s and '60s** |
| | | The reverse face reads "Goody Award for Skillful Twirling." This yo-yo is presumably a first place award yo-yo for Goody contests. It is unknown whether this yo-yo was ever retailed. |
| 633 | 25 | **WINNER (with jewel), paint seal, wood, tournament shape, one-piece, '50s and '60s** |
| | | This model was retailed both with and without jewels. See #634. |
| 634 | 25 | **WINNER, paint seal, wood, tournament shape, one-piece, '50s and '60s** |
| | | It is common to find Goody yo-yos with jewels and without jewels bearing the same seal. Models without jewels are slightly lower in value than the same models with jewels. |

Continued ...

## YO-YO LISTINGS

## HANNA BARBERA CHARACTERS

Hanna Barbera characters, like The Flintstones, Yogi Bear and The Jetsons, began ruling the TV cartoon world in the '60s. Thanks to reruns, their popularity continues to this day. Thousands of licensed products, including yo-yos, featuring Hanna Barbera characters have been released over the decades. In the early '80s, Festival (Union Wadding) held the license for Hanna Barbera yo-yos and has produced the largest variety of Hanna Barbera character yo-yos.

| | | |
|---|---|---|
| 635 | 26 | (Barney Rubble), paper sticker seal, plastic, tournament shape, 1981 |
| 636 | 26 | CAPTAIN CAVEMAN, paper sticker seal, plastic, Festival, butterfly shape, 1980 |

This yo-yo was retailed out of the same display box as Pebbles and Bam-Bam #650, and Scooby-Doo #653.

637  26  (Captain Caveman), plastic, Festival, tournament shape, view lens, 1980
For more information, see #645.

638  26  (Flintstones), paper insert seal, plastic, tournament shape, view lens, pinball, 1980
The "Flintstones Bedrock" was one of many Hanna Barbera pinball style game yo-yos produced by the Festival Company (Union Wadding). This toy combines both a yo-yo and a pinball game. It was introduced in 1980 as the Laff-A-Lympics game yo-yo. It retailed out of a counter display box with three other similar yo-yos; Yogi's "Space Race" #656, Scooby-Doo's "Haunted Ghost" #655, and "Team Play." The blister card from the Laff-A-Lympics series features Captain Caveman, but he does not appear on any of the yo-yo seals.

639  26  FRED FLINTSTONE (and Dino), lithograph seal, tin, tournament shape, '70s
This was from a series of tin character yo-yos sold in the '70s. They were retailed loose out of a display box marked "Metal Yo-Yo with Sleep Action." Others in this series include #640, #644, #647, #651, #652, #657, and #660. Some of these yo-yos can also be found on blister cards.

640  26  FRED FLINTSTONE (and Pebbles), lithograph seal, tin, tournament shape, mid-'70s
For more information, see #639.

641      (Fred Flintstone), paper sticker seal, wood, tournament shape, 1990
642  26  (Fred Flintstone), sculpted head, plastic, '70s
Fred and Yogi are the first of the sculpted character yo-yos to be produced. There are at least two size variations of the Fred Flintstones yo-yo. Both yo-yos were sold on cards and loose in a display box. Blister cards are marked Stock No. 6157 and carry a copyright date of either 1975 or 1976.

643  26  (Fred Flintstone), paper insert seal, plastic, miniature, view lens, '70s
This snap together yo-yo was a cereal premium from the '70s. For other similar snap together giveaways, see Keds Advertising #39 and Sugar Crisp Character #189.

644  26  FRED FLINTSTONE (in chair), lithograph seal, tin, tournament shape, '70s
For more information, see #639

645  26  (Fred Flintstone), paper insert seal, plastic, Festival, tournament shape, view lens, 1980
This was one of a series of four yo-yos released in 1980 by Festival, called "TV Stars Yo-Yo." These four yo-yos all retailed on blister cards. The other three characters were: Captain Caveman #637, Yogi Bear #658, and Scooby-Doo #654. The series was discontinued in 1981. These yo-yos are considered rare.

646  26  FRED FLINTSTONE, paper sticker seal, plastic, tournament shape, 1990
This was retailed on a blister card with a bar code, Stock No. 953.

647  26  HUCKLEBERRY HOUND, lithograph seal, tin, tournament shape, '70s
The Huckleberry Hound cartoon series first aired in 1958, but continues to be shown in reruns. This yo-yo was retailed out of a counter display box and on a blister card with a copyright date of 1975. For more information, see #639.

648  26  JETSONS, large disk insert seal, plastic, tournament shape, early '90s
This yo-yo was developed to promote the Jetsons animated movie released in 1990. The movie was based on the popular 1962 TV cartoon series.

•    Pac-Man - *See Duncan Miscellaneous Plastic section #427.*

650  26  PEBBLES AND BAM-BAM, paper sticker seal, wood, Festival, tournament shape, 1980
This yo-yo was originally introduced in 1980 by Festival. It retailed on a blister card out of a counter display box that read "Hanna-Barbera's Funtastic World of Yo-Yos." Two other yo-yos were sold out of the same counter display box: "Captain Caveman," a plastic butterfly model, and "Scooby-Doo," a plastic tournament shaped model. Compared to Pebbles and Captain Caveman, twice as many Scooby-Doo yo-yos were released.

651  26  PEBBLES (crawling), lithograph seal, tin, tournament, '70s
For more information see #639.

652  26  PEBBLES (standing), lithograph seal, tin, tournament shape, '70s
For more information, see #639.

653  26  SCOOBY-DOO, paper sticker seal, plastic, Festival, tournament shape, 1980
For the display card for this yo-yo, see Plate 21, #653.

*Continued ...* ▶

## YO-YO LISTINGS

● **HANNA BARBERA CHARACTERS** *Continued...*

654  26  **(Scooby-Doo), paper insert seal, plastic, Festival, tournament shape, view lens, 1980**
For more information, see #645.

655  **(Scooby-Doo's • haunted ghost), paper insert seal, plastic, tournament shape, view lens, 1980**
For more information on this style of view lens pinball game yo-yo, see #638.

656  **(Yogi's • space race), paper insert seal, plastic, Festival, tournament shape, view lens, 1980**
For more information on this style of view lens pinball game yo-yo, see #638.

657  26  **YOGI BEAR (and Boo-Boo), lithograph seal, tin, tournament shape, '70s**
For more information, see #639.

658  26  **(Yogi Bear), paper insert seal, plastic, Festival, tournament shape, view lens, 1980**
For more information, see #645.

659  26  **(Yogi Bear), sculpted head, plastic, '70s**
This yo-yo is similar to the sculpted Fred Flintstone yo-yo #642. The Yogi Bear yo-yo was produced at the same time and retailed out of the same counter display box. This yo-yo was also retailed on a blister card, Stock No. 7901.

660  26  **YOGI BEAR, lithograph seal, tin, tournament shape, '70s**
For more information, see #639. An example of this yo-yo is shown next to this section's title.

661  26  **(Yogi and Huckleberry Hound), paper insert seal, plastic, puck shape, view lens, early '60s**
This yo-yo is referred to as a "double puzzle game and whirling Hi-Lo." Each face has a pinball game. It is believed to be the first licensed Hanna Barbera character yo-yo. Huckleberry Hound was Hanna Barbera's first major TV hit from 1958 - 1962. Yogi Bear was introduced on this series and later got his own series in 1961. This yo-yo was retailed in a polybag with display card and header.

# HASBRO

Hasbro has been a toy giant for decades. Originally, Hasbro was a school supply company. In the late '30s, they expanded their product line to include toys. Their first major toy was Mr. Potato Head in 1952. In 1964 they released an even more popular toy, GI Joe.

662  45  **TOP-A-GO-GO, hot stamped seal, plastic, tournament shape, whistler holes, '60s**
This was a whistling yo-yo which could be used as a top, whizzer, and gyro. It came on a large 10-1/2" blister card with a trick sheet and a gyroscope stand. E.W. Frangos patented this toy in 1953, but Hasbro apparently did not produce it until 1964. An example of this yo-yo is shown next to this section's title.

663  32  **(Glow action yo-yo), no seal, plastic, tournament shape, late '60s**
This is the same design as the Top-A-Go-Go, but without the whistling hole and made from glow plastic. It does have the metal gyroscopic spindle coming out of the axle, but unlike the Top-A-Go-Go does not come with the gyroscope stand. This yo-yo was first introduced in 1968. The original blister card has a round paper sticker seal in the center which reads, "Glows in the Dark."

664  46  **(Pencil yo-yo), no seal, plastic, tournament shape, midget, '60s**
Hasbro was known as a pencil and school supply manufacturer in the '20s, so it is likely that they invented the pencil yo-yo. Two display card versions of the Hasbro Pencil Yo-Yo exist, both are shown on Plate 46. The rarest of the two yo-yos is on a long full-length card. The second variation has a small circular card that slips in the yo-yo string slot. Both Pencils have identical markings, "Pencil Yo-Yo By Hasbro." The plastic yo-yo holder slips over the eraser end of the pencil. The plastic yo-yo itself has no markings. It is believed these pencil yo-yos were all made in the early '60s. There is also a Duncan version, a foreign Coca-Cola version and another unnamed brand of yo-yo pencil. For any of these yo-yos to be considered complete, they should have the original pencil. If the pencil is sharpened, it decreases the value.

# YO-YO LISTINGS

## HI-KER

In early '30s, Wilfred H. Schlee began producing Hi-Ker yo-yos in Canada. These yo-yos were manufactured at the Kitchener Buttons factory owned by Schlee. These first yo-yos where called "Hi-Ker." Then, because of the intense popularity of the yo-yo in England, Schlee introduced the yo-yo named "Cheerio." This model became more popular than the Hi-Ker brand. Schlee continued to produce Hi-Ker in the '30s. These were marketed predominantly in Canada. It is believed Hi-Ker, as a brand, was largely ignored in the '40s and early '50s.

In 1946, Wilfred Schlee's son, Wilfred Schlee Jr., moved to the United States and began producing Cheerio yo-yos for Sam Dubiner, who had now acquired the Cheerio trademark. It is believed that the Hi-Ker yo-yos were not manufactured in the United States until around 1954. It was in 1954, that Sam Dubiner sold the Cheerio trademark to the Donald F. Duncan Co. Since Duncan had a factory which produced yo-yos, Schlee had to decide whether to stop making yo-yos or start his own line. He decided to go back to making the Hi-Ker Yo-Yo, the brand his father produced during the '30s. He hired his own promoters and began marketing the Hi-Ker brand in the United States. During this period, Schlee shadowed the highly promoted Duncan campaigns. In 1960, Schlee sold out to Duncan and became part of their organization. After acquiring the Hi-Ker Co., Duncan sold off the remaining inventory and dropped the Hi-Ker brand.

Several different styles of Hi-Ker yo-yos exist and all are wood. Hi-Ker yo-yos were retailed in the United States both on hanging cards and loose out of counter display boxes.

| | | |
|---|---|---|
| 666 | 56 | **55 FAMOUS, die stamped seal, '50s** |
| | | The item shown is the original die used to stamp the seal of this yo-yo. The model was only retailed in Canada. This die was from the Mastercraft Company in Canada which made at least some of the Hi-Ker yo-yos. |
| 667 | 15 | **59 (3 jewel), gold leaf stamped seal, tournament shape, '50s** |
| | | This Hi-Ker jewel was retailed in a polybag with a header card reading "Sparkle Master," but it does not have "Sparkle Master" lettered on the yo-yo seal. This yo-yo was first produced in 1955, and production continued into the late '50s. The original pre-printed price was 49 cents. The header card is marked #49R. |
| 668 | | **59 SPARKLE MASTER (4 jewel), gold leaf stamped seal, tournament shape, '50s** |
| | | This is perhaps the most desirable of the Hi-Ker yo-yos. It was first introduced in 1955 and continued to be produced into the late '50s. |
| 669 | 22 | **BEGINNER TOP, gold leaf stamped seal, tournament shape, miniature, pegged string, '50s** |
| | | The same seal was also used on a standard size model. |
| 670 | 22 | **FLAT TOP ORIGINAL, gold leaf stamped seal, butterfly shape, one-piece, '50s** |
| | | This is perhaps the best known yo-yo of the Hi-Ker line. An example of this yo-yo is shown next to this section's title. Duncan also briefly made a butterfly shaped yo-yo named the Flat Top. See #325. |
| 671 | 22 | **PROFESSIONAL, gold leaf stamped seal, tournament shape, one-piece, '50s** |
| | | This airbrush striped yo-yo retailed in a polybag with a header card and insert. The original pre-printed price of this yo-yo was 29 cents. The card is marked #29R. See Plate 32, #671. This was the standard line tournament model produced by Hi-Ker. The Professional was more common than either of the Spin Master. See #673 and #674. |
| 672 | | **SPARKLE MASTER (4 jewel), die stamped seal, tournament shape, '50s** |
| | | This version of the Sparkle Master does not have the number 59 on the seal. |
| 673 | 22 | **SPIN MASTER (metallic paint finish), die stamped seal, tournament shape, '50s** |
| | | This yo-yo is considered more rare than the Spin Master #674. |
| 674 | 22 | **SPIN MASTER (sparkle paint), gold leaf stamped seal, tournament shape, '50s** |
| | | When Duncan bought out Hi-Ker in 1960, their newly acquired inventory included quantities of this yo-yo. They turned them into Cheerios by adding the Cheerio 99 foil sticker seal. Duncan was marketing Cheerio as a non-promoted line at this time. When the foil seal is taken off some Cheerios, a Hi-Ker Spin Master seal can be found underneath. |

# YO-YO LISTINGS

## HOLIDAY: Christmas

Although Christmas yo-yos may be considered cross-collectible, there appears to be little demand for them by yo-yo collectors. Most are recently produced, inexpensive, tin lithograph yo-yos that carry little value. Only a few major yo-yo manufacturers have made Christmas yo-yos like the "Merry Christmas" yo-yo made by Festival in the '70s and the "Ho Ho" yo-yo by Playmaxx in the '90s. An unusual yo-yo related item was produced by Enesco in the '90s, a Christmas ornament yo-yo with a mouse climbing the string. Currently, there is no documentation of Christmas yo-yos produced prior to the 1980s. There are many styles of Christmas yo-yos available, most at a low cost, making this a good category for new collectors.

| | | |
|---|---|---|
| 675 | 27 | (Christmas series), lithograph seal, tin, O.T.C., tournament shape, '90s |
| | | This was a series of four Christmas yo-yos released in the middle '90s by O.T.C. The series includes Rudolph, Snowman, Santa and a Nutcracker soldier (shown). |
| 675.1 | | (Spanish Christmas series), lithograph seal, tin, O.T.C., tournament shape, late '90s |
| | | A series of four yo-yos released in 1997. All read "Feliz Navidad." |
| 675.2 | | (Snowflake series), lithograph seal, tin, O.T.C., tournament shape, late '90s |
| | | This is a series of three blue yo-yos released in 1997. The seals feature one, three or four snowflakes. Other than the words "Made In China," there is no lettering on the rim. |
| 675.3 | | (Christmas series), lithograph seal, tin, O.T.C., tournament shape, late '90s |
| | | This series of six yo-yos has snowflakes, but no lettering, on the rims. This series, released in 1997, includes a Polar bear with a present, a dog with a Christmas stocking, a penguin with a toboggan hat, a cat with a Santa hat, a snowman with red gloves and a broom, and Santa with a North Pole sign. The Santa with a North Pole sign version can be seen next to this section's title. |
| 675.4 | | (Hand-Painted Christmas yo-yos) |
| | | Because hand-painted yo-yos are not dated, their value is based purely on aesthetics. |
| 676 | | HO-HO ITSY BITSY YO-YO, sticker seal, wood, midget, '90s |
| | | For other Itsy-Bitsy yo-yos, see #1602. |
| 677 | | (Holly), lithograph seal, tin, tournament shape, '80s |
| | | There is no lettering on this yo-yo. |
| 678 | 27 | MERRY CHRISTMAS YO, YO, YO!, removable disk seal, plastic, Playmaxx, fly wheel shape, '90s |
| 679 | 27 | MERRY CHRISTMAS, paper insert seal, plastic, tournament shape, Festival, view lens, early '80s |
| | | This model retailed on a blister card that looks like a Christmas ornament. The card reads "My Christmas yo-yo" and was green on one side and red on the reverse. This yo-yo was produced by Union Wadding, the company that manufactured the Festival line. For an example of the blister card, see Plate 21, #679. |
| 680 | 31 | MERRY CHRISTMAS (yo-yo ornament), imprint seal, plastic, Enesco, tournament shape, 1990 |
| | | This is not a working yo-yo, but a special collectible Christmas ornament. A plastic mouse figurine is climbing the yo-yo string. The display box is shown on Plate 31. |
| 681 | 27 | (Mouse with candy cane), lithograph seal, tin, tournament shape, '80s |
| | | The seal features a mouse wearing a Santa hat. The copyright reads "© Kurt S. Adler, Inc." |
| • | | Raggedy Andy at Christmas - *See Character section #183.* |
| 682 | | (Reindeer), lithograph seal, tin, tournament shape, '90s |
| 683 | 27 | (Rudolph), paint seal, wood, puck shape, '90s |
| | | This is part of a hand-painted series released by O.T.C. Others in this series include a snowflake, snowman and Teddy bear with a ribbon. There is no lettering on these yo-yos. |
| 684 | 27 | (Santa face), lithograph seal, tin, tournament shape, '80s |
| | | "Made in Hong Kong" is printed on the rim. It retailed on a blister card, Stock No. 241. |
| 685 | 27 | (Santa face), lithograph seal, tin, tournament shape, '80s |
| | | Many tin Santa Face variations exist. |
| 686 | | (Santa face), paint seal, wood, tournament shape, plastic finger ring, '90s |
| 687 | | (Santa face), hand painted seal, wood, tournament shape, '90s |
| | | Yo-Yos with hand-painted seals carry value based on aesthetics only. There is no way to date them. |
| 688 | 27 | (Santa face), lithograph seal, tin, O.T.C., tournament shape, 1990 |
| 689 | | (Santa face • winking), lithograph seal, tin, O.T.C., tournament shape, 1989 |
| 690 | | (Santa kissing Mrs. Claus), lithograph seal, tin, tournament shape, '80s |
| 691 | | (Santa light-up), sticker seal, plastic, puck shape, '80s and '90s |
| | | Many Light-Up yo-yos with Christmas themes have been made. These were popular stocking stuffers. |
| 692 | 27 | (Santa hat on teddy bear), lithograph seal, tin, O.T.C., tournament shape, 1990 |

*Continued ...*

## YO-YO LISTINGS

| | | |
|---|---|---|
| 693 | 27 | **(Santa painting a toy soldier)**, lithograph seal, tin, tournament shape, '80s |
| | | The copyright reads "© Jamie Gardner Rehfeld." |
| 694 | 27 | **(Santa with horn)**, lithograph seal, tin, Shackman & Co., tournament shape, '80s |
| | | Imprinted on the rim is "© 1986." |
| 695 | | **(Snowman)**, lithograph seal, tin, tournament shape |
| | | Many Snowman styles exists and are popular on yo-yos with a Christmas theme. |
| 696 | | **(Teddy bear on rocking horse)**, lithograph seal, tin, tournament shape |
| 697 | 27 | **(Winking snowman)**, lithograph seal, tin, O.T.C., tournament shape, 1989 |
| 698 | 27 | **WINTER WONDERLAND**, lithograph seal, tin, tournament shape, '90s |
| | | This model has graphics of snowflakes and retailed in the '90s. The copyright reads "© C-P Inc." |

## HOLIDAY: Easter

Yo-Yos with an Easter theme were produced mainly in the '80s and '90s. They are typically low quality yo-yos, popular for stuffing in Easter baskets.

| | | |
|---|---|---|
| 699 | 27 | **(Baby bunny)**, lithograph seal, tin, tournament shape, '90s |
| | | This is a series of four Baby Bunny Easter yo-yos made by O.T.C. in the '90s. The series includes a baby bunny with chick (shown), a baby bunny with egg, and a baby bunny with flowers. |
| 700 | | **BUNNY & COMPANY**, sticker seal, plastic, tournament shape, '90s |
| 701 | 27 | **(Bunny jumping)**, sticker seal, wood, tournament shape, three-piece, '90s |
| | | This model was released by Applause. It has a drilled axle with a fixed string. |
| 702 | | **(Bunny multi-color)**, sculpted, plastic, 1998 |
| | | This yo-yo retailed on a blister card. An example of this yo-yo is shown next to this section's title. |
| 703 | 27 | **(Bunny riding the moon)**, paper sticker seal, wood, tournament shape, late '80s |
| | | This model was released by Applause, copyright 1988. This was part of a series of at least three different Easter related yo-yos. The series includes a bunny with fiddle and a bunny with a balloon. |
| 704 | 27 | **(Bunny with basket)**, lithograph seal, tin, tournament shape, '80s |
| | | This yo-yo was released by Midwest Importers Cannon Falls, Inc. |
| 705 | | **(Bunny with eggs)**, lithograph seal, tin, tournament shape, 1990 |
| 706 | 33 | **(Egg)**, painted, wood, sculpted, '90s |
| | | Sculpted Easter Egg yo-yos are common. More than one style exists. These typically don't have a manufacturer's seal. |
| 707 | | **(Egg)**, lithograph seal, tin, tournament shape, 1990 |
| | | The copyright reads "© O.T.C. 1990." |
| 708 | | **HAPPY EASTER**, lithograph seal, tin, U.S. Toy Company, tournament shape, 1990 |

## HOLIDAY: Halloween

Most Halloween yo-yos are inexpensive tin lithographed varieties. They are marketed as party favors and trick or treat gifts. Yo-Yos produced prior to '80s with a Halloween theme are poorly documented.

| | | |
|---|---|---|
| | • | **Eyeball** - *See Oriental Trading Co. section #1144.1.* |
| 709 | 27 | **(Ghost and pumpkin)**, lithograph seal, tin, tournament shape, '90s |
| | | This model was retailed by the U.S. Toy Company. The copyright reads "© C-P Inc." An example of this yo-yo is shown next to this section title. |

*Continued ...* ▶

## YO-YO LISTINGS

● **HOLIDAY: HALLOWEEN** *Continued ...*

710    27    **(Halloween series), lithograph seal, tin, tournament shape, '90s**
This is a series of four Halloween yo-yos released by O. T. C. in 1993. The series includes Frankenstein (shown), Witch, Ghost and Mummy.

711    27    **(Halloween series), lithograph seal, tin, tournament shape, '90s**
This is an O.T.C series dated 1990. The Skeleton (shown) and Pumpkin are known to exist, but there may be others.

712          **(Halloween series), lithograph seal, tin, tournament shape, '90s**
This O.T.C. series was released in 1997 and includes Dracula, Ghost and Pumpkin, Witch and Mummy.

713    27    **(Halloween candy series), paper sticker seal, plastic, puck shape, miniature, '90s**
This is a series of four different candy filled yo-yos released in the middle '90s. The four yo-yo variations are a Spider, Ghost, Wolfman, and Dracula (shown).

714    27    **(Hand-Painted Halloween series), paint seal, wood, puck shape, miniature, '90s**
This is a hand-painted yo-yo series retailed by O.T.C. in the middle '90s. The yo-yos in this series include a Scarecrow, Ghost (shown), and Black Cat with hat and cape.

715          **(Jack-O'-Lantern), lithograph seal, tin, tournament shape, '90s**
"Hong Kong" is imprinted on the rim.

716    27    **(Skull and crossbones), lithograph seal, tin, tournament shape, miniature, view lens, '60s**
This yo-yo has view lenses on both faces; one side has a mirror, the other a skull and crossbones pinball game. The yo-yo was retailed in polybag with a saddle header card reading "New Musical Top with Puzzle & Mirror."

717          **(Spider), lithograph seal, tin, O.T.C., tournament shape, 1989**

718          **(Witch on broom), lithograph seal, tin, tournament shape, '90s**

### HOLIDAY: New Year

719          **HAPPY NEW YEAR, lithograph seal, tin, tournament shape, 1997**
This model was distributed by the U.S. Toy Company. The copyright on the rim reads "© C-P." An example of this yo-yo is shown next to this section's title.

### HOLIDAY: St. Patrick's Day

720          **HAPPY ST. PATRICK'S DAY, lithograph seal, tin, tournament shape, 1997**
This model was distributed by the U.S. Toy Company. The copyright on the rim reads "© C-P." An example of this yo-yo is shown next to this section's title.

*Continued ...* ▶

## YO-YO LISTINGS

### HOLIDAY: Thanksgiving

The only Thanksgiving theme yo-yos currently known to exist are tin lithographed yo-yos from the '90s.

| | | | |
|---|---|---|---|
| 721 | 27 | **(Happy Thanksgiving series)**, lithograph seal, tin, O.T.C., tournament shape, '90s | |

This series of three yo-yos has a Thanksgiving theme. The series includes a turkey (shown on Plate 27), a turkey with falling leaves, and Happy Thanksgiving.

| | | |
|---|---|---|
| 722 | | **(Turkey)**, lithograph seal, tin, tournament shape, '90s |

This model was distributed by the U.S. Toy Company. The copyright on the rim reads "© C-P." An example of this yo-yo is shown next to this section's title.

### HOLIDAY: Valentine's Day

Listed below you will find yo-yos from the '80s and '90s with a Valentine's theme. Earlier yo-yos with a Valentine's theme may exist, but are not currently documented.

723    27    **(Cupid • with arrow quiver)**, lithograph seal, tin, tournament shape, '90s
The U.S. Toy Co retailed this yo-yo. An example of this yo-yo is shown next to this section's title.
724    27    **(Cupid • without arrow quiver)**, lithograph seal, tin, tournament shape, '80s
725    27    **(Heart)**, no seal, sculpted heart shape, plastic, Hallmark, metal axle, early '90s
Some of these yo-yos have a foil sticker identifying them as a Hallmark product. This yo-yo has a red string. One yo-yo half is red, the other black.
726    27    **I LOVE YOU**, lithograph seal, tin, tournament shape, '90s

### HUMMINGBIRD

Originally known as The Country Wood Shop Ltd., Hummingbird was the country's major producer of wooden yo-yos in the '80s and early '90s. Brad Countryman, a wood carver and furniture maker, founded the company. In 1988, The Country Wood Shop underwent a name change and became Hummingbird Toy Company. The first yo-yos were produced in 1982, as a method of cutting down on scrap wood waste. Original yo-yos were called the "Limited Yo-Yo" and were made of twelve different exotic woods. They were marketed to museum gift shops and other retailers. In 1983, the popular Hummingbird "Classic" was released. Several other mass-produced designs followed. Tournament shaped yo-yos made up Hummingbird's standard line, but butterfly shaped yo-yos were made exclusively for special orders. Hummingbird also produced hundreds of different ad and promotional yo-yos. Hummingbird closed its doors in 1995, but Countryman continues to produce yo-yos. (See "B.C. Yo-Yos.") All Hummingbird yo-yos are made of wood. This listing includes the Hummingbird standard line yo-yos and a sampling of some of the custom yo-yos produced.

*Continued ...*

## YO-YO LISTINGS

**HUMMINGBIRD** *Continued ...*

| 727 | | **ADULT YO-YO, die stamped seal, tournament shape, '90s** |
|---|---|---|

This is a natural wood yo-yo retailed on a blister card with Stock No. 82-000.

- **Brookstone** - *See Advertising Wood section #73.*

| 728 | *28* | **CLASSIC, die stamped seal, tournament shape, '80s and early '90s** |
|---|---|---|

This is the standard line Hummingbird Yo-Yo produced from 1983 - 1995. It was retailed in a shadow box. An example of this yo-yo is shown next to this section's title.

- **Church Street Station** - *See Souvenir section #1360.*

| 729 | *28* | **COMET, die stamped, butterfly shape, '80s** |
|---|---|---|

The Comet is made from the letters "c" "o" "m" "e" "t." This was the first butterfly shaped yo-yo made by Hummingbird.

| 730 | *28* | **DENNIS MCBRIDE JOHN 3:16 • YO-YOLOGIST, die stamped seal, butterfly shape, '90s** |
|---|---|---|

Dennis McBride is one of the best modern yo-yo players. Well-known for his two handed trick expertise, he has produced a series of instructional video tapes. The yo-yo was made in both solid colors and multi-colored laminates. Only 2000 of the McBride yo-yos were made. See Video section.

| 731 | *28* | **DALE MYRBERG WORLD CLASS CHAMPION, die stamped seal, tournament shape, '90s** |
|---|---|---|

Dale Myrberg is an entertainer and 1996 AYYA World Champion. Approximately 5000 of this tournament shaped yo-yo and 1200 of the butterfly style were produced.

| 732 | *07* | **DISNEYLAND BLAST TO THE PAST, laser diffraction seal, tournament shape, '80s** |
|---|---|---|

This yo-yo was retailed only at Disney outlets in 1988. There is a plastic version of Blast To The Past. See #262.

| 733 | *02* | **(Eddie Bauer), die stamped seal, tournament shape, '90s** |
|---|---|---|

This yo-yo does not have a Hummingbird logo. For the shadow display box, see Plate 31, #733.

| 734 | *31* | **(Executive), no seal, tournament shape, '80s** |
|---|---|---|

Like the limited, these yo-yos were made out of exotic woods. Executives do not have seals and should be collected in their original shadow box.

- **F.A.O. Schwarz** - *See Advertising Wood section #83.*

| 735 | *28* | **(Grateful Dead • Jerry Garcia), die stamped seal, tournament shape, '80s** |
|---|---|---|

This was a promotional yo-yo for the Grateful Dead band. The yo-yo has the image of the lead singer Jerry Garcia. It is reported that Hummingbird made at least three different Grateful Dead yo-yos.

| 736 | *28* | **HOL-YO-GRAM, laser hologram disks seal, tournament shape, '90s** |
|---|---|---|

The Hol-Yo-Gram, along with the Starburst #744 were called the "Rainbow Series." Both yo-yos had a laser diffraction seal. The Hol-Yo-Gram yo-yo has a disk that produces rainbow hues on both faces. It does not have a true hologram picture. This yo-yo is retailed both on blister cards and in shadow boxes.

| 737 | | **KNOTTS BERRY FARM, die stamped seal, tournament shape, '80s** |
|---|---|---|
| 738 | *28* | **LIMITED, die stamped seal, tournament shape, '80s** |

These yo-yos were made with and without seals. Limiteds without seals have little value unless they are in the original shadow display box. Twelve different exotic woods were used for this model: Bocote, Brazilian tulip, Satinwood, Purpleheart, Bolivian Rosewood, Bubinga, Padauk, Zircote, Zebrawood, Manderan, Rosewood, and Kingwood.

- **L.L. Bean** - *See Advertising Wood section #95.*

| 739 | *28* | **LUCKY & JACQUIE, die stamped seal, tournament shape, 1995** |
|---|---|---|

This is a wedding yo-yo created for the author and his wife. (Illustrated by T. Brown.) A total of 350 were made to be given as a gift to wedding guests. Hummingbird produced hundreds of promotional and advertising yo-yos like this. This model was one of the very last custom yo-yos produced by Hummingbird.

| 740 | *28* | **OH ZONE, paint seal, tournament shape, 1992** |
|---|---|---|

This yo-yo is one in a series of toys Hummingbird produced for the "Teenage Mutant Ninja Turtles II" movie released in 1991. The face of the yo-yo has a dripped paint surface. Other toys included a top, paddleball, and jump rope. Hummingbird also produced prop yo-yos used in the movie. Three paint splattered yo-yos of the following sizes were made: 3.5", 4.5" and 5.5" diameter. These yo-yos did not have seals.

| 741 | | **OPERATIONAL DESERT STORM • SAUDI ARABIA, die stamp seal, tournament shape, '90s** |
|---|---|---|

Whenever a customer purchased this model, a free yo-yo was sent to a serviceman fighting in Desert Storm. The Hummingbird logo appears on the reverse face.

| 742 | *04* | **ROY ACUFF, die stamped seal, tournament shape, early '90s** |
|---|---|---|

In the early '90s, this model retailed in the Grand Ole Opry gift shop. It sold in a polybag with a saddle header card. See #1395. Roy Acuff was a legend at the Grand Ole Opry for 50 years and was well known for using yo-yos in his act. Yo-Yos that bear his likeness are popular among collectors. The most money ever paid for a yo-yo was at the Acuff estate auction, where an autographed yo-yo from President Nixon given to Acuff sold for $16,029.00.

| 743 | | **SPLASH MOUNTAIN, die stamped seal, tournament shape, '90s** |
|---|---|---|

This a promotional yo-yo for the grand opening of Disney's Splash Mountain. Briar Rabbit is featured on the seal. Only 600 of these yo-yos were produced.

| 744 | | **STARBURST, die stamped seal, tournament shape, '80s** |
|---|---|---|

This was the first seal used on the Starburst. After a brief period, the metallic stamped seal was changed to the more common laser diffraction seal. See #746.

*Continued ...*

## YO-YO LISTINGS

745    *28*    **STARBURST (jewel • 5 jewels), die stamped seal, tournament shape, '80s**
This yo-yo has the same seal as #744, but with five jewels on the face. Linda Sengpiel purchased these Starbursts, added jewels and used them for her promotions. This yo-yo was not made by Hummingbird, but fifty ended up at the factory and were given out as gifts.

746    *28*    **STARBURST, laser diffraction seal, tournament shape, '80s and '90s**
This is a standard line yo-yo for Hummingbird, produced from 1988 to 1995. This has the same seal as #745, but is a laser diffraction version. The yo-yo was retailed both on a blister card and in a shadow box. See Plate 32, #746.

747    *28*    **TRICKSTER, decal seal, multi-colored maple laminate, tournament shape, '80s and '90s**
Early Tricksters did not have a seal. They were retailed in shadow display boxes that said "Trickster." More recent models have a decal seal with the word "Trickster." These were retailed on a blister card. All were made of brightly colored strips of dyed maple laminated together. Tricksters were made from 1988 until 1995.

       •    **The Nature Company -** *See Advertising Wood section #118.*

748    *07*    **WALT DISNEY STUDIOS, die stamped seal, tournament shape, '80s**

749    *39*    **YOYO MAN, gold leaf stamped seal, tournament shape, 1988**
This was a very popular yo-yo in the late '80s and early '90s. Hummingbird retailed this model on a blister card. It came in three different colors: red, white, and blue. The blue version is more difficult to find due to the small quantity produced. For an example of the blister card, see Plate, 32, #749.

750    *40*    **YOYO MAN, die stamped seal, tournament shape, late '80s**
"Nissan Los Angeles Open" is on the face of this model; the Smothers Brothers Yo-Yo Man logo is on the reverse side. Tommy Smothers is an avid golfer and celebrity sponsor for many golf tournaments.

751    *39*    **(YoYo Man) SMOTHERS BROTHERS KODAK, gold leaf stamped seal, tournament shape, late '80s**
This seal displays the Smothers Brothers' logo type. Kodak is die stamped on the reverse face. This yo-yo came with the Smothers Brothers' instructional video. See Plate #45. Approximately 210,000 of these yo-yos were produced.

752    *39*    **(YoYo Man) HUMMINGBIRD, die stamped seal, tournament shape, late '80s**
This is the same logo as #749, but also has "Hummingbird" on the seal. This version is less common than #749.

753    *28*    **YO-YO MASTER AWARD, gold leaf stamped seal, tournament shape, '90s**
This was an award yo-yo given out by Hummingbird in contest promotions. Three different Master Award yo-yos were made with the same seal, but different colored metallic lettering (gold, silver, and bronze). These yo-yos were not sold retail and only a small number, less than 100, were made.

754          **YO-YO PRINCESS • JENNIFER BAYBROOK, die stamped seal, tournament shape, '90s**
In 1997, Jennifer Baybrook became the first female to win the AYYA National Yo-Yo Championships. Two different variations of this yo-yo were made, one natural wood and one multi-colored laminate.

## HUMPHREY

Dave and Ken Humphrey started their company in the early 1970s. Originally the company produced the Humphrey Flyer, which was a Frisbee used in advertising promotions. In 1975, yo-yos were added to the promotion line.

Over the last two decades, the Humphrey Yo-Yo has been the king of the advertising yo-yos. The Humphrey Yo-Yo Company is the largest producer of ad yo-yos. Well over 1000 different types of Humphrey ad yo-yos are produced yearly. Total production numbers run into the millions. All types of companies use Humphrey yo-yos for promotions, from billion dollar corporations to mom and pop restaurants.

The Humphrey Corporation produces many promotional products for corporate advertising. One of the largest uses of ad yo-yos is as trade show promotional products. The Humphrey Company does not sell directly to buyers, but uses a network of over 2500 distributors throughout the United States.

The original Humphreys were produced in three different styles, All Pro (most common), Classic (tournament shape) and the Monarch (butterfly shape). The tournament and butterfly shapes are no longer in production. Currently, Humphrey only produces the "All Pro" style which yo-yo collectors refer to as the "Humphrey style."

*Continued ...* ▶

## YO-YO LISTINGS

**HUMPHREY** *Continued ...*

The Humphrey All Pro Yo-Yo is a fat, polystyrene plastic yo-yo with a wide face for advertising. It is rarely packaged or blister carded. Humphrey yo-yos that are considered collectible are always cross-collectible. Collectors should look for seals featuring major corporations, movie or character advertising, logos where a character is yo-yoing, or dated yo-yos. Yo-Yos with two or more colors imprinted on the seal are more desirable than single color graphics.

In the 1990s, Humphrey developed new molds. In addition to the bottle cap yo-yos produced for Pepsi and Coke, Humphrey created its Sports Ball line. This line includes a golf ball, soccer ball, baseball, and hockey puck. Since these yo-yos are created for advertising, a space is left on the face for advertising seals.

Below are some tips in determining the value of Humphrey yo-yos.

---

**TIPS ON COLLECTING HUMPHREY YO-YOS:**

1. Prime Rule:
   - The ad seal must be cross-collectible in some way.
2. Characteristics which add to the desirability of the yo-yo include:
   - Dates
   - Two or three color graphics
   - Cartoon characters
   - Characters playing a yo-yo
   - Slogans using the word "yo-yo"

---

| | | |
|---|---|---|
| 755 | | **3M, AIMS FAMILY DAY, 1980** |

Company picnics, reunions, etc. frequently provide Humphreys as souvenirs. Most often these are dated, but have little collectible value. The exception is any Humphrey with a 1975 or '76 date. These are some of the first Humphreys ever produced.

| | | |
|---|---|---|
| 756 | 30 | **(4-H logo)** |

Any nationally recognizable logo on a Humphrey has some slight cross-collectible value.

| | | |
|---|---|---|
| 757 | | **ACE DISPOSAL COMPANY** |
| 758 | | **ADMIRAL CRUISES** |
| 759 | | **A.E.A. 84** |
| 760 | | **AIRTECH AUTOMOTIVE DIVISION, 1935-1985** |

Dated commemorative yo-yos are collectible to some degree, but they need to be cross-collectible if they are expected to increase in value.

| | | |
|---|---|---|
| 761 | 30 | **(Airwalk)** |

The Airwalk logo is shown on the reverse face.

| | | |
|---|---|---|
| 762 | 28 | **ALL PRO • ONE OF THE INDUSTRY'S MOST POPULAR PREMIUMS** |

This is a promotional yo-yo from Humphrey that is advertising its own line of yo-yos. The All Pro is the name given to the standard Humphrey shaped yo-yo. All the yo-yos in this section are considered "All Pros." An example of this yo-yo is shown next to this section's title.

| | | |
|---|---|---|
| 763 | | **ALLERGAN** |
| 764 | | **ALLTECH • EVERYTHING FOR CHROMATOGRAPHY** |

High technology companies frequently give Humphreys out as promotional items on the trade show floor at conventions. Most have little, if any, collectible value.

*This is the cover of Humphrey's kit used to promote the initial release of their advertising yo-yo in 1975.*

*Continued ...*

## YO-YO LISTINGS

765　　　　AMERICAN CANCER SOCIETY • MAGIC IN THE STREETS
766　　　　AMERICAN GREETINGS
767　　　　AMERICAN HEART ASSOCIATION
768　　　　AMOCO
　　　　　　This yo-yo is made out of recycled plastic. Humphrey yo-yos made out of recycled plastic are black.
769　*30*　AMOCO, AT 100, 1989, CHALLENGE TOMORROW
770　*29*　ANGEL'S DINNER AND BAKERY
　　　　　　Unless small regional restaurants become multi-national chains, these yo-yos have little collectible value.
781　　　　ANHEUSER-BUSCH EMPLOYEES CREDIT UNION • MERRY CHRISTMAS
　　　　　　Banks, savings & loans, and credit unions have always loved giving out yo-yos.
782　*29*　ARKANSAS RAZORBACKS
　　　　　　Many colleges sell Humphrey yo-yos in the school book store. Few of them are dated and most have little value.
　　　　　　Yo-Yos typically feature the school mascot.
783　　　　ASTRO SAVERS, EMPORIA STATE BANK
784　　　　ATLAS POWDER COMPANY
　　　　　　"Block Buster Stress Reliever" is imprinted on the reverse face.
785　*29*　BABY RUTH
　　　　　　This is a promotion for the popular candy bar.
786　*29*　BALL STATE
787　　　　BANKERS TRUST COMPANY
788　*30*　BASF
　　　　　　"Agrichemicals for a Growing World" is imprinted on the face. Several variations of this yo-yo exist. Many companies use an
　　　　　　acronym, without graphics, for advertising on Humphrey yo-yos. These have little value unless the acronym is easily recognized
　　　　　　like AT&T or UPS.
789　*29*　BIG BOY
　　　　　　This is a good cross-collectible yo-yo, but it is overshadowed by its wooden predecessors, See #68 and #69.
　　•　　　Big Brutus - *See Souvenir Section #1349.*
790　　　　BLOCKBUSTER VIDEO
　　　　　　"Wow What a Yo-Yo" appears on the reverse face. Yo-Yos such as this have little value now. As technology changes, they may
　　　　　　have some nostalgic appeal. If cable TV or the internet ever wipe out the video rental market, these yo-yos will begin to have
　　　　　　collector appeal.
791　*30*　BLUE ANGELS
　　　　　　The reverse face has an imprinted logo of the Blue Angel jets.
792　　　　BLUE CROSS BLUE SHIELD • OFFICIAL SPONSOR 1992 OLYMPIC TEAM
793　*29*　BONANZA
　　　　　　Yo-Yos promoting national restaurant chains are somewhat collectible. They become more collectible if the chain releases
　　　　　　more than one style of promotional yo-yo. Few of these yo-yos are ever dated, but those with dates are more desirable.
794　*29*　BURGER KING
　　　　　　Big chains frequently re-release many styles of promotional yo-yo with similar seals. Since they are rarely dated, their values
　　　　　　remain low.
795　　　　BUSINESS & ESTATE ADVISORS, INC.
　　　　　　"We Want Yo Business" is imprinted on the reverse face. This slogan is very common on ad yo-yos produced for promo-
　　　　　　tions at trade show conventions.
796　　　　BUY AMERICAN UAW LOCAL 686
　　　　　　Having a catch phrase, like "Buy American," on a yo-yo may increase its cross-collectibility.
797　*01*　CABBAGE PATCH KIDS
　　　　　　This yo-yo advertises the famous line of dolls. Humphrey yo-yos with two-color graphics are more desirable than one-color
　　　　　　seal imprints. Because two-color yo-yos are more expensive to produce, there are more one-color models available.
798　*01*　CAP'N CRUNCH
　　　　　　Name-brand food products are usually cross-collectible especially if there is a character on the seal.
799　　　　CARNIVAL CRUISE LINES
800　*01*　CHEETOS (cheetah playing yo-yo)
　　　　　　Yo-Yos are more valuable if the seal features a character playing a yo-yo.
801　*01*　CHEETOS • PAWS
　　　　　　This yo-yo has two-color graphics and is made of glow plastic.
802　*29*　CHURCH'S CHICKEN
803　*30*　CLINTON AND GORE
　　　　　　Presidential campaign yo-yos have added value. Although there have been other presidential souvenir yo-yos, this is the first
　　　　　　campaign yo-yo that the author is aware of. See #840, #842, #871 and #881.
804　*30*　(Clydesdale logo)
　　　　　　This is a promotion for the Busch Clydesdales at Sea World.
805　　　　COMPUTER SCIENCE, MAGRAW HILL
806　　　　COUNSEL FOR SOLID WASTE SOLUTIONS • 100% RECYCLED POLYSTYRENE

*Continued ...*

The History and Values of Yo-Yos

### YO-YO LISTINGS

● HUMPHREY *Continued ...*

| | | |
|---|---|---|
| 807 | | (Cow) |

This yo-yo has two-color graphics.

| | | |
|---|---|---|
| 808 | | CRANBERRY WORLD WEST |
| 809 | *30* | CUB SCOUTS |
| 810 | *29* | DAIRY QUEEN |
| 811 | *03* | DARTH VADER |

This is a rare Dairy Queen promotion from the late '70s. It is the first time Star Wars characters appeared on yo-yos. Other characters from Star Wars may have been released in this series, but these have not been identified. This Darth Vader yo-yo is so far the highest valued of the Humphrey style yo-yos. For other Darth Vader yo-yos, see #1474.

| | | |
|---|---|---|
| 812 | | DELTA JUNCTION • ALASKA |

Many Humphrey yo-yos such as this one are made and sold as souvenirs.

| | | |
|---|---|---|
| 813 | *30* | DENNIS THE MENACE |

This was a promotional yo-yo from the 1993 movie starring Walter Matthau. There is a Crush logo on the reverse face. Movie promotion yo-yos are highly cross-collectible, especially if the movie was a success.

| | | |
|---|---|---|
| 814 | *30* | DISNEY'S TOY STORY ON VIDEO |

For other Toy Story yo-yos, see #1479.

| | | |
|---|---|---|
| 815 | *29* | DOMINOS PIZZA • NOBODY DELIVERS BETTER |

This yo-yo has two-color graphics.

| | | |
|---|---|---|
| 816 | *29* | DOMINOS, PIZZA DELIVERS FAST, FAST, FAST |

This yo-yo has two-color graphics. The reverse face is shown on Plate #1.

| | | |
|---|---|---|
| 817 | | DON'T BE A YO-YO AND MISS LAKE ARROW HEAD, SCI |

"Don't Be A Yo-Yo" is a common slogan on ad yo-yos.

| | | |
|---|---|---|
| 818 | *29* | (EKU) |

A friendly school with a swim coach but no swim team. You wouldn't understand, it's an Eel thing.

| | | |
|---|---|---|
| 819 | *29* | EAT AT ED DEBEVICS |
| 820 | | EUROPEAN PARTS INTERNATIONAL |
| 821 | | EXECUTIVE STRESS RELIEVER |

This is another common yo-yo slogan found on promotional yo-yos.

| | | |
|---|---|---|
| 822 | | EXIDE BATTERIES |
| 823 | | EXOTIC IMAGES |
| 824 | | FIRST WISCONSIN • MANY HAPPY RETURNS |

"Many Happy Returns" is a common yo-yo slogan for banks, and savings & loans.

| | | |
|---|---|---|
| 825 | | GALAXY • " THANKS FOR YO' BUSINESS" |

"Thanks for Yo' Business" is another common promotional slogan.

| | | |
|---|---|---|
| 826 | | GENERAL BICYCLE |
| 827 | *28* | GLO WITH HUMPHREY |

This is a promotional yo-yo for Humphrey's glow plastic line. Glow plastic Humphreys are less common than standard and recycled plastic yo-yos.

| | | |
|---|---|---|
| 828 | | GREAT TAKES VIDEO STORES |
| 829 | | GREATEST WALK ON EARTH WALK- A- THON |

A juggling clown is on the seal.

| | | |
|---|---|---|
| 830 | | GRAY BROTHERS • BURIAL VAULTS |

Unique ads such as this do make some yo-yos slightly more collectible.

| | | |
|---|---|---|
| 831 | | GREYBAR JACKSONVILLE FALL FESTIVAL • 1988 |

A squirrel carrying a large nut is on the reverse face.

| | | |
|---|---|---|
| 832 | | HERSHEY'S |
| 833 | *29* | HERSHEY'S • CHOCOLATE WORLD |
| 834 | *29* | HOLIDAY INN |

"Central... Tampa, FL" is imprinted on the reverse face.

| | | |
|---|---|---|
| 835 | | HONEY BEE |

There is a bee logo on the reverse face.

| | | |
|---|---|---|
| 836 | *30* | HOT ROD MAGAZINE |
| 837 | | I LOVE CAMP |

This yo-yo has two-color graphics.

| | | |
|---|---|---|
| 838 | | IN THE SWIM DISCOUNT POOL SUPPLIES |

This yo-yo has three-color graphics. Three-color graphics on Humphrey yo-yos are somewhat unusual because they are more expensive to produce. Because fewer were made, collectors concentrate only on those models with three-color graphics. Multiple colored graphics don't increase the value unless the yo-yo is cross-collectible.

| | | |
|---|---|---|
| 840 | *30* | INAUGURATION OF PRESIDENT AND VICE PRESIDENT • 1993 |

This yo-yo was sold as an inaugural souvenir for President Clinton's first term.

| | | |
|---|---|---|
| 841 | | INTERNATIONAL DIAMOND CORPORATION |

*Continued ...* ▶

# YO-YO LISTINGS

| | | |
|---|---|---|
| 842 | 30 | **JIMMY CARTER 39TH PRESIDENT** |
| | | See also #871. |
| 843 | 30 | **JOHN DEERE DAY** |
| 845 | | **JOSE FELICIANO** |
| 846 | 30 | **KIDDS MARSHMALLOWS • LAS VEGAS** |
| | | This yo-yo features a marshmallow playing a yo-yo. |
| 847 | | **KIDS SAVERS MID-AMERICA FEDERAL** |
| 848 | | **KODJ, glow plastic** |
| | | Glow plastic models of Humphreys are less common. They are more expensive to produce than standard plastic models. |
| 849 | 30 | **LEVI STRAUS AND CO., S.F. CAL.** |
| 850 | | **LIGHTEN UP WITH ORC (Optical Radiation Company)** |
| 851 | | **LITTON** |
| 852 | | **LONG JOHN SILVER'S • KD PLUS YOU** |
| 853 | | **M & Ms** |
| 854 | | **MAGRAW HILL • SCHAUM'S OUTLINE SERIES** |
| 855 | | **MAKE A WISH FOUNDATION** |
| 856 | 29 | **(Mars logo © 1989)** |
| 857 | | **MARTIN** |
| | | On the reverse face, a three-color graphic of stars and stripes is imprinted on a gear mechanism. |
| 858 | | **MARY'S DINER** |
| | | "Sam's Town" is imprinted on the reverse face. |
| 859 | | **MG DISPOSAL SYSTEMS** |
| 860 | 30 | **MINOLTA** |
| 861 | | **NATURE ISLAND** |
| | | "Kiwanis Riverview" is imprinted on the reverse face. |
| 862 | | **NORTHWEST BANKS** |
| 863 | | **NORTHWEST LUMBERMAN'S ASSOCIATION, "100 YEARS OF BUILDING," 1990** |
| 864 | | **OFF THE WALL • LA JOLLA • CA** |
| 865 | 30 | **ORKIN • "ONE CALL DESTROYS THEM ALL"** |
| | | The Orkin Man logo is imprinted on the reverse face. |
| 866 | | **PACIFIC STATES** |
| 867 | | **PARTS DEPOT** |
| 868 | 30 | **(Pegasus logo)** |
| 869 | 30 | **(Phillips 76 logo)** |
| 870 | 29 | **PIZZA HUT** |
| | | This has two-color graphics. The Pizza Man logo is imprinted on the reverse face. |
| 871 | 30 | **PLAINS GA., "HOME OF JIMMY CARTER"** |
| | | The image of the President is imprinted on the reverse face. For other Presidential yo-yos, see #803, #840, #842, and #881. |
| 872 | | **(Planters Peanuts)** |
| | | This was released in 1976. The seal shows Mr. Peanut. There are at least three different Planters Peanuts yo-yos. See #52 and #307.34. |
| 873 | | **PONTIAC WEST ASSEMBLY • 1 MILLION VEHICLES • SEPT. 12, 1990** |
| 874 | | **PRIME HEALTH** |
| | | This yo-yo has two-color graphics. |
| 875 | | **PURINA FARMS** |
| 876 | 29 | **RED LOBSTER** |
| 877 | | **REDUCE, RECYCLE, RE-USE** |
| | | Humphrey yo-yos such as this are molded from recycled plastic. Yo-Yos made from recycled plastic are black. |
| 878 | | **RESPIRATORY CARE, "YO' UR RIGHT CHOICE"** |
| | | University of Rochester is imprinted on the reverse face. |
| 879 | | **REVENGE OF THE NERDS II** |
| | | This is a promotional yo-yo for the movie. |
| 880 | 01 | **RON JON SURF SHOP** |
| | | This is a famous surf shop in Cocoa Beach, Florida. |
| 881 | | **RONALD REAGAN** |
| | | This yo-yo was sold at the Reagan Library. For other Presidential yo-yos, see #803, #840, #842, and #871. |
| 882 | 30 | **ROYAL NEIGHBORS OF AMERICA, glow plastic** |
| | | This is a life insurance company based in Illinois. They also produced a wooden promotional yo-yo. See #111. |
| 883 | 30 | **SATURN** |
| 884 | 30 | **SAY NO TO DRUGS** |
| 885 | | **ST. LOUIS, MO.** |
| | | The seal shows the start of a balloon race. It is dated 1986. |

*Continued ...*

## YO-YO LISTINGS

**HUMPHREY** Continued...

| | | |
|---|---|---|
| 886 | 29 | ST. LOUIS CARDINALS |

Many sports teams have used Humphrey yo-yos as souvenirs.

| | | |
|---|---|---|
| 887 | 30 | STATUE OF LIBERTY 1985 |

The reverse face reads "Comite Official Franco-American Du Centenaire 1886 - 1986." It commemorates the 100 year anniversary of the Statue of Liberty. This is a good Humphrey collectible for several reasons: it is a well known event and landmark, it is dated, and it is in three-colors.

| | | |
|---|---|---|
| 888 | 29 | STRAW HAT PIZZA |
| 889 | 29 | (Super chicken graphic) |

"The Hy Vee Deli" is imprinted on the reverse face.

| | | |
|---|---|---|
| 900 | 29 | TACO BELL |
| 901 | 29 | TAR HEELS |
| 902 | | TITANIC |

This yo-yo was retailed at the traveling Titanic exhibit in 1997. The logo reads "R.M.S. Titanic" and it shows the ship.

| | | |
|---|---|---|
| 903 | 29 | UCLA BRUINS |
| 904 | | UNC TAR HEELS |
| 905 | 30 | UNION PACIFIC (with train) |

Yo-Yos with graphics are usually more desirable than those without. See #906.

| | | |
|---|---|---|
| 906 | | UNION PACIFIC (without train) |
| 907 | 30 | UPS |
| 908 | | WINDOWS 3.0 ROLL-OUT |
| 909 | | YAMAHA 1985 GRAND NATIONALS |
| 1000 | 28 | YO-YO'S GET THE WORLD AROUND (cartoon of Yancy yo-yoing), '70s |

Yancy was the original promotional character associated with Humphrey yo-yos. Humphrey also released a small trick booklet with Yancy demonstrating tricks. See #1933.

## IMPERIAL TOYS

The Imperial Toy Company, founded in 1969, is currently the low price leader in toy marketing. Many of their products cost less than $9.00. Imperial Toys is known more for its marbles and bubble soap than its plastic yo-yos. Many, but not all, Imperial yo-yos have the highly detailed Imperial Crown logo somewhere on the seal. This Imperial Crown can help identify the origin of these yo-yos. Collectors should be aware that Duncan also produces a model called the Imperial. It is unrelated to the Imperial Toy Company Yo-Yo.

| | | |
|---|---|---|
| 1002 | 33 | CANDY CLUB, embossed seal, plastic, butterfly shape, late '90s |

This yo-yo was first introduced in 1997. This side plugs can be unscrewed to reveal compartments on two sides that hold Willy Wonka Tart N Tiny candies. This yo-yo is retailed on a blister card, Stock No. 7154.

- Barnyard Commando's - *See Character section #154.*

| | | |
|---|---|---|
| 1003 | 06 | CHAMPION YO-YO, gold leaf hot stamped seal, plastic, tournament shape, '80s |

These were retailed on blister cards, loose, and in party favor packs.

| | | |
|---|---|---|
| 1004 | | (Crown logo), gold leaf hot stamped seal, plastic, tournament shape, early '70s |

This yo-yo retailed on a blister card reading "Champion Yo-Yo." The card carried a pre-printed price of 59 cents.

- E-Z Go Ribbon Yo-Yo - *See Novelty section #1104.*

| | | |
|---|---|---|
| 1005 | 06 | GIANT TEENY, small disk insert seal, plastic, tournament shape, mid '70s |

This yo-yo was retailed loose out of a display box.

- Hawaiian Punch - *See Advertising Plastic section #32.*

| | | |
|---|---|---|
| 1006 | | HI-TECH, imprint seal, plastic, tournament shape, '90s |

This yo-yo was retailed on a blister card, Stock No. 8618. It has the same seal as #1007. An example of this yo-yo is shown next to this section's title.

| | | |
|---|---|---|
| 1007 | 06 | HI-TECH (Glo-Yo), imprint seal, glow plastic, tournament shape, '90s |

This yo-yo was retailed on a blister card, Stock No. 7249. The seal does not say "Glo-Yo," but "Glo-Yo" is imprinted on the blister card.

| | | |
|---|---|---|
| 1008 | | HI-TECH, imprint seal, plastic, tournament shape, '90s |

This is the checker board variation of the "Hi-Tech" seal. It was retailed on a blister card, Stock No. 7175.

Continued...

## YO-YO LISTINGS

1009           **HI-TECH, imprint seal, plastic, tournament shape, '90s**
This has an Imperial Crown logo on the reverse face. It was retailed on a blister card, Stock No. 7175.

1010           **HOT STUFF, imprint seal, plastic, butterfly shape, '90s**
This was retailed on blister card Stock No. 8389.

1011    *04*    **(Hot Wheel series), imprint seal, plastic, sculpted, '90s**
This was a series of four different metallic hub styles. They were retailed on a blister card, Stock No. 7012. These yo-yos are based on the famous toy car by Mattel.

1012           **IMPERIAL TOY, embossed seal, plastic, tournament shape, midget, '70s**
This yo-yo has a plastic insert disk seal. An Imperial Crown is in the center of the seal.

1013    *04*    **SUPER STOCK, imprint seal, plastic, sculpted, late '90s**
This yo-yo was first released in 1997. There is a button on the face of this wheel shaped yo-yo. When its pressed, it plays the sounds of a running race car.

1014           **SUPER SONIC SPACE YO-YO, imprint seal, plastic, bulge face shape, light and sound, '90s**
This yo-yo plays eight different space sounds and carries the copyright date of 1996.

1015           **TEENY YO-YO, insert disk seal, plastic, miniature, plastic finger ring, early '70s**
In the early '70s, this model was retailed two ways; it was sold loose, and on a blister card that held two yo-yos for the price of 29 cents.

1016           **(Totally colorful yo-yo), imprint seal, plastic, butterfly shape, '90s**

1017           **(Yo-Yo top twins), plastic, late '70s**
In the late '70s, this was retailed on a blister card for 69 cents. The two tops snap together to make the yo-yo.

## JA-RU

Often mistaken by collectors as an abbreviation for the unrelated Jack Russell Company. JA-RU is a toy company based in Jacksonville, Florida.

1018    *43*    **ALL STAR, imprint seal, plastic, bulge face shape, '90s**
This model retailed on a blister card, Stock No. 966.

1019    *43*    **AMERICAN CHAMP, imprint seal, plastic, tournament shape, '90s**
This yo-yo has a small insert disk seal.

1020           **JA-RU, imprint seal, plastic, bulge face shape, '90s**
Even though the blister card reads "All Star Yo-Yo," (Stock No. 966), All Star is not imprinted on the yo-yo itself. An example of this yo-yo is shown next to this section's title.

1021           **TODAY'S TOYS, imprint seal, plastic, bulge face shape, '90s**
JA-RU is imprinted on the reverse face. This was retailed on a blister card with Stock No. 1986.

1021.1          **(Super Sports Yo-Yo) FESTIVAL, molded seals, plastic, sculpted ball shapes, '90s**
In the mid-'90s, JA-RU released three models from Festival's "Be A Sport" series on their own cards. They were the Football, Basketball, and Baseball. See the Festival section for a discussion on how to tell original Festival sports balls from re-releases.

*These are examples of yo-yo reproductions made from the original Festival molds and distributed by JA-RU. The original Festival yo-yos, from the '70s, carry a much higher value than these re-released models.*

# YO-YO LISTINGS

## KAYSONS

Kaysons Novelty Company began producing yo-yos, called "Streamline Tops," in the '30s. It is unclear whether Kaysons continued production after WWII. There are three different yo-yos known, all made of wood. Kaysons also manufactured a display box and a string pack.

| | | |
|---|---|---|
| 1022 | 39 | **KAYSONS, gold leaf stamped seal, tournament shape, three-piece** |
| | | This is believed to be Kaysons most recently produced yo-yo. |
| 1023 | 39 | **KAYSONS, decal seal, tournament shape, three-piece, '30s** |
| | | Like all decal seals, this had a tendency to chip easily. An example of this yo-yo is shown next to this section's title. |
| 1023.1 | 39 | **KAYSONS, decal seal, tournament shape, three-piece, '30s** |
| | | This model is easily distinguished from the more common #1023 because the word "Genuine" is written in a script font. This model is believed to be the earliest Kaysons Yo-Yo and the most difficult to find. Although "Patent App. For" is on the seal, there is no evidence that Kaysons ever tried to patent their yo-yo. Marking yo-yos this way was a technique used to dissuade other companies from entering the market. It created the illusion that the idea was already patented. |

## KA-YO (CAYO)

Julius N. Cayo founded the Cayo Manufacturing Co. as a metal stamping business. Long before the production of yo-yos, Cayo was a leading producer of metal wastepaper baskets and dustpans. At its peak, Cayo was producing 30,000 wastepaper baskets a day. In the early '30s, Cayo began manufacturing tin yo-yos in Benton Harbor, Michigan. The Cayo Company also supplied Duncan with its metal yo-yo blanks for its whistling yo-yo line. It is unknown whether Cayo introduced his line of Ka-Yo yo-yos before or after he was contracted by Don Duncan Sr. to produce Duncan's whistling line. Either way, J.N. Cayo is credited for inventing the whistling yo-yo.

To bypass Duncan's trademark rights to the word "yo-yo," Cayo called his model the "Ka-Yo." In the '30s the Ka-Yos had multi-colored lithographed designs. Like the Duncan line of whistlers, there are two pairs of air holes on the front rim and one pair on the reverse rim. In 1941, due to WWII, production stopped on both Ka-Yos and Duncan's tin whistling yo-yos. It didn't start back up again until 1946. Following the war, J.N. Cayo retired, but his son Robert Cayo resumed yo-yo production. The original lithographed lines were not produced after the war, but the simulated wood grain style was made for a few years.

Robert Cayo also continued to produce Duncan whistling yo-yos after the war, but only in the solid colors. For more information on the Duncan tin whistling yo-yos, see the Duncan Tin section.

Ka-Yo yo-yos, like other tin yo-yos, are hard to find in near mint condition. They are prone to rusting within the string gap.

| | | |
|---|---|---|
| 1024 | 19 | **KA-YO MUSICAL, lithograph seal, tin, tournament shape, '30s** |
| | | Collectors call this lithograph pattern the "Captain's Wheel." |
| 1025 | 19 | **KA-YO WHISTLING, lithograph seal, tin, tournament shape, '30s** |
| | | This version is considered more rare than #1024. Some collectors call this lithograph pattern the "Aztec Design." An example of this yo-yo is shown next to this section's title. |
| 1026 | 19 | **KA YO WHISTLING, paper sticker seal, tin, tournament shape, '40s** |
| | | Although this pattern was produced briefly before WWII, it is the only Ka-Yo yo-yo made after the war. This yo-yo is known as the "wood grain Ka-Yo," because the plastic cover has a simulated wood grain pattern. |
| 1026.1 | | **KA YO, paper sticker seal, tin, tournament shape, '40s** |
| | | Similar to #1026, except this yo-yo is missing the word "whistling" on the seal and has a lithograph wood grain pattern, instead of the plastic cover. |

# YO-YO LISTINGS

 **KUSAN**

Kusan was a toy producer in Nashville, Tennessee best known for its plastic blocks and other molded plastic toys. In the early '60s, they entered into a manufacturing and marketing agreement with a man named Coleman who held the combination yo-yo and top patent. The toy was introduced in the early '60s at the New York Toy Fair. The Kusan booth featured a two by three foot giant spinning Twin Twirler. The toys were briefly promoted by demonstrators such as Bob Baab and Eddie Leader. Ad literature referred to the twin top not as a yo-yo, but as a "Spinning Top" and "Return Top." All Kusan yo-yos are made of plastic and carry the Patent Number 2,614,364. This number places the patent date in the middle to early '50s, but it was not the first top yo-yo combination patented. J.L. Spencer patented the first top yo-yo combination in 1951.

Contests required a combination of both yo-yo and top tricks. Although contests presented prizes such as bicycles and portable radios, there were no award patches. Some areas of the country did provide award sweatshirts that said "Twin Twirler Champions." The Ohio champion won a free trip to New York. Contests were typically run in shopping centers and local merchants put up the prizes. Like most yo-yo contests of this period, there were two age groups, 12 and under and 13 to 15. There were different divisions for boys and girls.

Kusan did run one world championship at the Singer Bowl during the New York World's Fair on June 27, 1964. Local elimination contests leading up to the "World Championship" awarded winners tickets to the World's Fair so they could participate in the championship contest. Grand prizes for the Championship included television sets, movie cameras and projectors, and bicycles. Although sales were good in many areas of the country, the Twin Twirler line was marketed and produced for less than two years. Kusan ultimately went out of business in the early '80s and never produced a yo-yo line after the Twin Twirlers.

Kusan yo-yos have a dual function, both as a yo-yo and as a top. The sides of the Kusan are pointed which allows the user to perform top tricks. For other top and yo-yo combinations, see Top-A-Go-Go by Hasbro #662, Gemini Gyro by Yomega #1666, and Gyro yo-yo by Charmore #1112.

*Ed Leader, shown (above) in photo, along with Bob Baab, were the two major demonstrators of the Kusan Twin Twirler.*

| | | |
|---|---|---|
| 1027 | 34 | **TWIN TWIRLER (falcon model), molded seal, marbleized plastic, '60s** |
| | | The face of this yo-yo has external fins which form a point on which the yo-yo can be spun as a top. It was sold on a blister card, Stock No. K-239, with a pre-printed price of 39 cents. For an example of the blister card, see Plate 32, #1027. |
| 1028 | 34 | **TWIN TWIRLER (bat model), molded seal, plastic, '60s** |
| | | This model was produced in both marbleized and translucent plastic. The translucent style is shown. It was sold on a blister card with a pre-printed price of 59 cents. An example of this yo-yo is shown next to this section's title. |
| 1029 | 34 | **TWIN TWIRLER (flying eagle model), molded seal, plastic, '60s** |
| | | This was sold on blister card, Stock No. K-211, with a pre-printed price of $1.00. |

## YO-YO LISTINGS

### LIGHT-UP

Light-Up yo-yos work by centrifugal force. When the yo-yo spins, it causes an electric switch to complete a circuit which lights the yo-yo. Typically these yo-yos use one AA penlight battery on each side. Light-Ups are commonly sold at fairs and theme parks. Those without seals are not considered collectible unless packaged in an unusual display box. Duncan's "Electric Lighted Yo-Yo" is the most collectible Light-Up Yo-Yo, see #381.

In 1952, D.J. Testino patented the first Light-Up Yo-Yo. This yo-yo had only one battery that ran through the axle. At least a half dozen other Light-Up yo-yo patents have been issued since the original patent. In 1989, a new patent was issued for a light up yo-yo that used a centrifugal switch and button batteries. See Duncan Light and Glow, and Disney sections for other Light-Up yo-yos.

| | | |
|---|---|---|
| 1030 | 01 | **(Alvin)**, paper sticker seal, plastic, puck shape, early '90s |
| | | In 1991, this promotional yo-yo for Del Monte Snack Cups was given away free in exchange for five proof of purchase seals from the product. The yo-yo also produces music when played. |
| 1031 | | **BRAVES**, plastic sticker seal, plastic, puck shape, '90s |
| | | The sticker seal is designed like a baseball. This yo-yo comes with a plastic finger ring at the string end. |
| | • | **Duncan Satellite** - *See Duncan Light-Up and Glow section #382.* |
| | • | **Duncan Big "G" Light-Up** - *See Duncan Light-Up and Glow section #381.* |
| 1032 | 31 | **ELEYO**, embossed seal, plastic, tournament shape, '60s |
| | | This model was released in 1967. It reads "Nighter Yo-Yo" on the box. The original yo-yo came with a thin brass finger ring and has a Patent Number of 375471. This is one of the few Light-Ups that has a name on the seal. It is considered one of the more collectible of the Light-Up yo-yos. |
| | • | **El Comerc de Barcelona** - *See Souvenir section #1394.1.* |
| 1033 | 31 | **EXPO 74 WORLD'S FAIR**, imprint seal, plastic, puck shape, 1974 |
| | | The box describes this model as the "Electric yo-yo." The original price was $3.95. Expo '74 was held in Spokane, WA. |
| 1034 | | **FIREFLY**, embossed seal, plastic, tournament shape, '50s |
| | | Rare early light-up yo-yo made by the Novel Products Corp. The battery and light is placed only on one side of the yo-yo. |
| 1035 | | **(Louie Anderson)**, plastic sticker seal, plastic, puck shape, 1997 |
| | | This was given out as a promotion during a Louie Anderson appearance in 1997. The seal on this yo-yo depicts Louie, the title character from the "Life with Louie" cartoon show. |
| 1036 | 03 | **MIGHTY MORPHIN POWER RANGERS**, paper sticker seal, plastic, puck shape, '90s |
| | | This yo-yo was first released in 1994. |
| 1037 | 03 | **(No seal)**, plastic |
| | | Generic Light-Up yo-yos without any identifying marks or seals have little, if any, collectible value. These yo-yos typically use AA batteries. Some Light-Ups are retailed in boxes with interesting graphics which can add value to the yo-yo, see Plate 3, 1037. An example of a generic Light-Up Yo-Yo is shown next to this section's title. |
| 1038 | 01 | **OREO**, imprint seal, plastic, sculpted, '80s |
| | | Sculpted and advertising Light-Up yo-yos have more collectible value than standard Light-Up yo-yos. For other Oreo yo-yos, see #53. |
| 1039 | 34 | **(Planet earth)**, paper sticker seal, plastic, puck shape, '90s |
| | | The seal shows view of Earth in space. This light-up model uses two AA batteries. |

### ROCK AND ROLL YO-YO

Introduced in 1996, these are light and sound yo-yos with button batteries. The yo-yos are retailed on six different cards. Each plays a different tune:

| | | |
|---|---|---|
| 1040 | 33 | #06110 "Wipe Out," no seal, plastic, bulge face shape, late '90s |
| 1040.1 | | #06120 "La Bamba," no seal, plastic, bulge face shape, late '90s |
| 1040.2 | | #0615 "Can't Buy Me Love," no seal, plastic, bulge face shape, late '90s |
| 1040.3 | | #06125 "Tequila," no seal, plastic, bulge face shape, late '90s |
| 1040.4 | | #6135 "The Lion Sleeps Tonight," no seal, plastic, bulge face shape, late '90s |
| 1040.5 | | #6140 "California Girls," no seal, plastic, bulge face shape, late '90s |

| | | |
|---|---|---|
| 1041 | | **SHAKEY'S PIZZA PARLOR**, imprint seal, plastic, puck shape, '70s |

*Continued ...* ▶

## YO-YO LISTINGS

1042    34     **(U.F.O. yo-yo ball), paper sticker seal, plastic, ball shape, late '90s**
                      First released in 1997, this was retailed in a box that reads "U.F.O. Yo-Yo Ball." The seal has a graphic of a frog's face. What makes this yo-yo unusual is that as the ball bearings inside move, the colors of light change. The yo-yo uses AA batteries.

1043    33     **(Yo tune), no seal, plastic, sculpted, early '90s**
                      This light and sound yo-yo retailed on a blister card and uses a penlight battery. It was one of the first combination Light-Up and sound yo-yos.

## MCDONALD'S

McDonald's yo-yos are very cross-collectible. They are generally inexpensive plastic yo-yos that can be "played with," but only in the most limited sense. The majority of McDonald's yo-yos were produced for local or regional giveaways. One series of yo-yos was part of the first Happy Meal that was test marketed in March of 1979. (The first official, nationally promoted Happy Meal campaign was in June of 1979.) These were "generic" premium yo-yos, which means the Happy Meal box art did not correspond to the giveaway. The yo-yos were a last minute substitute for the French Fry Flute that had production delays. The test market was limited to just a few cities across the U.S. These yo-yos are the most collectible of the McDonald's yo-yo promotions. There has not been a national Happy Meal promotion with yo-yos...yet.

1047          **(Golden arches logo), imprint seal, plastic, Humphrey shape, '90s**
                Humphrey yo-yos with McDonald logos are very common. An example of this yo-yo is shown next to this section's title.

1048          **(Hamburglar), embossed seal, plastic, tournament shape, early '90s**
                This was a 1992 Canadian giveaway called a "McSpinner." It has a metal axle and internal fins. The seal has the Hamburglar's head embossed on the face. An European promotion used a similar yo-yo (Jo Jo) in 1993, which included the Hamburglar, Ronald, and Grimace. None of these yo-yos were released in the United States.

### HAPPY MEAL TEST MARKET SERIES

This was a series of miniature plastic yo-yos with McDonald's characters imprinted on paper sticker seals. The stickers were affixed to both faces. The logos carry a copyright date of 1978. These came with a polystyrene finger ring at the string end that slipped over the yo-yo's body and into the string slot. This held the string in place when the yo-yo was not in use. These are considered the most collectible of the McDonald's plastic yo-yos.

1049    04     **(Big Mac • blue), paper sticker seal, tournament shape, miniature, 1979**
1050         **(Fries • blue), paper sticker seal, tournament shape, miniature, 1979**
1051         **(Grimace • purple), paper sticker seal, tournament shape, miniature, 1979**
1052         **(Hamburglar • yellow), paper sticker seal, tournament shape, miniature, 1979**
1053         **(Ronald • red), paper sticker seal, tournament shape, miniature, 1979**

1054    04     **MCDONALD'S, gold leaf stamped seal, wood, tournament shape, miniature, '60s**
                This is a pegged string beginners model from the late '50s or early '60s. This is one of the first McDonald's advertising yo-yos produced. The yo-yo's colors are white and red, the colors associated with early McDonald's restaurants.

1055         **MCDONALD'S (logo), sticker seal, wood, tournament shape, '80s**
                The sticker seal has the standard golden arches logo.

1056    04     **MCDONALD'S (logo), imprint seal, plastic, slimline shape, '80s**
                There is a Coke ad on the reverse face.

1057         **MCDONALD'S (logo), imprint seal, plastic, tournament shape, '90s**
1058    04     **MCDONALD'S (logo), embossed seal, plastic, miniature, '90s**
                Several different sizes of yo-yos with this same seal can be found.

1059    04     **(Ronald McDonald), sculpted, plastic, tournament shape, '90s**
                This was a premium yo-yo used in Australia.

1060    04     **RONALD MCDONALD CHAMPIONSHIP, molded seal, plastic, slimline shape, '60s**
                It is believed that this yo-yo was used in the one McDonald's promotion that was done by the Jack Russell Company. This promotion was run in Florida during the middle '60s.

1061    04     **(Ronald McDonald), paper insert seal, plastic, slimline shape, view lens, '80s**
1062         **(Ronald McDonald), paper insert seal, plastic, miniature, view lens, 1989**
                This was a Canadian release.

# YO-YO LISTINGS

## MEDALIST

Fred Strombeck released the Medalist Yo-Yo line in 1967. Strombeck purchased Duncan's wood turning lathes at the 1966 liquidation auction following Duncan's bankruptcy. Strombeck set up a new yo-yo plant near the old Duncan factory and briefly produced two wood lines and one plastic line. Strombeck, the owner of the company died in the early '70s, at which time the Medalist series of yo-yos ceased production.

| | | |
|---|---|---|
| 1063 | | **CADET, gold leaf die stamped seal, wood, tournament shape, late '60s** |
| | | This is a beginners style yo-yo. Each side of the yo-yo "half" is in a different color, either red, blue, or black. |
| 1064 | *22* | **TRICKMASTER, die stamped seal, wood, tournament shape, late '60s** |
| | | The Trickmaster is considered the most collectible of the Medallist series. An example of this yo-yo is shown next to this section's title. |
| 1064.1 | | **TROPHY, hot stamped seal, plastic, tournament shape, late '60s** |
| | | This is the only Medalist Yo-Yo made of plastic. |

## METAL (MISCELLANEOUS MANUFACTURERS)

This section contains a variety of metal yo-yos made by miscellaneous manufacturers.

| | | |
|---|---|---|
| 1065 | | **ALVIN STERLING, engraved rim seal, sculpted, metal axle, '70s** |
| | | This yo-yo is similar to the Gorham Sterling #1070, but with a different ornate pattern. It is more rare than the Gorham Sterling. |
| 1066 | | **BAL-YO, engraved seal, tournament shape, riveted axle** |
| | | This yo-yo is made by riveting two convex metal disks together. The date and maker are unknown. |
| 1067 | | **(Cartier), gold, 1932** |
| | | This is a very rare miniature gold yo-yo sold in Paris by Cartier during the 1932 yo-yo craze. The original price for the yo-yo was 280 francs. It was small enough to be worn as a bracelet bangle. Do not confuse this with the gold plated Charmore "Pocket Yo-Yo." See #1075. |
| 1068 | | **(Comstock Yo-Yo), engraved seal, riveted disks** |
| | | This yo-yo was made in Kansas City, though its production date is unknown. Two versions of this yo-yo exist. |
| 1069 | *19* | **EVERSTICK ANCHOR • ST. LOUIS, engraved seal, flywheel shape, riveted disk axle** |
| | | A bead inside of the yo-yo produces a bell sound during spinning. |
| 1070 | *37* | **GORHAM STERLING, engraved rim seal, sculpted, metal axle, early '70s** |
| | | Gorham introduced this yo-yo in 1971 with an original price of ten dollars. The Gorham yo-yo is metal and has a sterling silver shell with a highly ornate pattern. The central portion is left blank for engraving. Yo-Yos that have been engraved carry a lower value. Bob Rule, the Festival demonstrator, did several New York promotions associated with the release of this yo-yo. |
| | | Because the yo-yo has a sterling silver shell, it is seldom thrown out and invariably winds up in antique shops. In January of 1980, when silver broke the sixty dollar an ounce mark, the yo-yo had a silver value of over forty dollars. At current silver prices, the value is less than five dollars. This is one of the few yo-yos to drop in value over the last fifteen years. |
| 1071 | *19* | **HILL BRASS CO., paper sticker seal, tournament shape, internal bell** |
| 1072 | *19* | **HY LO, molded seal, riveted axle** |
| | | Many collectors believe this yo-yo to be a Bandalore (Pre-Flores), but this has not been confirmed. This yo-yo comes with a metal finger ring attached to the string end. |
| 1072.1 | | **HY-LO, ink stamped seal, riveted disk axle** |
| | • | **Lumar** - *See Foreign section.* |
| 1073 | | **MEGASPIN, engraved seal, aluminum, flywheel shape, '90s** |
| | | This is a heavy aluminum yo-yo with a wooden axle, produced by J.D. Ryan Enterprises. Only 80 of these yo-yos were made. The original retail price was $80.00. An example of this yo-yo is shown next to this section's title. |

*Continued ...* ▶

## YO-YO LISTINGS

1073.1      **MOONSTAR, engraved seal, aluminum, flywheel shape, transaxle, late '90s**
This is a ball bearing transaxle yo-yo made in Germany, but retailed in the U.S.A. The seal reads "CAME-YO" and the design features a chameleon with a yo-yo on the end of the its tongue. The yo-yo is retailed in a cardboard display box.

1074      **(O-Hi-O climbing top), no seal, tournament shape, '30s**
This unmarked yo-yo is one of the very few that can be identified by its color pattern and shape. It appears in an old Gibbs Manufacturing Company advertising flyer from the '30s.

1075     *34*     **(Pocket yo-yo), no seal, gold plate finish, midget, jewelry clasp mounted to face, '50s and '60s**
Duncan granted Charmore a license to use the "yo-yo" trademark in 1948. Around 1950, Duncan purchased all the remaining Pocket yo-yos and used them as gifts and as awards up to the early '60s. Some were also sold retail. This yo-yo could be attached by a jewelry clasp to a necklace or key chain. One of these yo-yos was an exhibit during the Duncan vs. Royal trademark trial.

1076     *35*     **REAL LUCKY CHAMPION YO-YO, paint seal, tin, tournament shape, '70s**
This yo-yo has an internal bell.

1076.1      **(Sterling necklace yo-yo), no seal, sterling silver, tournament shape, '90s**
This was a sterling silver yo-yo that slides into a U-clip which attaches to a necklace. It was made by the Reo Company.

1077     *37*     **TIFFANY & CO., engraved rim seal, wood body, sterling silver shell, tournament shape, '90s**
This is a sterling silver gift yo-yo with a wooden axle produced by Tiffany in the '90s. It originally retailed for $80.00. This yo-yo came in a Tiffany box with a felt pouch.

1078     *37*     **TOWLE SILVER PLATE, engraved rim seal, silver plated shell, tournament shape, metal axle, '70s**
The pattern on this silver plated yo-yo is identical to the Gorham #1070, but unlike the Gorham version, it is not in a sterling silver shell. "Towle Silver Plate" is engraved on the rim. Towle also produced a combination wood and silver yo-yo. See Wood Miscellaneous #1637.

## NATIONAL CHAMPIONSHIP

The modern day National Yo-Yo Championships began in 1993, in Chico, California. Previously, Chico was the site for several California State Championships. The town of Chico held its first modern yo-yo contest in 1988. The driving force behind these contests has been Bob Malowney, owner of Bird In Hand, and curator of the National Yo-Yo Museum. The contest attracts top players from around the country competing for the title of National Champion. In conjunction with the National Competition, contests are also held for novice and advanced players. The competition is open to all players and any brand of yo-yo can be used. This event recognizes the yo-yo as a sport rather than a promotional event for selling a particular brand. The winner is recognized by the AYYA as the official National Champion.

A variety of yo-yos, pins, patches, and posters have been produced for these competitions. Typically, this event is held at the beginning of October. It is a great opportunity to meet other collectors and to do some trading.

### 1993 NATIONAL CHAMPIONSHIP

1079      **1993 NATIONAL CHAMPIONSHIP, silver leaf seal, wood, Bird in Hand, tournament shape**
This yo-yo was produced with only two color variations, blue and black.

1080     *28*     **1993 NATIONAL CHAMPIONSHIP, removable disk seal, plastic, Playmaxx, fly wheel shape**
There is a championship seal insert on both faces.

### 1994 NATIONAL CHAMPIONSHIP

1081     *28*     **1994 NATIONAL CHAMPIONSHIP, gold leaf seal, wood, Bird in Hand, tournament shape**
This yo-yo was only produced in red.

1082     *28*     **1994 NATIONAL CHAMPIONSHIP, removable disk seal, plastic, Playmaxx, fly wheel shape**
This has the championship seal on the front face and the Bird in Hand logo on the reverse.

### 1995 NATIONAL CHAMPIONSHIP

No wooden yo-yos were made this year. Hummingbird, who previously made the wood Championship yo-yos, had gone out of business.

1083     *28*     **1995 NATIONAL CHAMPIONSHIP, paper insert seal, plastic, Duncan, slimline shape**
This has the championship seal on the front face, and the Bird in Hand logo on the reverse.

*Continued ...* ▶

## YO-YO LISTINGS

● **NATIONAL CHAMPIONSHIP** *Continued ...*

### 1996 NATIONAL CHAMPIONSHIP

1084    *28*    **1996 NATIONAL CHAMPIONSHIP, gold leaf seal, wood, Bird in Hand, tournament shape**
The championship seal is on the front face and the Bird in Hand logo on the reverse. See Plate 39, #1084 for the Bird in Hand logo on the reverse face. This yo-yo was produced in only one color, dark blue.

1085    **1996 NATIONAL CHAMPIONSHIP, removable disk seal, plastic, Playmaxx, flywheel shape**
The Championship seal is on the front face and the Bird in Hand logo on the reverse.

### 1997 NATIONAL CHAMPIONSHIP

1086    **1997 NATIONAL CHAMPIONSHIP, Gold leaf seal, wood, BC, tournament shape**
This yo-yo, made by BC, is blue with a white stripe.

1087    **1997 NATIONAL CHAMPIONSHIP, removable disk seal, plastic, Playmaxx, flywheel shape**
This yo-yo is a Pro Yo II. An example of this yo-yo is shown next to this section's title.

### 1998 NATIONAL CHAMPIONSHIP

1087.1    **1998 NATIONAL CHAMPIONSHIP, die stamped seal, wood, BC, tournament shape**
1087.2    **1998 NATIONAL CHAMPIONSHIP, imprint seal, plastic, Yomega, tournament shape**
1087.3    **1998 NATIONAL CHAMPIONSHIP, removable disk seal, plastic, Playmaxx, flywheel shape**

## NOVELTY

To the general public, all yo-yos are viewed as novelties. To the yo-yo enthusiast, novelty yo-yos are those with a unique decoration, shape, color, or function. Bells and figures were added to Bandalores long before Pedro Flores ever set foot in America. In 1882, Chas Wurst patented the first yo-yo with bells attached to the outside. Other unusual patents granted for yo-yos with special attachments include propellers, combustible wicks, and flint sparklers.

1088    **3 IN 1 SPACE YO-YO, embossed seal, plastic, view lens, pinball game, '70s**
This yo-yo uses the same model design and graphics as the Keds promotional space yo-yo #39. This yo-yo is more rare than the Keds version.

1089    **(Antique car series), paper insert seal, plastic, slimline shape, view lens, '80s**
The seals show antique cars. A 1900 Wolseley is on one face; a 1920 Bull Nose Morris is on the reverse.

•    **Avon Soap Yo-Yo Set -** *See Wood Advertising section #66.*

1090    **(Bat key chain yo-yo), sticker seal, plastic, Arco, tournament shape, miniature, '90s**

1091    **(Bilboquet-yo-yo), no seal, wood**
This is a combination toy and yo-yo. This toy is somewhat reminiscent of the ball and cup game. Instead of a ball on the string, there is a yo-yo. The yo-yo has a hole in the center, like a doughnut. The object of the game is to flip the yo-yo and have it land on the wooden peg at the end of the handle. This toy can also be used as a yo-yo. It is doubtful that a named toy manufacturer ever made this game. These generally fall in the classification of folk art toys, and were more popular in South America than North America.

1092    *33*    **(Bracelet yo-yo), no seal, brass, tournament shape, miniature**
This solid brass yo-yo clips to a brass bracelet. The date and maker of this yo-yo are unknown.

1093    *45*    **BUBBLE YO, molded seal, plastic, jumbo, '90s**
This unique yo-yo produces bubbles while spinning. It was retailed on a blister card with bubble soap, yo-yo holder, and soap dish. It was copyrighted in 1984 by Shure products.

1094    **(Burger Yo), no seal, plastic, sculpted hamburger shape, '90s**
This can be distinguished from the Spectra Star hamburger yo-yo, #1439, by the absence of the Spectra Star markings on the inside edge of the string slot lip.

# YO-YO LISTINGS

|  |  |  |
|---|---|---|
| | • | **Candy Club** - *See Imperial Toys section #1002.* |
| 1095 | | **(Candy yo-yo with light and sound)**, plastic, puck shape, '90s |

This yo-yo has candy filled compartments, a push button light on one side and a sound mechanism on the other. This was first released in 1996.

1096  *33*  **(Candy filled)**

Several styles of candy filled yo-yos have been made. Most are plastic, miniature in size and contain candy compartments. The yo-yo shown is from the '80s. It is opaque, however, most candy filled yo-yos have transparent faces so the candy can be seen.

1097  *06*  **(Cat yo-yo)**, plastic, sculpted face with moveable eyes, '90s

1098  **CAUTION P.M.S.**, medallion seal, wood, tournament shape, '90s

A clear acrylic medallion seal is affixed to the face of the yo-yo.

1099  *46*  **(Cocktail mixer yo-yo)**, plastic, jumbo, drilled plastic axle, '70s

Two clear plastic glasses screw into the face of this yo-yo. When it is played, it mixes the drinks. The yo-yo comes packaged in a cardboard box with cartoon graphics.

1100  **(Condom yo-yo)**, condom inserts, plastic, view lens, late '90s

This yo-yo holds an Aladan condom on each face under a snap out view lens.

1101  *34*  **(Doll)**, no seal, celluloid, '20s

This is arguably not a true yo-yo, although it does have a yo-yo-like mechanism. A string is attached to a bow on the doll's head. The string passes through the hollow body to the base where it winds around an axle connected to two wheels. Pulling the string causes the wheels to turn making the doll walk. The doll can also be "yo-yoed" up and down the string.

1102  **(Dog series)**, paper insert seal, plastic, slimline shape, view lens, '80s

This series is similar to the Antique Car Series #1089, but features dogs. This model has a carriage dog on one face and a Tibetan prayer on the reverse.

1103  *33*  **(Eraser)**, no seal, rubber erasers, '80s

More than one style of eraser yo-yo exists. The one shown in this book is believed to be from the '80s, however, there are no seals or identifying markings.

1104  *33*  **(E-Z go ribbon yo-yo)**, imprint seal, plastic, Imperial Toy Co., early '90s

This has a one inch wide ribbon instead of a string. The seal reads "Hi Tech" and has the Patent Number 4,290,224. This yo-yo was retailed on a blister card. The first ribbon yo-yo patent was granted to Patrick MacCarthy in 1981. See also #1129.

1105  **(Game • chain yo-yo)**, paper insert seal, plastic, bulge face shape, view lens

This is a combination game toy. The yo-yo features a cartoon face with a chain for a nose. The shape of the nose changes as the yo-yo is shaken.

1106  **(Glow yo-yo)** no seal, glow plastic, tournament shape

Several companies have retailed glow-in-the-dark yo-yos without seals. These are not considered valuable unless they are on a unique blister card.

1107  *15*  **(Golden yo-yo • 19 jewels)**, imitation gold finish, metal, tournament shape, miniature

This yo-yo retailed in a jewelry box. "For The Girl Who Has Everything" is imprinted on the jewelry box. The yo-yo came with colored string. At least three variations of this yo-yo are known.

1108  *33*  **(Go-Go ring)**, no seal, plastic, tournament shape, midget, '60s

This is a detachable yo-yo mounted on a plastic ring. It is a rare premium found in Nabisco's Rice Honeys cereal. Assembly was required. The user had to supply the "string," in this case sewing thread was recommended. Instructions on how to assemble and use this model, were printed on the back of the cereal box. For other ring yo-yos, see #1130.

1109  **(Goofus)**, HO HO YOYO, imprint seal, Kenner, tournament shape, early '70s

This is believed to be the first "soft" yo-yo produced. It was retailed in a cardboard box with a copyright date of 1971.

1110  **(Gumball machine yo-yo)**, no seal, plastic, tournament shape, midget, '80s and '90s

This is a yo-yo prize from a gumball machine that comes in a plastic prize capsule. This yo-yo was made in Hong Kong. Most gumball prize yo-yos do not have seals.

1111  *04*  **GRAND PRIX WHEEL YO-YO**, imprint seal, plastic, sculpted shape, '90s

This yo-yo was made from the same mold as the Imperial - Mattel Hot Wheel series. These were retailed in cardboard boxes and on blister cards.

1112  *31*  **(Gyro yo-yo)**, no seal, metal and plastic, yo-yo top combination, '50s

This is a high quality constructed yo-yo/top/gyroscope combination. The Charmore Company made this toy, Patent Number 2337334. The axle and top end are stainless steel. The plastic yo-yo body has internal fins. This was retailed in a box with a trick sheet and a special string clip.

1113  *33*  **(Half dollar yo-yo)**, metal, '70s

This yo-yo was retailed as a magic trick. It is made out of two Kennedy half dollars. The yo-yo came with a magic trick instruction sheet.

1114  **(His and hers yo-yos)**, foil sticker seal, plastic, tournament shape, '80s

This is a set of gag yo-yos that came in a gift box reading "I'll let you play with mine, if you let me play with yours."

1115  **(Jumbo yo-yos)**

A jumbo yo-yo is any yo-yo with a diameter greater than 2-3/4". Over the years, many different oversized yo-yos have been produced. Some have interesting graphics, but no other identifying marks. When a jumbo has an unidentified maker,

*Continued ...*

The History and Values of Yo-Yos

## YO-YO LISTINGS

● **NOVELTY** Continued ...

the value is based only on the aesthetics of the graphics. Jumbos have been made out of wood, plastic, and metal.

| | | |
|---|---|---|
| 1116 | | (Lady bug • 4 dots), paint seal, wood, tournament shape, miniature, three-piece, pegged string, '90s |
| 1117 | | (Lady bug • 6 dots), wood, paint seal, tournament shape, three-piece, pegged string, '90s |
| 1118 | | (Lady bug • 7 dots), wood, paint seal, tournament shape, jumbo, three-piece, pegged string, '90s |
| 1119 | 33 | (Lady bug • 18 dots), lithograph seal, tin, tournament shape |

For unknown reasons, ladybug designed yo-yos are popular. Several more styles exist than are listed in this book.

1120      (Large cookie yo), no seal, plastic, sculpted, jumbo, metal string slot axle, early '90s
This yo-yo retailed in a cellophane wrapper reading "Cookie-Yo."

1121    33  (Lipstick yo-yo), imprint seal, plastic, red and pink lipstick
This yo-yo is made out of two lipstick pots. It was called "Funshine Lip Pots" and was distributed by Dorothy Grey Ltd.

•     **Music and light-up yo-yos** - See *Light-Up section.*

1122    34  (Music yo-yo), decal seal, plastic, slimline shape
Many styles of music playing yo-yos exist. Musical yo-yos should not be confused with whistling yo-yos that make sound via air currents passing through holes in the yo-yo. Musical yo-yos have a battery and a centrifugal switch that starts the music. Most play an entire tune once the switch makes contact. Early models used AA batteries, but more recent models have changed to button batteries. See also Duncan Mel-Yo-Dee #422 and the Light-Up section.

1123      (Nut yo-yo), no seal, bulge face shape, midget, mid '90s
This yo-yo may look like ivory, but it was turned from a hard nut. It was retailed by The Nature Company.

1124    35  (Orange Slice), lithograph seal, tin, tournament shape, jumbo, bell attachments, '70s - '90s
There are several varieties of these yo-yos, most sold in souvenir shops in Florida. The word "Florida" may be marked on the seal. Both tin and wooden versions exist.

1125    34  PAMELA, imprinted seal, plastic, slimline shape, late '90s
The seal reads, "The World's First Silicone Implanted Glow Yo-Yo."

1126      (Penny yo-yo), sculpted, plastic, miniature, one-piece
The ink stamped seal on the reverse face reads "First National Bank." An example of this yo-yo is shown next to this section's title.

•     **Pencil Yo-Yo** - See *Duncan Miscellaneous #409 and Hasbro section #664.*

1127    34  POOF, embossed seal, plastic, tournament shape, late '90s
This plastic foam yo-yo, released in 1997, is made to play indoors. More than one Poof variation exists. The sides and rim of the yo-yo are slightly compressible.

1128    04  RACE WHEELS SUPER RETURN, imprint seal, plastic, butterfly shape, '90s
This is a knock off of the Duncan Long Spin Wheels style yo-yo.

1129      (Ribbon spinning yo-yo), no seal, plastic, '90s
A nylon ribbon replaces the string on this yo-yo. There is no seal on the face just paint swirls. The yo-yo was retailed on blister card by Marchon in the mid '90s, Stock No. 60322. This yo-yo is similar to the Imperial Ribbon yo-yo #1104

1130    56  (Siren ring and yo-yo), no seal, tournament shape, midget
This is a rare Kellogg's Pep premium from the late '40s or early '50s. The ring, as shown in the advertisement, was 25 cents with one Kellogg's Pep box top. This yo-yo was attached to a siren on the ring. When the yo-yo was removed, it exposed the siren which, when blown, made a whistling noise. For other yo-yo rings see # 1108.

•     **Soccer Ball** - See also *Spectra Star and Sports Ball section.*

1131      SPARKLING GLO YO, die stamped seal, tin, tournament shape, '30s
In 1934, Irving C. Brown received Patent Number 1,949,858 for this yo-yo. The yo-yo has a flint mounted through the face. When the flint contacts the spinning inner metal disk, sparks are generated.

1132    33  (Stinger Yo-Yo), no seal, plastic, puck shape, embedment, late '90s
This blister carded yo-yo was released in 1996. There is an actual scorpion embedded in the acrylic plastic. The Scorpion fluoresces under ultraviolet light. There is also a glow version of this yo-yo.

1133    33  SQUISH YO, dyed cloth seal, tournament shape, early '90s
This yo-yo was not made like traditional models. Its form was made of a soft foam, then covered with cloth. The seal was dyed on to the same fabric which covers the yo-yo form. These were retailed on hanging display cards. There were four cloth patterns produced for this yo-yo.

1134      THE BIGGEST YO-YO IN ____, paper sticker seal, plastic, jumbo, '90s
This 4-3/4" diameter plastic yo-yo with internal fins is used as a souvenir in many states. The name of the state can be inserted in the blank. This yo-yo is made in Taiwan and distributed by Four Star International Trading Company. It is retailed loose or in polybag with a saddle header card.

•     **Transformer Yo-Yo** - See *#1139.*

1135      (Watch yo-yo), working LCD clock display, plastic, tournament shape, midget size, early '90s
The watch snaps out of the watch band holder to make a yo-yo.

1136      (World yo-yo), imprint seal, plastic, sculpted globe shape, '80s
This is similar to the Spectra Star Earth Zone yo-yo, but does not have the Spectra Star imprint inside the string slot lip. The success of the Spectra Star "Earth Zone" #1482 inspired this knock off. It was made before the breakup of the Soviet Union. This model was retailed on a blister card.

*Continued ...* ▶

## YO-YO LISTINGS

1137  33  YO BALL, molded seal, plastic, Knots, ball shape, '70s
          This yo-yo has an internal spring-loaded rewind mechanism on the axle.
1137.1  45  YO-GUN, early '60s
          This is not a true yo-yo. The toy gun has a string attached to it and a plastic ball attached to the string. It has a rewind mechanism, so when the trigger is pulled, the ball returns to the tip of the gun. The toy was made by Ideal.
1138  33  YO-YO BALL, imprint and molded seal, plastic, ball shape, '90s
          This model is very similar to Knot's Yo-Ball #1137. The rewind mechanism has colorful graphics, several styles of which were produced. This yo-yo retailed on a Marchon blister card, Stock No. 60300.
1139  45  (Yo-Yo bike • Yamaha), foil sticker seal, plastic, sculpted, '90s
          Like a Transformer toy, this motorcycle can be turned into a yo-yo.
1140  34  (Yo-Yo top), early '90s
          Many yo-yo/top combinations exist. See also Kusan section.

## OLIVER TOYS

Dale Oliver, a former Duncan and Playmaxx demonstrator and 1992 World Champion, founded Oliver Toys in 1995. Dale Oliver was also the first president of the American Yo-Yo Association. In 1996, when Oliver Toys began selling parts to Spintastics, the yo-yo seal design changed. A seal change occurred again, in 1998, dropping the word "Terminator."

1141  37  TERMINATOR, insert disk seal, plastic, flywheel shape, '90s
          The earliest Terminator models read "Oliver Toys" on the seal. More recent versions of this yo-yo read "Spintastics." There is also a glow plastic version of the Terminator. An example of the original Terminator Yo-Yo is shown next to this section's title.
1141.1    TERMINATOR TORCH, insert disk seal, plastic, fly wheel shape, light-up, late '90s
          The light-up version of the Terminator uses button batteries.
1141.2  37  TERMINATOR TORNADO, insert disk seal, plastic, fly wheel shape, '90s
          This yo-yo is similar to the Terminator #1141, but has a ball bearing transaxle. Earlier models read "Oliver Toys" on the seal; more recent models read "Spintastics."

## ORIENTAL TRADING COMPANY

Oriental Trading Company (O.T.C) is a large distributor of novelty toys. O.T.C. markets inexpensive toys for party gifts, treasure chests for doctors' and dentists' offices, etc. It is best known for inexpensive lithographed tin yo-yos made in China. These are marked O.T.C. Some O.T.C. yo-yos may be marked with copyright dates for the graphics. These yo-yos are largely from the '80s and '90s. O.T.C. also distributes other yo-yos in addition to tin models. Many of these are generic yo-yos without manufacturer seals and have little collectible value. (See also Holiday sections.)

1142  35  (Birthday series), lithograph seal, tin, tournament shape, '90s
          These are birthday party favor yo-yos. This is a series of three yo-yos including the Elephant (shown), Clown, and Unicorn.
1143  35  (Dinosaur series), lithograph seal, tournament shape, '90s
          This was produced by O.T.C in the middle '90s. There are three dinosaur variations in this series.
1144  35  (Dinosaur series • baby), lithograph seal, tournament shape, '90s
          This is a series of three baby dinosaur yo-yos. The series features a triceratops (shown), a stegosaurus, and a brontosaurus with a boom box.
1144.1  33  (Eyeball), lithograph seal, tin, tournament shape, late '90s
1145  35  (Geometric pattern series), lithograph seal, tin, tournament shape, '90s
          This series of four yo-yos was release in the mid-'90s. They feature such designs as arrows and paper airplanes.
1146  35  (Goofy face series), lithograph seal, tin, tournament shape, early '90s
          This is a series of Goofy faces, all date 1990. Of the four faces, the winking female (shown) and the bucktooth versions are slightly harder to find.

*Continued ...*

## YO-YO LISTINGS

● **ORIENTAL TRADING COMPANY** *Continued ...*

| | | |
|---|---|---|
| 1146.1 | *3.5* | (Insect series), lithograph seal, tin, tournament shape, late '90s |
| | | This was released in 1997 and features four insects, bee (shown), ladybug, moth, and lighting bug. |
| 1147 | *3.5* | (Laughing face series), lithograph seal, tin, tournament shape, '90s |
| | | This is similar to the Goofy faces series #1146. It was released in 1993. These yo-yos do not have a date. There are four different faces. The face with eyeglasses (shown) is the most difficult to find. |
| 1148 | *3.5* | (Rainbow pattern series), lithograph seal, tin, tournament shape, late '80s |
| | | This is a series of three yo-yos from 1989. Each has a different multi-colored pattern. |
| 1149 | | (Religious series), lithograph seal, tin, tournament shape, early '90s |
| | | This is a series of three yo-yos released in 1990. They feature the sayings: Share It, Love One Another, and Jesus Loves You. |
| 1149.1 | | (Religious series), lithograph seal, tin, tournament shape, '90s |
| | | New patterns for series #1149 were released in 1997. These feature: Jesus ❤ You, God's Promise and a smiling sun on a rainbow. |
| 1149.2 | | (Smile face), lithograph seal, tin, tournament shape, late '90s |
| | | This yo-yo, released in 1997, features the typical smile face. This yo-yo comes in four colors: yellow, red, blue and green. |
| 1150 | *3.5* | (Sports Ball series), lithograph seal, tin, tournament shape, '90s |
| | | This series of four Sports Ball yo-yos was released in the mid-'90s. The series features a football (shown), a soccer ball, a basketball and a baseball. |
| 1150.1 | | (Star of David series), lithograph seal, tin, tournament shape, '90s |
| | | This is a series of two yo-yos, one with a single star and one with multiple stars. |

# PARKER - NATIONAL - CANADA GAMES

In Canada the word "yo-yo" is not generic, as it is in the United States. Only the company owning the trademark "Yo-Yo" may call it a yo-yo. All other yo-yo manufacturers must use other names for their toy or be licensed. A recent example is the Yomega yo-yos that were marketed in Canada. In the U.S. they are known as the "Yo-Yo With A Brain." In Canada they were marketed as a "Bandalore With A Brain."

Beginning in the '30s, Cheerio held the first Canadian yo-yo trademark. Sam Dubiner, an early Cheerio promotions director, registered both the Cheerio and yo-yo trademarks in 1938. Dubiner sold the Canadian Cheerio name to Al Krangle in 1954, but retained the rights to license the "Yo-Yo" trademark. In the early '60s, there was a dispute between Dubiner and Krangle resulting in the loss of Cheerio's license to use the word "Yo-Yo." In 1968, Al Gallo started the National Yo-Yo Company in Canada and purchased the "Yo-Yo" trademark from Dubiner. In 1978, Gallo sold the trademark to Parker Brothers. In 1986, Parker Brothers sold the trademark to Canada Games.

In the '50s, the Canadian Cheerio Company purchased their yo-yos from the Mastercraft Wood Company based in Canada. Sometime during the early '60s, Cheerio began buying yo-yos from Elfversons out of Sweden, but continued to buy from Mastercraft. Any Cheerio yo-yo manufactured in Sweden has "Sweden" die stamped on the reverse face. Al Gallo also bought yo-yos for the National Yo-Yo Company from Mastercraft and Sweden. As late as 1971, Mastercraft was producing 100,000 yo-yos a year for National. In 1974, Mastercraft discontinued making yo-yos. Any National Yo-Yo that does not have "Sweden" die stamped on the reverse face was made prior to 1974.

*This is an original display box from the early '70s containing Pro Tournament yo-yos from the National Yo-Yo Company.*

*Continued ...*

## YO-YO LISTINGS

| | | |
|---|---|---|
| 1151 | 15 | (3 jewel) SATELLITE, paper sticker seal, wood, National, tournament shape, '70s |
| 1151.1 | | (4 jewel) SATELLITE, paper sticker seal, wood, National, tournament shape, late '60s |
| 1152 | 37 | CALIFORNIA YO-YO TOP, imprint seal, plastic, Canada Games, tournament shape, '90s |
| 1153 | 37 | GLOW YO-YO, imprint seal, glow plastic, Canada Games, tournament shape, '90s |
| 1154 | 37 | LIGHT-UP YO-YO, imprint seal, plastic, Canada Games, bulge face shape, '90s |
| 1155 | | OLYMPIC, gold leaf stamped seal, wood, Canada Games, tournament shape, '90s |
| 1156 | 37 | OLYMPIC • ORIGINAL GENUINE, die stamped seal, wood, National, tournament shape, '60s - '70s |

Al Gallo trademarked the name "Olympic" for the National Co. in the '60s. This trademark was later passed on with the yo-yo trademark to Parker and then to Canada Games. During the 1976 Olympics held in Montreal, the International Olympic Committee challenged Gallo's "Olympic" trademark on this yo-yo. The IOC backed down when they discovered that the trademark was a legitimate Canadian yo-yo trademark. Olympic yo-yos were painted either silver or gold. The yo-yos were retailed both on a blister cards and loose out of a display box which held one dozen yo-yos.

1156.1   OLYMPIC • ORIGINAL GENUINE, foil sticker seal, wood, National, tournament shape, '60s - '70s
This model is similar to #1156, except it has a foil sticker seal. The Olympic yo-yos and the foil seals were either silver or gold. This version is less common than the die stamped seal.

1156.2   PARKER JUNIOR, paper sticker seal, plastic, tournament shape, '80s
1157  37  PARKER OLYMPIC, die stamped seal, wood, tournament shape, '80s
1157.1   PARKER OLYMPIC, foil sticker seal, wood, tournament shape, '80s
1158  37  PARKER PRO TOURNAMENT, gold leaf stamped seal, wood, tournament shape, '80s
An example of this yo-yo is shown next to this section's title. For an example of the blister card, see Plate 32, #1158.

1159  37  PRO CANADA GAMES, gold leaf stamped seal, wood, tournament shape, '90s
1160  37  PRO TOURNAMENT SPINNER, gold leaf stamped seal, wood, Parker, tournament shape, '80s
1161  37  PRO TOURNAMENT GENUINE, gold leaf stamped seal, wood, Parker, tournament shape, '80s
1161.1   PRO TOURNAMENT GENUINE, foil sticker seal, wood, National, tournament shape, '70s
1162     SHRIEKER YO-YO TOP, lithograph seal, tin, National, tournament shape, whistler holes, '70s
National made this model for only a few years. It was a whistling yo-yo with air holes in the rim. National demonstrators did not like this yo-yo, and the sales were slow compared to other yo-yos in the National line.

# PARTY FAVOR

Party favor yo-yos are sold in multi packs, generally four to six yo-yos per pack. These yo-yos are low quality and don't play well. With the exception of the Duncan "Happy Birthday" Yo-Yo, this category is not considered to have much collectible value. Some party favor yo-yos may be found retailed individually on blister cards.

1163   BETA, hot stamped seal, plastic, tournament shape, '90s
The yo-yos are all stamped "Made in Hong Kong." These were retailed in a six pack on a blister card, Stock No. 2006.

1164   CHAMPION YO-YO, hot stamped seal, plastic, tournament shape, '90s
Imperial Toy Corporation distributed this six pack on a blister card. This yo-yo has the same design style as the Beta yo-yo. See #1163.

•   **Duncan Happy Birthday** - *See Duncan Miscellaneous section #399.*

1165   HAPPY BIRTHDAY, paper sticker seal, plastic, tournament shape, '90s
The seal shows a cake with one candle. "Made in China" is hot stamped in plastic. The yo-yo was distributed by Unique Industries, six on a blister card.

1166   (Pin wheel), paper sticker seal, plastic, tournament shape, '90s
These yo-yos were distributed, six to a blister card, by Unique Industries.

1167   (Space yo-yos), paper insert seal, plastic, miniature, view lens, '90s
This was a series of five yo-yos with paper insert seals of space ships. These were distributed on a blister card by Beta Toy and Party Favors.

1168   (Unique party favors), paper sticker seal, soft plastic, miniature, '90s
The seal shows a pinwheel with "China" written on it. These yo-yos were retailed either four on a blister card or six in a polybag with a saddle header card. An example of this party favor pack is shown next to this section's title.

1169   WINNER YO-YO, gold leaf hot stamped seal, plastic, tournament shape, '90s
This yo-yo may or may not have "Taiwan" stamped in the center. This was distributed by Unique Industries, Stock No. 4018.

*The History and Values of Yo-Yos*

## YO-YO LISTINGS

 **PEANUTS**

Peanuts, created in the '50s by cartoonist Charles Schulz, has been a cartoon icon for the last four decades. Originally a popular cartoon strip, it later evolved into TV specials, movies and the source of hundreds of licensed products.

The United Feature Syndicate owns all Peanuts character rights. Hallmark has been the licensee for yo-yos since the '70s. Almost all Peanuts yo-yos have been produced for Hallmark. Festival (Union Wadding) manufactured Hallmark's first Peanuts yo-yos. In the '80s and '90s, Monogram Products bought out Festival and began producing the Hallmark Peanuts yo-yos.

Peanuts character yo-yos, like Disney character yo-yos have increased in value because they are cross-collectible. Snoopy is the most common Peanuts character found on yo-yos.

| | | |
|---|---|---|
| 1170 | 36 | (Charlie Brown), imprint seal, plastic, Hallmark, tournament shape, early '80s |
| | | This yo-yo was retailed on a Hallmark blister card with a pre-printed price of $1.00. The original yo-yo has a colored string. |
| 1171 | 36 | (Linus), imprint seal, plastic, Hallmark, tournament shape, early '80s |
| | | This yo-yo was retailed on a Hallmark blister card with a pre-printed price of $1.00. The original yo-yo has a colored string. |
| 1172 | 36 | (Lucy), imprint seal, plastic, Hallmark, tournament shape, early '80s |
| | | This yo-yo was retailed on a Hallmark blister card with a pre-printed price of $1.00. The original yo-yo has a colored string, |
| | • | **Snoopy** - *See also Tom Kuhn section, #1592 and #1593.* |
| 1173 | 36 | (Snoopy and Woodstock), paper sticker seal, plastic, Monogram, butterfly shape, '90s |
| | | This was a Knott's Berry Farm promotional yo-yo. |
| 1174 | 36 | (Snoopy balloons), paper insert seal, plastic, Festival, tournament shape, view lens, '70s |
| | | This is one in a series of three Peanuts pinball game yo-yos made by Festival (Union Wadding), see #1177. |
| 1175 | 36 | (Snoopy laying on yo-yo), imprint seal, plastic, butterfly shape |
| 1176 | | (Snoopy tangled in yo-yo string) KNOTT'S BERRY FARM, imprint seal, plastic, butterfly shape |
| 1177 | 36 | (Snoopy tennis), paper insert seal, plastic, Festival, tournament shape, view lens, '70s |
| | | This is one in a series of three Peanuts pinball game yo-yos made by Festival (Union Wadding). This was retailed on a Hallmark card with a pre-printed price of $2.00. On the blister card this was named "Peanuts Puzzler Yo-Yo." |
| 1178 | 36 | (Snoopy one yo-yo), imprint seal, plastic, Monogram (Hallmark), tournament shape, '80s |
| 1179 | 36 | (Snoopy walking the dog), imprint seal, plastic, tournament shape |
| | | An example of this yo-yo is shown next to this section's title. |
| 1180 | 36 | (Snoopy playing three yo-yos), paper sticker seal, wood, tournament shape, '80s |
| | | This was retailed on a Hallmark blister card as a beginners yo-yo. The card had a pre-printed price of $1.00. |
| 1181 | 36 | (Snoopy colonial hat), imprint seal, plastic, Hallmark, butterfly shape, 1976 |
| | | This yo-yo was distributed by Hallmark. The yo-yo retailed on a blister card with a pre-printed price of $1.59. The display card was labeled "Snoopy Hummingbird Yo-Yo." |
| 1182 | 36 | (Snoopy Joe Cool), imprint seal, plastic, Monogram (Hallmark), tournament shape, '80s |
| | | Snoopy first appeared in the cartoon strip as Joe Cool in 1971. |
| 1183 | 36 | (Snoopy football), imprint seal, plastic, Monogram (Hallmark), tournament shape, '80s |
| 1184 | 36 | (Snoopy sport fan), imprint seal, plastic, Monogram (Hallmark), tournament shape, '80s |
| 1185 | 36 | (Snoopy stars and stripes top hat), imprint seal, plastic, tournament shape. |
| 1186 | 36 | (Snoopy Woodstock), imprint seal, plastic, Monogram (Hallmark), tournament shape, '80s |
| 1187 | 36 | (Snoopy playing one yo-yo), paper sticker seal, wood, tournament shape, '80s |
| | | This was made by Monogram for Hallmark. |
| 1188 | 36 | (Snoopy WWI pilot), imprint seal, Hallmark plastic, tournament shape, early '80s |
| | | Snoopy first appeared in the cartoon strip as a Flying Ace in 1965. This yo-yo retailed on a Hallmark blister card with a pre-printed price of $1.00. See Plate 21, #1188. The original yo-yo has a colored string, |
| 1189 | | (Snoopy WWI ace with handlebar moustache), imprint seal, plastic, tournament shape, '80s |
| | | This yo-yo was produced by Monogram for Hallmark. The yo-yo was retailed on a blister card which said "Snoopy Champion Yo-Yo." The blister card has a pre-printed price of $1.25. |
| 1190 | 36 | (Woodstock), imprint seal, plastic, tournament shape, 1976 |
| 1191 | 36 | (Woodstock), paper insert seal, plastic, Festival, tournament shape, view lens, '70s |
| | | This is one in a series of three Peanuts pinball game yo-yos made by Festival (Union Wadding), see #1177. |

# YO-YO LISTINGS

## PLASTIC MISCELLANEOUS

Prior to WWII almost all yo-yos were made of wood or metal. Plastic yo-yos did not appear on the scene in any appreciable numbers until the early '50s. Now, over 40 years later, the majority of yo-yos being produced are made of plastic. Over the years, a variety of companies have produced plastic yo-yos. The ones listed in this section all have seals or identifying marks and are from smaller companies. Most of these yo-yos have limited collectible value, but there are some exceptions.

| | | |
|---|---|---|
| 1192 | *3* | **(Alien face)**, foil sticker seal, bulge face shape, '90s |
| | | This yo-yo preceded the release of the popular Duncan Alien Yo-Yo #411. |
| 1193 | *32* | **ALL AMERICAN**, embossed seal, metallic gold finish, tournament shape, **1971** |
| | | This gold and silver plated All American Yo-Yo, produced by the All American Yo-Yo Co., forced Duncan to produce its own line of silver and gold plated diamond tournament yo-yos in 1971. This yo-yo was retailed on a blister card with a pre-printed price of $1.19. |
| 1194 | | **BIG CHIEF**, molded seal, tournament shape, late '50s |
| | | This variegated plastic yo-yo is believed to have been produced in the late '50s, but the exact date and manufacturer have not been documented. Do not confuse this yo-yo with the Duncan Chief #437 or the Cheerio Big Chief #203. |
| 1195 | *43* | **BLACK MAMBA**, imprint seal, tournament shape, late '90s |
| | | Two versions of this yo-yo were released in the United States; the Five Star which has a ball bearing transaxle, and the Three Star which has an internal clutch. |
| 1196 | | **CHAMPION YO-YO**, hot stamped seal, tournament shape, '90s |
| | | This yo-yo is imported out of China and distributed by more than one U.S. company. |
| 1197 | | **CHARMORE YO-YO**, hot stamped seal, plastic, tournament shape, '50s |
| | | Duncan granted Charmore a license to use the "yo-yo" trademark in the '50s. They also produced the Pocket Yo-Yo #1075 and the Gyro Yo-Yo #1112. |
| 1198 | | **COASTER YO-YO**, imprint seal, bulge face shape, '80s |
| | | "Championship Contest Use" is written on the face. These were retailed loose out of a display box of one dozen. |
| 1199 | | **(Cockeyed finger tops)**, paper sticker seal, miniature, plastic finger ring, early '70s |
| | | There were two yo-yos on a blister card with a pre-printed price of 39 cents. The design on the yo-yos and card make the yo-yos look like they are cockeyed. Jak Pak distributed these yo-yos. |
| 1200 | | **DA BOMB**, imprint seal, tournament shape, ball bearing axle, late '90s |
| | | Only one out of a hundred are made of black plastic. |
| 1201 | | **DON'T WORRY**, paper sticker seal, tournament shape, '90s |
| | | This has a smile face sticker. The yo-yo is retailed in a polybag with a saddle header card. The card is marked with Stock No. 2650 by Louis Greenberg & Son, Inc. |
| 1202 | | **DYNA YO**, imprint seal, tournament shape, late '90s |
| | | This yo-yo has an internal clutch mechanism, that allows for automatic returns. |
| 1203 | | **GIANT JOLLY HO-HO**, paper sticker seal, tournament shape, '70s |
| | | Jak Pak retailed this yo-yo on a blister card in the early '70s. The card had a pre-printed price of 69 cents. |
| 1204 | | **KNIBB DYNO GLOW SKIL TOP**, hot stamp seal, tournament shape, '50s |
| | | This is considered the first glow plastic yo-yo ever produced. It is made from a combination of glow and standard plastics. |
| 1205 | *32* | **LIDO PERFECT SPINNING TOP PLASTIC**, hot stamped seal, tournament shape, '50s |
| | | This was retailed on a card with polybag, Stock No. 613. The display card had a pre-printed price of 59 cents. |
| 1206 | | **LONG SPIN YO-YO**, paper sticker seal, glow plastic, butterfly shape, early '90s |
| | | This is the same design as #1208, but in glow plastic. |
| 1207 | | **LONG SPIN YO-YO** (factory error) |
| | | This is the same as Long Spin #1206, except the blister card reads "growids in dark" instead of "glows in the dark." |
| 1208 | *04* | **LONG SPIN YO-YO**, imprint seal, butterfly shape, early '90s |
| | | This is another Duncan Wheels knock off. It has a wheel shape, a fan shaped hub molded on the face and it retailed on a blister card. The blister card is marked with the Stock No. YC039. |
| 1209 | *06* | **MAGIC MARXIE MAJESTIC**, hot stamped seal, tournament shape, '60s |
| | | This yo-yo retailed on a blister card. The blister card has the Magic Marxie character on the card, see Plate 32, #1209. In 1960, Marx introduced Magic Marxie as its trademark character. For more about Louis Marx and his line of yo-yos, see Lumar in the Foreign Yo-Yo section. |
| 1210 | *06* | **MARX YO-YO**, embossed seal, plastic, midget, '70s |
| | | The faces of this yo-yo have insert disk seals painted with metallic gold paint. The rims are painted with a metallic blue paint. |

*Continued ...* ▶

*The History and Values of Yo-Yos*

## YO-YO LISTINGS

**PLASTIC MISCELLANEOUS** *Continued ...*

1211    *06*    **MARX TOYS (Magic Marxie character), imprint seal, butterfly shape, '90s**
In 1995, when Marx Toys re-entered the toy market, they released this yo-yo featuring the Magic Marxie character. This yo-yo was produced in England. The blister card reads "Vale Royal Winged Yo-Yo."

1212    *06*    **MONARCH SLEEPER, hot stamped seal, butterfly shape, '70s**
This yo-yo was retailed on a display card. It appears to be one of the earliest attempts at making a functional take-apart yo-yo. It has a long screw as an axle.

1213    *06*    **OLYMPIC, several companies distribute yo-yos with the name "Olympic"**
These "Olympic" yo-yos are made in China or Hong Kong and are distributed by more than one U.S. company. Some have no seals, but are found on Olympic blister cards. The following are variations from the '80s and '90s. They are all tournament shaped.
- Champion Yo-Yo, made in China, hot stamped seal, Olympic blister card Stock No. T-8056
- Olympic made in China, hot stamped seal, Kidmark, Stock No. 12522
- Olympic made in Hong Kong, hot stamped seal, DC International, Stock No. 12522
- Olympic Yo-Yo, imprint seal, Magic Stock No. 8056
- Olympic, imprint seal, glow plastic (shown), BC International, Stock No. 12522
- No Seal, Olympic blister card Stock No. 12522

1214        **PLAY BALL, paper insert seal, puck shape, view lens, pinball baseball game, '50s**
This model was made by the Knibb Company of Chicago.

1215    *37*    **PARK'S TWIRLEE, hot stamped seal, tournament shape, '60s**
This yo-yo is believed to have been produced in the early '60s. It was retailed in a polybag with a header card and a pre-printed price of 49 cents. The card is marked with Stock No. 300-B. The manufacturer was Park Plastics, out of Linden, N.J.

1216    *43*    **PRO SPIN OLYMPIAN, insert seal, plastic, view lens, tournament shape**
The yo-yo is similar in appearance to the Royal TV yo-yos. The maker and date of this yo-yo is not known.

1216.1    *43*    **PRO SPIN STINGER, plastic sticker seal, butterfly shape, '80s**
The bee on the seal is playing a yo-yo.

1217    *06*    **PROFESSIONAL EAGLE RETURN TOP, hot stamped seal, butterfly shape, '70s**
This appears to be a knock off of the Duncan style butterfly from the early '70s.

1217.1       **PROFESSIONAL YO-YO MARK IV, hot stamp seal, tournament shape, '60s**
In the '60s, Joe Radovan, the owner of Royal, contacted Al Gallo, the owner of National in Canada. Radovan told Gallo about a U.S. plastic company he was using to make some Royal yo-yos. Gallo decided to try making a plastic model himself and registered the trademark "Mark IV" in Canada. The yo-yos were molded and sent to Gallo, but the majority were ruined during shipping. The order was sent back and the model was never released by National. The mold was apparently used later to make yo-yos for release in the United States. The font for the word "yo-yo" on these later models is the same as the font used for the National yo-yos during that earlier period. It is unclear how many of these yo-yos were released and what years they were produced. An example of this yo-yo is shown next to this section's title.

1218    *37*    **SAYCO, molded seal, slimline shape, wood axle, '60s and '90s**
This is a plastic yo-yo with a wooden axle. L.J. Sayegh applied for this patent in 1966 and received it, and Patent Number 3,444,6444, in 1969. Sayegh, who performs under the name Larry Sayco, had his first job as a demonstrator with Duncan in 1951. Initially, he performed as a field demonstrator in the United States, but later moved to D.I.R., Duncan's international promotion division.

In the early '60s, when his stint with Duncan ended, Sayegh took a job at a plastic factory working with injection molding machines. At this time, he developed his plans for the Sayco Yo-Yo. Sayco not only promotes and demonstrates his yo-yos, but continues to make each one himself using the machinery he developed. He manufactures as many as needed to meet the demand. Some years he produced as many as 100,000. He does not sell his yo-yos through stores, only to sponsors of his demonstrations. This yo-yo has been produced for three decades without a seal change. It is made in a view lens style and an opaque style (shown). See #1715 and #1906.

1219    *06*    **SLEEPER KING, hot stamped seal, tournament shape, '90s**
This yo-yo was marketed by C & C Fun Co. on a blister card. Multi-colored metallic paint is used for the stars. Earlier yo-yos by this company used only one color of paint. See #1228

1220    *30*    **SPINMASTER, imprint seal, plastic, Humphrey shape, '80s**

1221    *37*    **SUPERYO CRUISER, imprint seal, butterfly shape, late '90s**
This seal design was introduced in 1997, and changed in 1998.

1222    *37*    **SUPERYO TURBO TREAD, imprint seal, sculpted tire shape, late '90s**
This yo-yo was introduced in 1997. There is more than one variation.

1223    *37*    **SUPERYO SONIC SPIN, imprint seal, flywheel shape, late '90s**

1223.1       **SUPERYO SPINBOSS, imprint seal, butterfly shape, late '90s**
This take-apart yo-yo has a modular design. It is made out of temperature sensitive plastic that changes color.

1224        **SPORT DELUXE, hot stamped seal, slimline shape, view lens, '80s**
A bear character is holding blocks that read "Barrbola." Inside the view lens is a molded design that looks like a mag wheel. The axle is wooden.

1225    *29*    **SUPER SLEEPER, imprint seal on diffraction disk, Humphrey Shape, '90s**

1226    *37*    **TOPS EM ALL • IDEAL TWIRLER (on reverse), embossed seal, marbleized plastic, flywheel shape, metal axle, '50s**

*Continued ...*

## YO-YO LISTINGS

1227    *06*    **THE RATTLER, hot stamped seal, tournament shape, '50s**
         The seal reads "Plastic Perfect Spinning Top." Internal beads make a rattlesnake sound when the yo-yo is played.

1228    *06*    **TRICKSTER, hot stamped seal, plastic, tournament shape, glitter embedments, '80s**
         This was marketed in a polybag by the C&C Fun Co.

1229    *29*    **TRICK MASTER, imprint on diffraction disk, Humphrey shape, mid '90s**

1230    *37*    **WHIRL-E-GIG FREE WHEELING PLASTIC, hot stamped seal, tournament shape, metal axle, '40s**
         Some consider this to be the first post-WWII plastic yo-yo produced. Production began in 1946 by Decker Products.

1231          **WINNER, imprint seal, tournament shape, '90s**
         This brittle plastic yo-yo is a common Taiwan production. It has been distributed by many different companies such as Jak Pak and Lucky Star. See the Party Favor section.

1232          **YO-BONIC, paper insert seal, fly wheel shape, late '90s**
         In 1998, Yo-Bonic released several different yo-yos. They appear to be from an old Playmaxx mold.

1233          **(Yo-Top)**
         This combination top and yo-yo was released in 1990. It was made by Aaron Morierity of Cool Toys.

1234    *43*    **YO-YO M.A.C, hot stamped seal, tournament shape, '70s**

## PLASTIC (MISCELLANEOUS - UNNAMED)

Most plastic yo-yos that do not carry a manufacturer's seal, advertising, or a known character have limited collector interest. Listed below are a few examples that collectors may come across.

1235    *18*    **(Bats), imprint seal, tournament shape**
         "Patent Applied For" is embossed on the central face. This was apparently an attempt to capitalize on the Batman craze. The date and origin for this yo-yo are unknown.

1236          **(Composition plastic), no seal, tournament shape**
         This yo-yo has multiple plastic chips embedded in clear acrylic. It is similar to the Duncan Mardi Gras. Various styles of this yo-yo may be found. The collectible value of a yo-yo without a seal is based purely on its uniqueness.

1237    *34*    **(Concentric rings), bulge face shape**
         This unusual acrylic plastic yo-yo has a frosted appearance and concentric rings painted inside of the string slot. There are rivets on the outside that hold the wooden axle in place. The maker and date of this yo-yo are unknown. An example of this yo-yo is shown next to this section's title.

1238    *06*    **(Military aircraft series), paper insert seal, bulge face shape, view lens, '90s**
         This is a series of four photo insert seals with pinball games on both faces. Different photos are on each face. These were retailed loose out of a counter display box labeled "Sleep'n Return Top." The series includes: a Single Jet, Double Jet, Bomber, and a Helicopter.

1239    *27*    **(Mythological animal series), paper insert seal, bulge face shape, view lens, '90s**
         This is a series of four animals with pinball games on both faces. The yo-yo has a different animal on each face. Sometimes a military aircraft from series #1238 will be mixed in with this series. These were retailed loose out of a counter display box labeled "Sleep'n Return Top." The series includes: a Winged Unicorn, Eagle Leopard, Easter Bunny, and an Ugly Duckling.

1240          **(Spinner top series), imprint seal, bulge face shape, early '90s**
         This is a series of four animal characters, Bat, Bug, Caterpillar, and Turtle. These were retailed on blister cards reading "Spinner Top." They were made in Hong Kong by Toythy Plastic MFY and distributed by Famus Co.

1241    *34*    **(World map), imprint seal, tournament shape, metal string slot axle, '80s**
         This is one of many world map yo-yos.

# YO-YO LISTINGS

## PLAYMAXX

In 1974, after several years hiatus from manufacturing yo-yos, Don Duncan Jr. re-entered the market with his new patented yo-yo, the "Pro Yo." Duracraft was licensed to manufacture and sell this new model. When it was introduced, Duncan claimed the Pro Yo had the longest sleep time of any model ever produced. Two design improvements allowed the Pro Yo to achieve this long spin time. First, 80% of the weight of the flywheel shaped yo-yo was shifted to the rim. Second, the axle was made out of brass and reduced in diameter. The removable disk inserts allow the Pro Yo to easily be made into an advertising yo-yo. Pro Yos with clear mylar lenses are considered earlier models. In 1987, Duncan ended his agreement with Duracraft and formed his own company, Playmaxx.

1242        **CALIFORNIA STATE CHAMPIONSHIPS, removable disk seal, plastic, flywheel shape, '80s and '90s**
                 All yo-yos made for the California State Yo-Yo Championships were produced by Playmaxx.

1243   *37*   **DON DUNCAN SIGNATURE, removable disk seal, plastic, flywheel shape**
                 This is a Pro Yo style yo-yo.

1244        **DONALD F. DUNCAN CORPORATE SEAL, removable disk seal, plastic, flywheel shape, 1992**
                 The seal on this Pro Yo features a reproduction of the original 1930s Duncan corporate seal. This came with a numbered statement of authenticity and a small, felt yo-yo champion pennant. Only 250 of these were made in 1992.

1245        **DRAGONFLY • PRO SPIN, imprint seal, plastic, butterfly shape, 1990**
                 This butterfly style yo-yo was briefly produced in Mexico. It reads "Pro Spin" on the blister card.

1246        **GENUINE DUNCAN FAMILY COLLECTION, removable disk seal, plastic, flywheel shape, 1992**
                 This Pro Yo was sold at the Chico Museum. It commemorates the Duncan Family Collection on display there from October 17, 1992 to February 28, 1993.

1247        **MR. PRO YO, removable disk seal, plastic, flywheel shape, '90s**
                 This standard Pro Yo features a yo-yo headed character playing a yo-yo.

      •    **National Championship** - *See the National Championship section.*

1248        **OLYMPIAN PRO SPIN, imprint seal, plastic, slimline shape, 1990**
                 This slimline style yo-yo was briefly produced in Mexico. This yo-yo is called Pro Spin on the blister card.

1249        **(Pog Yo), removable disk seal, plastic, flywheel shape, '90s**
                 This yo-yo was marketed to capitalize on the Pog phenomenon that began in 1996. The yo-yo is designed to hold a child's favorite Pog.

1251   *37*   **PRO YO BY PLAYMAXX, removable disk seal, plastic, flywheel shape, '90s**
                 This is a standard Pro Yo with the old style Duncan Mr. Yo-Yo on the seal.

1252        **PRO YO BRONZE AWARD CHAMPION, removable disk seal, plastic, flywheel shape, '90s**

1253   *37*   **PRO YO GOLD AWARD CHAMPION, removable disk seal, plastic, flywheel shape, '90s**
                 This is an award yo-yo used by Playmaxx.

1254        **PRO YO SILVER AWARD CHAMPION, removable disk seal, plastic, flywheel shape, '90s**

1255   *37*   **PRO YO II, holographic insert seal, plastic, flywheel shape, wood axle, '90s**
                 Production of this yo-yo began in 1996. Several colors are available: Amber, Red, Blue, Purple, Green, Pink, Turquoise, and Clear. This Pro Yo looks very similar to the original model, but it is slightly heavier. The yo-yo is a take-apart style.

### PRO YO (Rainbow)

The Pro Yo was first produced in 1977. The original Pro Yo had a rainbow seal on a removable paper disk insert that fit under a mylar lens. These were the two versions of the original Pro Yo retailed, both on blister cards. They were produced by Duracraft. In the early '80s, Trag E. Toys also marketed the Pro Yo.

1256   *32*   **(Wheeler), paper insert seal, plastic, flywheel shape, brass axle, late '70s**
                 The Wheeler is made of opaque polystyrene. The display card has a pre-printed price of $1.29 and is marked, Stock No. #1290.

1257        **(Deluxe), paper insert seal, plastic, flywheel shape, brass axle, late '70s**
                 The Deluxe is made of translucent co-polyester. The display card has a pre-printed price of $1.79 and is marked, Stock No. #1790.

1258   *37*   **PRO YO (Ultimate Duracraft version), removable disk seal, plastic, flywheel shape '80s**
                 This yo-yo has a snap out clear plastic lens on the face. It does not say "ultimate" on the seal. For a yo-yo in the original display box, see Plate 31, #1258. These were retailed in a clear plastic display case with a trick book and string.

1259        **PRO YO ULTIMATE (Playmaxx version), removable disk seal, plastic, flywheel shape, '80s and '90s**
                 There is no plastic lens. The word "ultimate" appears on the seal. These were retailed in a clear plastic display case with a trick book and string.

*Continued ...* ▷

## YO-YO LISTINGS

1259.1     **PRO YO ULTIMATE (II), removable disk seal, plastic, flywheel shape, late '90s**
This version of the Ultimate comes in black or turquoise and was released in 1996. The yo-yo was packaged in a gift box with a trick sheet, spare axles and strings.

### RETURN OF THE YO-YO TRAVELING SHOW
Even though this Playmaxx series was recently produced, it is still considered collectible. These yo-yos were retailed at the "Return of the Yo-Yo" shows at Taubman Malls across the country. A different yo-yo was produced for each site. The Duncan Family Collection, featuring several classic yo-yos, was on display at these shows. Yo-Yo contests and performances by demonstrators were held in conjunction with these shows. A wooden, gold finish yo-yo was used as an award for these contests. See #148. Following the tour, the Duncan display was moved to the National Yo-Yo Museum in Chico, California. The display is responsible for cultivating many people's interest in yo-yo collecting.

| | | |
|---|---|---|
| 1260 | *37* | **BELLEVUE, removable disk seal, plastic, flywheel shape** |
| 1261 | | **FAIR OAKS, removable disk seal, plastic, flywheel shape, 1990** |
| 1262 | | **HILLTOP, removable disk seal, plastic, flywheel shape, 1990** |
| 1262.1 | | **LAKE FOREST, removable disk seal, plastic, flywheel shape, 1990** |
| 1263 | | **MARLEY STATION, removable disk seal, plastic, flywheel shape, 1991** |
| 1264 | | **TWELVE OAKS, removable disk seal, plastic, flywheel shape, 1990** |

Two variations of the Twelve Oaks were produced.

1265     **WOODLAND, removable disk seal, plastic, flywheel shape, 1990**

1266     **PRO SPIN • SPIN BREAKER, imprint seal, plastic, tournament shape, early '90s**
This model yo-yo was briefly produced in Mexico. It was called a Pro Spin on the blister card when it was first introduced in 1990.

1267     **TURBO BUMBLE BEE, removable disk seal, plastic, flywheel shape, late '90s**
This is Playmaxx's first ball bearing transaxle yo-yo. It was first released in 1997, and the seal changed in 1998.

1268   *39*   **YOYO MAN (Smothers Brothers), removable disk seal, plastic, flywheel shape, '90s**
This is one of many Smothers Brothers related yo-yos. It was the only plastic Smothers Brothers yo-yo produced. See also Smothers Brothers section.

## ROYAL

Joe Radovan founded the Royal Tops Manufacturing Company in 1937. Radovan was born in the Philippines in 1909. He arrived in Seattle in 1930 and immediately went to Los Angeles looking for fellow countrymen. The first Filipinos he came across were in a pool hall and they were playing with a Flores Yo-Yo. That year, he bought his first American manufactured yo-yo, a Flores, in a candy store. Radovan toured the U.S., Europe, Brazil, and Portugal as a Duncan demonstrator. During these tours Duncan promoted him as the holder of the World Champion title. (This title was bestowed on almost all the early Duncan demonstrators in the '30s.)

Radovan was one of Duncan's earliest demonstrators. This was the time of The Depression and the yo-yo men led a charmed life. They made good money when many people were unemployed. They also achieved a kind of celebrity status. They performed before packed crowds in theaters and frequently rubbed elbows with the rich and famous. Nice suits, big cars, and frequent trips to the nightclubs were standard fare for the early demonstrators. Radovan himself gave a command performance before the Prince of Spain.

After leaving Duncan, Radovan became one of Duncan's fiercest competitors. In 1951, he filed a court action for a declamatory judgment to determine the validity of Duncan's trademark "Yo-Yo." Before any action was taken, Duncan approached Royal and offered an exclusive license to the trademark "Yo-Yo" if the action was dropped. The action was dropped, and in 1955 Duncan gave Royal an "exclusive license."

In 1955, Radovan entered a partnership with Rueben Delagana and changed the name of his company to Royal Tops Mfg. Co. Inc. Delagana ran the factory, shipping, and accounting; Radovan designed the yo-yos, ran promotions, and was in charge of sales. Compared to Duncan, Radovan had few demonstrators and put little money into the promotion of products. Radovan, at peak times, would have four to five demonstrators working for him. These demonstrators did not work full-time. Most of

*Continued ...* ▶

● ROYAL *Continued ...*

them had other jobs. Like many other yo-yo companies of that time period, Royal chose to shadow the large, heavily promoted Duncan campaigns. Regarding Duncan's high profile campaigns, Radovan is quoted as saying, "You grease them up and I'll slide a yo-yo to 'em."

Royal Tops was one of the few Duncan competitors to use the word "yo-yo" on their toys. Radovan used the word "yo-yo" on his Royal and Chico lines. (See the Chico section.) Duncan owned the trademark on the names. "Genuine Duncan Yo-Yo" (1932), "Yo-Yo" (1933), and "Flores Yo-Yo" (1932). Radovan felt the word "yo-yo," from his native Filipino language, was a generic term and refused to honor the trademark.

Duncan finally brought legal action against Royal in 1961. After a four year battle, the courts judged that the word "yo-yo" was a Malayo-Polynesian word of Philippine origin and a generic word in the islands. The court also ruled that "yo-yo" had become descriptive in this country by continued usage and was no longer entitled to trademark protection. Duncan lost its "yo-yo" trademark and in that same year filed for bankruptcy.

Although Royal never rebounded completely from the expense of legal costs incurred during the court battle, Radovan continued to produce Royal yo-yos into the early '80s. Following the trial, Radovan seldom employed demonstrators electing to do most of the promotions himself. Like many other companies in the '60s, Royal began producing more and more plastic yo-yos. Radovan owned the molds and had several different plastic companies produce his yo-yos. These were stored in warehouses and shipped to areas where sales were hot. Duncan promotions in the '70s tended to be centered on specific large chain stores, so Radovan gave his product to the toy jobbers in each city that supplied the mom and pop toy stores. He frequently offered these small stores a money back guarantee, taking back any unsold yo-yos. By the '70s, because the profit margin was so much higher with plastic, all newly produced Royals were plastic.

The most familiar logos on Royal seals are crowns and chevrons. Royal yo-yos are categorized by the type of crown and by the absence or presence of crowns and chevrons. Jewel Royals are listed in their own category.

*Joe Radovan in the early '30s. Notice the yo-yo labeled megaphone. He used this while organizing large contests.*

*This is a Royal postcard from the '60s used for advertising.*

# YO-YO LISTINGS

## CROWNLESS ROYAL

1269    *38*    **CHAMPION PLASTIC**, hot stamped seal, tournament shape, early '50s
This was the first plastic yo-yo made by Royal. The word "Plastic" is hot stamped on the seal. Companies that made early plastic yo-yos often imprinted the word "Plastic" on the seal. In the '50s, plastic was considered high tech and unique. Early plastic yo-yos were typically sold at a higher price than wooden models. This is considered the rarest and most desirable of the plastic Royal yo-yos.

1270       **ROYAL MUSICAL YO-YO TOP**, foil sticker seal, marbleized plastic, tournament shape, whistler holes, '50s
This model was retailed on a blister card. Royal also made whistling plastic yo-yos with standard seals. See #1275.1.

1271    *38*    **THUNDERBIRD**, foil seal, wood, tournament shape, three-piece, pegged string, '60s
This yo-yo is considered a beginners model.

1271.1    *38*    **THUNDERBIRD**, decal seal, wood, butterfly shape, one-piece, '60s
The triangular decal Thunderbird seal is considered more collectible than the round foil sticker seal. An example of this yo-yo is shown next to this section's title.

1272    *38*    **(TV yo-yo)**, plastic, view lens, '60s and '70s
This yo-yo carries the Patent Number 3,256,635. The patent was applied for in 1962, and granted to Joe Radovan in 1966. This yo-yo came unassembled in a cardboard box. The box is pre-printed with a price of $1.00. Photographs could be inserted beneath the plastic view lens. This model was used as both a premium and an ad yo-yo. "Royal Tops Mfg. Co. Inc." can be found embossed on the yo-yo's face, under the view lens.

## FIVE POINT CROWN WITH CHEVRON ROYAL

The "Five Point Crown with a Chevron" became the logo most frequently associated with Royal yo-yos. Its use began in 1956 and continued through the '60s. This chevron is very similar to the one used on Duncan's Imperial Yo-Yo that began production in 1954. (See #347.) Many believe that the design was chosen to capitalize on the success of the Imperial Chevron that had become one of Duncan's most popular yo-yos. All Royal yo-yos that have a chevron on the seal also have a five point crown. The yo-yos listed below all bear this logo on the seal.

1273    *38*    **500**, die stamped seal, wood, tournament shape, three-piece, pegged string, '50s and '60s

1274    *38*    **BUTTERFLY**, gold leaf stamped seal, wood, butterfly shape, one-piece, early '60s
This yo-yo was used as an exhibit during the Duncan vs. Royal yo-yo trademark trial. In addition to the yo-yo trademark suit, Duncan was also trying to protect its Butterfly trademark. Duncan eventually won the Butterfly trademark battle with Royal. Duncan Flambeau is still the only yo-yo maker allowed to use the name Butterfly in the United States. The Royal "Butterfly" was only produced briefly and is considered highly collectible.

1275    *38*    Die stamped seal, wood, tournament shape, three-piece, pegged string, '50s and '60s
This is a standard red and black beginners yo-yo.

1275.1    *38*    Foil sticker seal, plastic, tournament shape, wood axle, whistler holes, early '60s
See also #1270.

1276    *38*    Hot stamped seal, plastic, Humphrey shape, wood axle, late '70s
This is one of the last yo-yos made by Royal.

1276.1    *38*    Hot stamped seal, plastic, tournament shape, wood axle, pegged string, early '60s

1277    *38*    **MONARCH**, imprint seal, plastic, tournament shape, metal axle, '70s

1278    *38*    **OFFICIAL TOURNAMENT**, decal seal, wood, tournament shape, '50s and '60s
There are two different versions of this seal, one round and one with a scalloped edge. An example of this yo-yo is next to this section's title.

1279    *38*    **OFFICIAL TOURNAMENT**, gold leaf stamped seal, wood, tournament shape, one-piece, '50s and '60s

1280    *38*    **SPECIAL**, hot stamped seal, plastic, tournament shape, early '60s

1280.1    *38*    **TOURNAMENT**, foil sticker seal, wood, butterfly shape, early '60s

The History and Values of Yo-Yos

# YO-YO LISTINGS

### FIVE POINT CROWN WITHOUT CHEVRON ROYAL

These seals have five point crowns of different styles. They do not have a chevron on the seal.

| | | |
|---|---|---|
| 1281 | | **CHAMPION FILIPINO SPINNING TOP**, decal seal, wood, tournament shape, '50s |
| | | This is a one-piece yo-yo believed to have been sold in the early '50s. |
| 1282 | 38 | **CHAMPION JUNIOR TOP**, gold leaf stamped seal, wood, tournament shape, '40s and '50s |
| | | In 1954, the original retail price of this three-piece yo-yo was 10 cents. This model had a Stock No. 233. |
| 1283 | | **CHAMPION SPORT TOP**, silver leaf stamped seal, wood, tournament shape, pegged string, '40s and '50s |
| | | An example of this yo-yo is shown next to the Royal main section's title. |
| 1284 | 38 | **CHAMPION STANDARD TOP**, gold leaf stamped seal, wood, tournament shape, '40s and '50s |
| | | In 1954, the original price of this three-piece yo-yo was 15 cents. This model had a Stock No. 255. An example of this yo-yo is shown next to this section's title. |
| 1285 | 38 | **CHAMPION OFFICIAL TOURNAMENT TOP**, decal seal, wood, tournament shape, '30s - '50s |
| | | This one-piece yo-yo was the standard line tournament model first used in the late '30s. |
| 1286 | 38 | **MASTER CHAMPIONSHIP OFFICIAL TOP**, decal seal, wood, tournament shape, '50s |
| | | This yo-yo was first produced in 1950. The "o" in the word "Royal" is in the shape of a yo-yo. This is an unusual font style. This font was also used by Duncan in the '50s. |
| 1287 | 38 | **RADOVAN CHAMPION MASTER OF ALL TRICKS**, decal seal, wood, tournament shape, miniature, '50s |
| | | This is one of the rarest Royal yo-yos. It is not believed to have been retailed. |
| 1288 | 38 | **SPECIAL OFFICIAL TOURNAMENT TOP**, decal seal, wood, tournament shape, '50s |
| 1289 | 38 | **TOURNAMENT**, gold leaf stamped seal, wood, butterfly shape, '50s |
| | | This has "Duncan Tops" lettered with the Mr. Yo-Yo logo on the reverse face. Royal bought yo-yos from Duncan in the '50s before the lawsuit. Rare yo-yos may have seals from both companies, one on each face. |
| 1290 | 38 | **YO-YO**, paper sticker seal, plastic, butterfly shape, '70s |

### ONE POINT CROWN ROYAL

The Monarch-one point crown logo was introduced in 1960. Although a few other one point crown seals exist, the majority of these seals were named Monarch. The Monarchs from the early '60s may carry the registration mark ® with the word "yo-yo." Later Monarchs have no registration mark.

| | | |
|---|---|---|
| 1291 | 38 | **MASTER**, foil sticker seal, wood, tournament shape, early '60s |
| | | An example of this yo-yo is shown next to this section's title. |
| 1292 | 12 | **MONARCH YO-YO ®**, hot stamped seal, plastic, tournament shape, metal axle, early '60s |
| | | The registration mark ® is located after the word "yo-yo." At the time this yo-yo was made, Duncan owned the trademark for the word "yo-yo." |
| 1293 | | **MONARCH**, hot stamped seal, plastic, butterfly shape, metal axle, '70s |
| 1294 | 38 | **MONARCH**, hot stamped seal, plastic, tournament shape, metal axle, '70s |
| 1295 | | **MONARCH**, imprint seal, plastic, tournament shape, metal axle, late '70s |
| 1296 | | **MONARCH**, gold leaf stamped seal, wood, tournament shape, early '60s |
| 1297 | | **OFFICIAL TOURNAMENT**, gold leaf stamped seal, wood, tournament shape, early '60s |

## CROWN OF ENGLAND ROYAL

This is another crown logo variation used by Royal. The logo resembles the crown of England. It is the most detailed of the Royal crowns and is found only on decal seals.

1298  38  **KING SIZE**, decal seal, wood, tournament shape, jumbo, '50s
This oversized (3" in diameter) model is the largest yo-yo retailed by Royal. It is one of the most collectible.

1299  38  **MASTER OFFICIAL**, decal seal, wood, tournament shape, '50s
An example of this yo-yo is shown next to this section's title.

1300  38  **SPECIAL OFFICIAL TOURNAMENT**, decal seal, wood, tournament shape, '50s
This yo-yo was released in 1947 and was used into the early '50s.

## JEWEL ROYAL

Adding rhinestones to the face of the yo-yo was common practice in the '50s and '60s. Yo-Yos with rhinestones are referred to as "jewels" and are favorites among collectors. Royal took many of their standard line yo-yos and made them into jewel models. Unlike Duncan, they did not have a model or a specific seal that would indicate the yo-yo had rhinestones added. Royal placed jewels on the reverse face of some models. Models that have jewels on the seal side are considered more desirable. Like Duncan, Royal drilled the jewel holes prior to painting the yo-yo. Since the rhinestones were added after painting, paint can be found on the inside edge of the jewel holes.

In addition to the yo-yos listed below, Royal produced some jewel models without seals. In 1954, a three-jewel model, Stock No. 288, was retailed without a seal. This yo-yo had rhinestones on one face and retailed for 39 cents. A four-jewel model without a seal was produced at the same time, Stock No. 299. It had rhinestones on both faces and retailed for 69 cents. Because these models do not have seals and were not retailed on blister cards, they carry little value.

1301       **(3 jewel • reverse face)**, 5 point crown, decal seal, wood, tournament shape, one-piece, '50s
The seal reads "Royal Special Official Tournament." This is the same seal as #1288, but it has three jewels in a row on the reverse face. It carries only a slightly higher value than #1288 because the jewels are on the reverse face.

1302       **(3 jewel • 5 point crown)**, decal seal, wood, tournament shape, '50s and '60s

1303  15  **(3 jewel • 5 point crown no chevron)**, gold leaf stamped seal, wood, tournament shape, '50s
The jewels are placed in a row on both faces. The seal may be on one or both faces. The two-sided seal carries a slightly higher value.

1304  15  **(3 jewel • 5 point crown with chevron)**, gold leaf stamped seal, wood, tournament shape, '50s and '60s
The jewels on this yo-yo are placed on each chevron point.

1305  12  **(3 jewel) MONARCH (5 point crown with chevron)**, gold leaf stamped seal, wood, tournament, shape, '60s
The jewels on this yo-yo are placed on each chevron point. The yo-yo was retailed on a blister card with a pre-printed price of 59 cents.

1306  15  **(4 jewel) DELUXE TOURNAMENT**, gold leaf stamped seal, wood, tournament shape, '50s
An example of this yo-yo is shown next to this section's title.

1307  15  **(4 jewel • 5 point crown with chevron)**, gold leaf stamped seal, wood, tournament shape, '50s and '60s

1308  15  **(4 jewel) CHAMPION SUPER DELUXE**, decal seal, wood, tournament shape, one-piece, '40s and '50s
This was the first of the post-WWII yo-yos made by Royal Tops Mfg. Company. The rhinestones are slightly larger (6mm) than the standard Royal Jewel (4mm). This yo-yo was known as model #400, and in 1954 had an original price of 65 cents.

1309  12  **(4 jewel)**, no seal, wood, tournament shape, '60s
Because there is no seal, this yo-yo is only considered collectible if it is still on the original blister card, as is shown in this book. The blister card has a pre-printed price of $1.00.

# YO-YO LISTINGS

## RUSS BERRIE AND COMPANY INC.

Russ Berrie and Company is a large producer of gifts and toys.

| | | |
|---|---|---|
| 1310 | | **ALL STAR DAD**, paint seal, wood, tournament shape, pegged string, plastic finger ring, early '90s |
| 1311 | | **(Bear with snowcap)**, paper sticker seal, plastic, miniature, '90s |
| | | This was retailed loose in 1996. |
| 1312 | | **(Light-Up)**, no seal, plastic, puck shape, '90s |
| | | The only way to tell if a Light-Up is a Russ is if the foil sticker seal, usually placed near the string slot, is still attached. |
| 1313 | | **(Number One) DAD**, paint seal, wood, tournament shape, pegged string, plastic finger ring, early '90s |
| 1314 | | **(Number One) GRAD**, paint seal, wood, tournament shape, pegged string, plastic finger ring, early '90s |
| | | An example of this yo-yo is shown next to this section's title. |
| 1315 | | **(Penguin with scarf)**, paper sticker seal, plastic, miniature |
| | | This was retailed loose in 1996. |
| 1316 | 43 | **SUPER DAD**, paint seal, wood, tournament shape, pegged string, plastic finger ring, early '90s |
| 1317 | | **(Trolls)**, paper sticker seal, plastic, tournament shape, miniature, '90s |
| | | This was a series of Troll yo-yos. |

## RUSSELL

The Jack Russell Company is based out of Stuart, Florida. Like many other yo-yo brands, Russell can trace its roots back to Duncan. Jack Russell, the company founder, began his yo-yo career as a Duncan demonstrator in the '40s. In the '50s, Duncan put together a foreign promotion division called D.R.I., an acronym for directors Duncan, Russell, and Ives. Duncan's brother-in-law, Tom Ives, unexpectedly died in 1958 and Duncan decided to pull out of foreign operations. Russell elected to buy D.R.I. and form his own company continuing with foreign promotions and sales.

Most Russell yo-yos are associated with foreign Coca-Cola promotions. From the 1960s through the 1980s, the Jack Russell Company worked closely with Coca-Cola to promote their product internationally. Coke would arrange for the dealers and the plastic companies to produce the yo-yos. The Russell Company would then provide the plastic company with the Russell yo-yo molds and Russell was reimbursed a service fee based upon the number of yo-yos sold during each promotion.

Coca-Cola Yo-Yo promotions lasted approximately eight weeks. During promotions, demonstrators would typically visit each store twice, the shopping malls more frequently. Small stores usually received the first box of yo-yos on consignment. Each additional box was purchased. These large promotions were very expensive to run. To consider the promotion a success, approximately 10% of the targeted con-

*Russell Coca-Cola Yo-Yo contest winners, New Guinea, 1978. Dan Volk, a Jack Russell demonstrator, is standing on the right.*

Continued ...

sumer had to purchase a yo-yo. Contests were heavily promoted, but typically less than 2% of the children who bought yo-yos actually participated in the contests. Most of the promotions are actually provided by the children themselves. What child wouldn't want a yo-yo after seeing other children playing with one? This has been the backbone of all contest-based promotions since Flores ran his first contest in the late '20s.

The majority of Russell yo-yos are plastic and most advertise Coca-Cola related products. Prior to 1978, Russell yo-yos either had molded seals on opaque convex faces or convex clear lenses. After 1978, the yo-yos were manufactured as flat bulge face shapes with advertising imprinted on the faces. In the '50s, Russell made one wood model and one tin. These are the most collectible of the Russell yo-yos. For Russell yo-yos retailed in the U.S., see the Coke and Soft Drink sections. The list below includes two examples of Russell yo-yos distributed to foreign markets. Many other styles do exist.

1318   24   **RUSSELL PROFESSIONAL**, molded seal, plastic, slimline shape, '60s
             This yo-yo is from one of the rare non-Coca-Cola promotions. This promotion was for Kolynos toothpaste and was held in Peru in 1966.
1319   24   **COCA-COLA SUPER**, paper insert seal, plastic, slimline shape, view lens, pinball, '60s
   • **Coca-Cola** - *See Coca-Cola section.*
   • **Soft Drink** - *See Soft Drink (Non-Coke) section.*
   • **McDonald's** - *See McDonald's #1060.*

## SMOTHERS BROTHERS

Tom and Dick Smothers have been in the entertainment business as a comedy team for over 35 years. Their first performance was reportedly at The Purple Onion in San Francisco in 1959. Their music and popular comedy act, featuring their famous "Mom always liked you best" routine, led to repeated show business successes. In 1967, they received their own variety series on CBS, "The Smothers Brothers Comedy Hour." The show was controversial, but highly rated. Their outspoken views on politics lead to censorship of some sketches and ultimately to the show's cancellation.

In the early '80s, Rick Cunha gave Tommy Smothers the song, "(I'm A) Yo-Yo Man." Smothers remembered a few yo-yo tricks from his childhood and added them to his performance of the song. The "Yo-Yo Man" was born. In 1984, with the help of Dan Volk and some other top demonstrators, Smothers incorporated the yo-yo more completely into his act. In 1986, the YoYo Man made his television debut on The Tonight Show with Johnny Carson. This was followed by the development of the Smothers Brothers YoYo Man Instructional Video, released in 1988. (See #751.) It has been the most successful yo-yo instructional video, reportedly selling over 200,000 copies. In 1988 and 1989, the Smothers Brothers were back on television again with their own show.

In 1982, Tommy Smothers bought the Remik Ridge Winery. Yo-Yo related items have been sold in the winery gift shop. The Smothers Brothers and the "Yo-Yo Man" remain Las Vegas headliners and continue to perform at conventions and fairs throughout the country.

   • **Hummingbird YoYo Man** - *See Hummingbird #752.*
   • **Kodak YoYo Man** - *See Hummingbird #751.*
   • **Nissan Open YoYo Man** - *See Hummingbird #750.*
   • **Playmaxx YoYo Man** - *See Playmaxx #1268.*
   • **Smothers Brothers Silver Bullet** - *See Tom Kuhn #1589.*
   • **Smothers Brothers Smo•Bro** - *See Tom Kuhn #1590.*
   • **Smothers Fine Wines** - *See Tom Kuhn #1591 and 1591.1.*
   • **YoYo Man** - *See Hummingbird #749.*
   • **YoYo Man Wine Stopper** - *See Miscellaneous section #1741.*

# YO-YO LISTINGS

## SOFT DRINK (NON-COKE)

Soft Drink advertising yo-yos are considered cross-collectible. This influences their value. Yo-Yo collectors are most interested in wooden Soft Drink advertising yo-yos produced in the '50s. The yo-yos listed below are a sampling of various Soft Drink advertising yo-yos. (See also the Coca-Cola section.)

1320   0.5   **ANNUAL PEPSI CHAMPIONSHIP TAMPA 1962**, paper insert seal, plastic, tournament shape, view lens, '60s
  This appears to be a Royal TV yo-yo. See #1272.

1321   0.5   **DR PEPPER 10-2-4**, gold leaf stamped seal, wood, tournament shape, three-piece, '50s
  These were given free with the purchase of a carton of Dr Pepper. The yo-yo came in a polybag with a cardboard saddle header which fit over the neck of a bottle. In the '40s and '50s, 10-2-4 was the ad campaign for Dr. Pepper. The 10-2-4 numbers represented the time of day you should drink a Dr Pepper. There are several different seal variations for the Dr. Pepper yo-yo. An example of this yo-yo is shown next to this section's title.

1322         **DIET PEPSI**, imprint seal, plastic, Humphrey shape, metal axle, '90s
1323         **DRINK PEPSI COLA**, decal seal, wood, tournament shape, '50s
1324   0.5   **FANTA GALAXY 200**, imprint seal, plastic, Russell, bulge face shape, wood axle, early '80s
  This yo-yo was used in a Canadian Jack Russell promotion in 1981. It does not have the word "yo-yo" on the seal because it was a Canadian promotion. Instead, it says "Return Top."

1325         **FANTA SUPER ENJOY**, imprint seal, plastic, Russell, bulge face shape, wood axle, '80s
  This was retailed loose out of a Russell counter display box of 12. All yo-yos in the box were marked Super. The box included six Coke, two Fanta, two Mello Yellow, and two Sprite yo-yos. This is from an Australian promotion in 1984.

1326         **FANTA SUPER**, imprint seal, plastic, Russell, bulge face shape, wood axle
  •     **Kist** - *See Duncan Advertising section #307.29.*

1327         **MELLOW YELLOW SUPER**, imprint seal, plastic, Russell, bulge face shape, wood axle, '80s
  This was retailed loose out of a Russell counter display box of 12. All yo-yos in the box were marked Super. The box included six Coke, two Fanta, two Mello Yellow, and two Sprite yo-yos. This is from an Australian promotion in 1984.

1328   0.5   **MOUNTAIN DEW**, paper sticker seal, plastic, sculpted, '90s
  This sculpted bottle cap shaped yo-yo was made by Imperial Toy Co. It was released in 1996 on a blister card, Stock No. 8525B.

1329         **OVALTINE**, paper insert seal, plastic, slimline shape, view lens, '60s
  This appears to have been made from the old Duncan "Eagle" mold. It is likely a foreign Duncan or a Russell promotional yo-yo from the late '50s or early '60s.

1330         **PEPSI**, embossed seal, plastic, midget, plastic axle, '80s
1331   0.5   **PEPSI**, hot stamped seal, plastic, sculpted, metal axle
  This is the earlier style of bottle cap shaped yo-yo. It is similar to the Kooky Kaps mold used by Coke. See #234.

1332   0.5   **PEPSI**, paint seal, wood, tournament shape, three-piece, '90s
1333   0.5   **PEPSI**, paper sticker seal, plastic, sculpted, metal axle, '90s
  This is a bottle cap shape produced by Humphrey and distributed by Imperial Toy Co. It was produced in two colors, gold and silver, and retailed on a blister card, Stock No. 8525.

1334         **PEPSI**, imprint seal, plastic, Humphrey shape, metal axle, '90s
1335         **PEPSI**, paper insert seal, plastic, slimline shape, view lens, plastic axle, '80s
  This is from a foreign promotion.

1336         **PEPSI LIGHT**, imprint seal, plastic, Humphrey shape, metal axle, '80s
1337   0.5   **ROYAL TRUE-ORANGE**, imprint seal, plastic, Russell, bulge face shape, wood axle, 1984
  This was from a Philippines promotion in 1984.

1338         **SPRITE • GALAXY 200**, imprint seal, plastic, Russell, bulge face shape, 1981
  This yo-yo is from a Canadian promotion in 1981. The seal reads "Return Top." It could not be called a "yo-yo" because Parker owned the Canadian trademark at that time.

1339   0.5   **SPRITE • GALAXY S 300**, imprint seal, plastic, Russell, bulge face shape, wood axle, 1981
  This yo-yo is from a Canadian promotion in 1981. The seal reads "Return Top." It could not be called a "yo-yo" because Parker owned the Canadian trademark at that time.

1340         **SPRITE SUPER**, imprint seal, plastic, Russell, bulge face shape, wood axle, 1984
  This was retailed loose out of a Russell counter display box of 12. All yo-yos in the box were marked Super. The box included six Coke, two Fanta, two Mello Yellow, and two Sprite yo-yos. This is from an Australian promotion in 1984.

# YO-YO LISTINGS

## SOUVENIR

The value of a souvenir yo-yo depends upon its age and its importance to the collector. Even for advanced collectors, collecting souvenir yo-yos can be fun. They not only make great additions to any collection, they also serve as personal mementos. These yo-yos are perfect for new collectors because they are inexpensive and easy to find. Many souvenir yo-yos are not packaged or blister carded. A souvenir yo-yo in its original packaging is not much more valuable than one found loose. An exception to this would be a yo-yo with packaging that advertises the attraction.

Thousands of different souvenir yo-yos have been produced. Favorites among collectors are pre-'70s wood yo-yos with decal seals, and tin yo-yos with photo lithographs. The following is a small sampling of several styles that may be found by collectors. For other souvenir yo-yos, see also World's Fair, Disney, Sports, Light-Up, Humphrey, and Novelty sections.

| # | Value | Description |
|---|---|---|
| 1342 | | ADVENTURELAND, paper seal, plastic, tournament shape, '80s |
| | | A paper seal affixed under a clear plastic disk is attached to the face. |
| 1343 | | ATLANTA ZOO, large disk insert seal, plastic, tournament shape, '90s |
| | | The seal shows "Willie B" the gorilla. |
| 1344 | 36 | ALASKA, laser carved seal, wood, tournament shape, three-piece, late '90s |
| | | This is from a set of seven different laser-carved natural wood yo-yos produced by Laser Tech. of Alaska. Other designs include polar bear cubs, polar bear, walrus, wolf cubs, mosquito, and a beagle in a seaplane. |
| 1345 | 40 | ARIZONA (lithograph seal of stage coach), tin, tournament shape, '70s |
| 1346 | | ATLANTA THE DOGWOOD CITY, foil sticker seal, plastic, tournament shape |
| 1347 | | AUSTRALIA, ink stamp seal, wood, tournament shape, '90s |
| 1348 | | BERMUDA, imprint seal, plastic, slimline shape, '90s |
| 1349 | 30 | BIG BRUTUS, imprint seal, plastic, Humphrey shape, '90s |
| | | At 16 stories tall, Big Brutus is the second largest electric shovel in the world. It is currently a museum in West Mineral, Kansas. |
| 1350 | | (Bowling ball • twin cities), imprint seal, plastic, ball shape, '70s |
| 1351 | | BOSTON • OLD NORTH CHURCH, ink stamp seal, wood, tournament shape, three-piece, '90s |
| 1352 | | BUFFALO BILL CODY HOMESTEAD, ink stamped seal, glow plastic, tournament shape, '80s |
| 1353 | | BUSCH GARDENS • TAMPA • FLORIDA, small disk insert seal, plastic, tournament shape, '90s |
| 1354 | | CALIFORNIA, imprint seal, plastic, tournament shape, '90s |
| 1355 | 40 | CANADA, decal seal, rainbow rings, wood, tournament shape, three-piece, pegged string, '70s |
| | | This yo-yo has a plastic finger ring. |
| 1356 | | (Cedar series), decal seal, wood, tournament shape, '80s and '90s |
| | | These are very common souvenir yo-yos. This is a line of cedar yo-yos that have the names of different tourist attractions on them. The yo-yos are sealed in polyurethane. |
| 1357 | | CELEBRATE WISCONSIN, imprint seal, plastic, Flambeau, tournament shape, '80s |
| | | The Duncan Imperial logo is on the reverse face. |
| 1358 | | CHEYENNE SALOON, hot stamped seal, wood, Hummingbird, tournament shape, '80s |
| 1359 | | CHINATOWN • NEW YORK (rainbow rings), paper sticker seal, wood, jumbo, pegged string, '70s |
| 1360 | 28 | CHURCH STREET STATION, die stamped seal, wood, Hummingbird, tournament shape, '80s |
| | | This is a popular tourist attraction in downtown Orlando, Florida. |
| 1361 | | CIRCUS CIRCUS CASINO, large disk insert seal, plastic, tournament shape, '90s |
| 1362 | | COLONIAL WILLIAMSBURG, paint seal, wood, tournament shape, '80s |
| | | This was the original capitol of Virginia. It has been restored and is now a popular tourist attraction. |
| 1363 | 36 | CORN PALACE, photo lithograph seal, tin, tournament shape, early '70s |
| | | The Corn Palace was established in 1892 and has been re-built three times. Giant murals made of South Dakota agricultural products are on the building exterior. |
| 1364 | 40 | CORN PALACE (rainbow rings), decal seal, wood, three-piece, pegged string, '70s |
| 1365 | | COUNTRY MUSIC HALL OF FAME, paper insert seal, plastic, slimline shape, view lens, '80s |
| | | The copyright date on the yo-yo is 1981. |
| 1366 | 36 | EVERGLADES HOLIDAY PARK, paint seal, wood, tournament shape, '60s |
| 1367 | 40 | FIELD OF DREAMS • DYERSVILLE • IOWA, removable disk seal, plastic, flywheel shape, '90s |
| | | This yo-yo is sold as a souvenir at the baseball field from the movie, "Field of Dreams." This is a Pro Yo. See also #1367.1. |
| 1367.1 | 40 | FIELD OF DREAMS, hot stamped seal, plastic, sculpted, '90s |
| | | This is a souvenir sold at the "Field of Dreams" baseball field. This is made from the Humphrey baseball mold. See #1367. |

*Continued ...*

## YO-YO LISTINGS

### SOUVENIR Continued ...

**1368**   *36*   **FLORIDA "SOUVENIR OF,"** decal seal, wood, tournament shape, 1959
Standard size, wood, tournament models with decal seals from the '50s and '60s are very rare. Not only were very few produced, but the seals do not age well and frequently chip off. Souvenir yo-yo collectors would consider yo-yos like this one highly desirable.

**1369**   *40*   **GATEWAY ARCH**, photo lithograph seal, tin, tournament shape, '60s
The St. Louis Gateway Arch National Monument was built in 1965 to commemorate westward expansion in the United States. This yo-yo retailed in a polybag with a saddle header card labeled "Return Top."

**1370**   *40*   **HAWAII**, paint seal, wood, tournament shape, jumbo, '60s
The yo-yo has rings painted in the rainbow colors and has external bells attached.

**1371**   *40*   **HAWAII • 50TH STATE**, small disk insert seal, plastic, tournament shape, '80s

**1372**   *40*   **HAWAII**, paint seal, wood, tournament shape, '90s
This is a series of three yo-yos: Whale, Helmet, and Hieroglyphics (shown). They are made by Lanakila Crafts of Honolulu. These yo-yos are retailed in a polybag with saddle header card.

**1373**   *40*   **HEARST CASTLE**, large disk insert seal, plastic, tournament shape, '80s
See #1374 for details of William Randolph Hearst.

**1374**   *40*   **HEARST CASTLE**, paint seal, wood, tournament shape, three-piece, pegged string, '60s
This is the most desirable of the Hearst yo-yos. See also #1373. William Randolph Hearst was an American publishing giant. He owned eighteen newspapers and numerous magazines. Hearst had an important part in yo-yo history. In the early '30s, Don Duncan, Sr. approached him with a promotional idea to stimulate newspaper circulation. To enter a Duncan yo-yo contest, kids were required to sell a three month subscription to the local newspaper. This was part of the deal Duncan made to get free advertising space in the newspaper.

The Herald Examiner in Chicago was the first newspaper to take advantage of Duncan's idea. When a reported 50,000 new subscriptions were sold, other Hearst newspapers jumped on the bandwagon. The Hearst Castle pictured on this yo-yo is now a state museum in San Simeon, California. An example of this yo-yo is shown next to this section's title.

**1375**     **INDIANA • THE HOOSIER STATE**, medallion seal, plastic, tournament shape, '80s
**1376**   *40*   **INDIANAPOLIS ZOO**, imprint seal, plastic, tournament shape, early '90s
**1377**     **IOWA**, ink stamped seal, wood, tournament shape, '90s
**1378**     **KEY BISCAYNE • FL**, imprint seal, plastic, tournament shape, '90s
**1379**     **KINGS ISLAND**, large disk insert seal, plastic, tournament shape, '90s
**1380**     **KITES OF BOSTON**, paint seal, wood, tournament shape, '90s
**1381**   *40*   **LAS VEGAS (rainbow rings)**, decal seal, wood, tournament shape, pegged string, '70s
This yo-yo has a plastic finger ring attachment.

**1382**     **LAS VEGAS**, lithograph seal, tin, tournament shape, '80s
**1383**     **LOCH NESS MONSTER**, paper sticker seal, plastic, tournament shape, '90s
The Loch Ness Monster is a 1500 year old Scottish legend. The movie "Loch Ness" was released in 1995.

**1384**   *40*   **LOOKOUT MOUNTAIN • LOVERS LEAP**, paper sticker seal, wood, tournament shape, '50s
This is a rare '50s style souvenir yo-yo from Rock City Gardens in Tennessee. Although this yo-yo does not have a Duncan seal, it does have the half and half paint style briefly used by Duncan in the late '50s.

**1385**   *35*   **MACKINAC ISLAND MICHIGAN**, lithograph seal, tin, tournament shape, '90s
The lithograph seal shows a painting of a horse drawn carriage.

**1386**   *36*   **MARDI GRAS**, lithograph seal, tin, tournament shape, late '80s
This yo-yo features the comedy and tragedy masks. For other Mardi Gras yo-yos, see #328 and #420.

**1387**   *36*   **MARINELAND**, lithograph seal, tin, tournament shape, '70s
**1388**     **MARINE WORLD**, plastic, coaster shape, '80s
**1389**     **MOVIELAND WAX MUSEUM**, paper insert seal, plastic, slimline shape, view lens, '80s
**1390**   *40*   **MUNICIPAL AIRPORT DES MOINES**, lithograph seal, tin, tournament shape, '70s
**1391**     **NEW HAMPSHIRE (picture of moose)**, paint seal, wood, tournament shape, '60s
**1392**   *40*   **NEW MEXICO**, photo lithograph seal, tin, tournament shape, '60s
**1393**   *40*   **NICKELODEON**, imprint seal, plastic, tournament shape, '90s
This is a promotional yo-yo given out at Nickelodeon Studios in Orlando, Florida.

**1394**     **OLYMPICS ATLANTA**, laser carved, wood, tournament shape, 1996
This was sold at the 1996 Atlanta Olympics. The seal pictures the Olympic rings on the torch.

**1394 .1**   *24*   **(Olympics, Barcelona)**, paper sticker seal, plastic, Light-Up, 1992
The seal pictures Cobi the dog, the 1992 Olympic mascot.

**1394.2**     **(Olympics Los Angeles)**, removable disk seal, plastic, flywheel shape, 1984
This yo-yo was sold at the 1984 Olympics. The seal of the yo-yo features the 1984 Olympic mascot, Sam the Eagle.

**1395**   *04*   **OPRYLAND (Roy Acuff on reverse)**, paper insert seal, plastic, slimline shape, view lens, '70s
Roy Acuff was the King of Country Music and a regular performer at the Grand Ole Opry for over 50 years. His signature performance is known to have included a yo-yo. See also #742.

**1396**   *40*   **OPRYLAND**, paper insert seal, plastic, slimline shape, view lens, '80s
This yo-yo is from the amusement park in Nashville, Tennessee. The same seal is on both faces.

*Continued ...*

## YO-YO LISTINGS

| | | |
|---|---|---|
| 1397 | | **ORIENT EXPRESS**, large disk insert seal, plastic, tournament shape, '90s |
| | | An Asian mask design is on the seal. |
| 1398 | | **PROJECT EARTH • TEXAS**, paint seal, wood, tournament shape, '80s |
| 1399 | | **PERU • IND.**, decal seal, marbleized plastic, tournament shape, '60s |
| | | This yo-yo has an internal sound generating device. |
| 1400 | | **PUERTO RICO**, imprint seal, plastic, tournament shape, '90s |
| 1401 | *40* | **ROCK ISLAND**, decal seal, wood, tournament shape, miniature, 1952 |
| | | The seal also reads "One Hundred Years of Progress, 1852 - 1952." |
| 1402 | | **SAN DIEGO ZOO**, imprint seal, plastic, tournament shape, miniature, '80s |
| | | This yo-yo has Koala bears on the seal and a metal string slot axle. |
| 1403 | *35* | **SAN FRANCISCO (cable car)**, lithograph seal, tin, tournament shape, '90s |
| 1404 | *35* | **SAN FRANCISCO (Golden Gate Bridge)**, lithograph seal, tin, tournament shape, '90s |
| 1404.1 | | **SAN FRANCISCO (Golden Gate Bridge)**, imprint seal, plastic, tournament shape, '90s |
| 1405 | | **SEA WORLD**, large disk insert seal, plastic, tournament shape, '90s |
| | | The yo-yo features Shamu on the seal. |
| 1406 | | **SIX FLAGS GREAT ADVENTURE • JACKSON • NEW JERSEY**, imprint seal, plastic, tournament shape, '90s |
| 1407 | *36* | **SMITHSONIAN INSTITUTION**, lithograph seal, tin, tournament shape, '90s |
| | | This yo-yo has a copyright of 1991. It was sold for several years in the Smithsonian Institution gift shops. |
| 1408 | | **STONE MOUNTAIN**, large disk insert seal, plastic, tournament shape, '80s |
| | | Stone Mountain is a Confederate Memorial Carving in Georgia. |
| 1409 | | **TARONGA ZOO**, paper insert seal, slimline shape, view lens, '80s |
| 1410 | | **TASMANIA**, carved seal, wood, tournament shape, '90s |
| 1411 | | **TENNESSEE VALLEY R.R.**, large disk insert seal, plastic, tournament shape, '90s |
| | | A locomotive design is on the seal |
| 1412 | | **TEXAS GIANT**, plastic sticker seal, wood, tournament shape, jumbo, early '90s |
| | | This was retailed in a polybag with a saddle header card. |
| 1413 | *40* | **TEXAS LONE STAR STATE (map)**, lithograph seal, tin, tournament shape, '70s |
| 1414 | *23* | **UNDERWOOD PARK • CALIFORNIA (rainbow rings)**, decal seal, wood, jumbo, '60s |
| 1415 | *40* | **WAHPETON • N.D. (rainbow rings)**, paper sticker seal, wood, jumbo, pegged string, '60s |
| 1416 | *36* | **YELLOW STONE PARK**, photo lithograph seal, tin, tournament shape, '60s |
| | | Yellow Stone is a National Park in Wyoming. |

## SPECTRA STAR

Frank Alonso founded Spectra Star in 1973 as a kite company. Alonso saw how yo-yo faces could be sculpted. This sparked his interest and around 1982 - '83, Spectra Star began manufacturing sculpted yo-yos.

Originally, Spectra Star started with a non-licensed series of yo-yos: the Globe "Earth Zone," Leaping Lizard, Hamburger "Fast Food," etc. These were called Radical Yos. The success of these early models lead to the manufacturing of several licensed yo-yos: "Ghostbusters," "Pee-Wee Herman," "Donald Duck," "G.I. Joe," etc. These licensed character yo-yos did extremely well and later defined the type of yo-yos that Spectra Star was best known for.

When Spectra Star produced a new yo-yo, between ten and 50-thousand units were made. After this initial run, more yo-yos were produced based upon the demand. If a yo-yo is successful, as many as two or three-hundred-thousand could be produced before the molds were retired. When a style is retired, the molds are recycled and the hand-made spray masks used for painting each yo-yo are destroyed.

Although Spectra Star primarily produced sculpted yo-yos, they have released other variations. One was the Ninja Turtles Yo-Yo. Three designs were made: one for Avon, another for Spectra Star, and one non-sculpted yo-yo painted with an imprinted process. The demand for the Ninja Turtles yo-yos was so high that the imprinted versions were made briefly to fill the orders. The imprinted version is harder to find than the two sculpted versions.

There was also more than one style of the Pee-Wee Herman Yo-Yo. Again there was an imprinted version and a sculpted version. Pee-Wee did not approve of the initial sculpted style so it was dropped. In this case, the sculpted yo-yo is a rare find.

*Continued ...*

## YO-YO LISTINGS

● **SPECTRA STAR** *Continued ...*

Spectra Star also produced yo-yos with photo stickers on the faces. These stickers feature such images as the "New Kids on the Block," "Star Trek: The Next Generation," and "Star Trek: Deep Space Nine."

Spectra Star yo-yos can be identified by the name indented on the inside of the plastic edge of the string slot. All Spectra Star yo-yos have metal string slot axles.

Spectra Star packaging has changed over the years. In the '80s and early '90s, yo-yos were on large nine inch hanging blister cards. The blister formed a stand at the base of the card so the yo-yo could be displayed standing upright. In the mid-'90s, the card height was reduced to seven inches and the standing base was eliminated. The cards for the original series have white backgrounds. More recent cards have full color backgrounds. The model numbers start at 1501 with the "Have a Blast" model and are now in the 1600s. Some numbers have not been assigned to specific yo-yos. It is unclear whether these numbers were deliberately left out or if some models were dropped and never released. The Stock Numbers refer to the characters on the yo-yos, not the different models produced. For example, all three Ninja Turtle yo-yos have the same stock number and both G.I. Joe yo-yos have the same stock number even though they are on different cards.

1417    *41*    **57 CHEVY, sculpted**
This yo-yo was retailed on a blister card, Stock No. 1531.

1418    *08*    **(Genie), sculpted**
This yo-yo is from the 1992 animated Disney movie "Aladdin." This was retailed on a blister card, Stock No. 1589.

1419    *41*    **BARBIE, imprinted seal**
This was the first Spectra Star Barbie yo-yo. It was retailed on a blister card, Stock No. 1509. For a Skipper related yo-yo, see #1618.

1420    *41*    **(Barbie), paper sticker seal**
This yo-yo was retailed on a blister card, Stock No. 1509.

1421    *41*    **(Bart Simpson), imprinted seal**
Bart is the lead character in the animated prime-time cartoon series "The Simpsons." This yo-yo was retailed on a blister card, Stock No. 1544.

1422    *41*    **(Baseball), sculpted baseball shape**
"Curve Ball," a baseball shaped yo-yo, is part of the original Spectra Star series. It was retailed on a blister card, Stock No. 1502.

1423    *41*    **(Basketball), sculpted basketball shape**
"Slam Dunk," a basketball shaped yo-yo, is part of the original Spectra Star series. This was retailed on a blister card, Stock No. 1523.

1424    *18*    **(Batman the movie), paper sticker seal**
This yo-yo is based on the 1989 film "Batman." It was retailed on a blister card, Stock No. 1529.

1425    *18*    **(Batman • bat signal), imprinted seal**
This is the first Batman yo-yo released by Spectra Star. The seal is made to look like the bat signal used to summon Batman when there was trouble. It was retailed on a blister card, Stock No. 1529.

1426    *18*    **(Batman), paper sticker seal**
This yo-yo is based on the animated television series. It was retailed on a blister card, Stock No. 1529. See Character and Duncan Super Heroes for other Batman yo-yos.

1427    *18*    **(Batman), sculpted**
This yo-yo was retailed on a blister card, Stock No. 1602.

1428    *41*    **(Bowling ball)**
This yo-yo was retailed on a blister card, Stock No. 1526. For other bowling ball yo-yos, see #459, #460 and #531.

1429    *18*    **(Cyclops), paper sticker seal**
Cyclops is a character from Marvel Comics' "X-Men." This yo-yo was retailed on a blister card, Stock No. 1599.

1430    *41*    **(Dick Tracy), imprinted seal**
The yo-yo is based on the 1990 Disney movie "Dick Tracy" starring Warren Beatty. It was retailed on a blister card, Stock No. 1545.

1430.1    *45*    **(Dick Tracy • wrist yo-yo), paper sticker seal, midget**
The yo-yo is based on the 1990 movie "Dick Tracy." This model was designed to resemble Dick Tracy's wrist communicator/wrist watch. It pops out of the wristband. This yo-yo was retailed on a blister card, Stock No. 1550.

1431    *41*    **(Dinosaur), sculpted**
This sculpted dinosaur yo-yo was marketed as "Dino." It was retailed on a blister card, Stock No. 1532.

1432    *41*    **(Freddy Krueger), sculpted**
This is based on the famous horror film character from the "Nightmare on Elm Street" series. The yo-yo was retailed on a blister card, Stock No. 1519.

1432.1    *08*    **(Donald Duck), sculpted**
This model, known as "Quazy Quacky," was one of Spectra Star's original releases. It was retailed on a blister card, Stock No. 1510.

1432.2    *41*    **DOUBLE DARE, imprinted seal**
This is based on the Nickelodeon TV show for kids. This was retailed on a blister card, Stock No. 1517.

*Continued ...* ▶

## YO-YO LISTINGS

**1433** *04* **GARFIELD, sculpted**
Garfield is a popular cartoon strip by Jim Davis. This yo-yo was retailed on a blister card, Stock No. 1516. Avon also released this same yo-yo on a 4-1/2" x 4-1/2" blister card. For other Garfield yo-yos, see #174.

**1434** *41* **(Ghostbusters), sculpted**
This yo-yo bears the logo from the 1984 blockbuster movie starring Bill Murray and Dan Aykroyd. This was retailed on a blister card, Stock No. 1513.

**1435** *41* **G.I. JOE, sculpted**
This yo-yo is based on Hasbro's 11-1/2" action figure. This was the first of the Spectra Star G.I. Joe yo-yos. The yo-yo was retailed on a blister card, Stock No. 1512.

**1436** *41* **G.I. JOE, paper sticker seal**
This yo-yo was retailed on a blister card, Stock No. 1512.

**1437** *41* **(Golf), sculpted**
This is a golf ball shaped yo-yo. It was retailed on a blister card, Stock No. 1525.

**1438** *41* **(Gumby), sculpted**
This model is based on "Gumby," the claymation character featured in movies and television. The yo-yo was retailed on a blister card, Stock No. 1518.

**1439** *41* **(Hamburger), sculpted**
"Fast Food," a hamburger shaped yo-yo, is part of the original Spectra Star series. It was retailed on a blister card, Stock No. 1504.

**1440** *41* **(Hand grenade) BOOM, sculpted**
"Have a Blast," a grenade shaped yo-yo, is part of the original Spectra Star series. It was retailed on a blister card, Stock No. 1501.

**1441** *41* **HOME ALONE 2, photo sticker seal**
This yo-yo is from the 1992 film "Home Alone 2: Lost in New York." It was retailed on a blister card, Stock No. 1576.

**1442** *41* **HULK RULES, imprinted seal**
This yo-yo features World Wrestling Federation Superstar Terry "Hulk" Hogan. Two versions of this yo-yo exist. The yo-yo was retailed on a blister card, Stock No. 1537.

**1442.1** **HUNCHBACK OF NOTRE DAME, paper sticker seal**
This yo-yo is from the 1996 Disney animated film. This yo-yo was retailed on a blister card, Stock No. 1633. An example of this yo-yo is shown next to this section's title.

**1443** *18* **(Joker) HA HA HA, imprinted seal**
The Joker is Batman's arch rival. See also #175. This yo-yo was retailed on a blister card, Stock No. 1535.

**1443.1** **JURASSIC PARK, photo sticker seal**
This yo-yo is from the 1993 blockbuster film. It was retailed on a blister card, Stock No. 1583.

**1444** *41* **(Leaping lizards), sculpted**
This is part of the original Spectra Star series. This yo-yo was retailed on a blister card, Stock No. 1506.

**1445** *08* **(Lion King), sculpted**
This yo-yo features young Simba from the 1994 Disney animated film, "The Lion King." This yo-yo was retailed on a blister card, Stock No. 1595.

**1446** **LITTLE MERMAID, imprinted seal**
This yo-yo features Ariel from the 1989 Disney animated film. This yo-yo was retailed on a blister card, Stock No. 1549.

**1447** *08* **(Mickey Mouse), sculpted**
For other Mickey Mouse yo-yos, see Disney section. This yo-yo was retailed on a blister card, Stock No. 1569.

**1448** *41* **(New Kids on the Block), paper sticker seal**
This yo-yo was retailed on a blister card, Stock No. 1546.

**1449** *41* **(Nightmare Before Christmas), sculpted**
This yo-yo is from the 1993 stop action animated film by Tim Burton. This was the first glow yo-yo produced by Spectra Star. The yo-yo features the character Barrel from the movie. This yo-yo was retailed on a blister card, Stock No. 1584.

**1450** *18* **(Ninja Turtle • Raphael), sculpted**
This yo-yo is based on the "Teenage Mutant Ninja Turtle" comic book and movie series. This yo-yo was retailed on a blister card, Stock No. 1515.

**1450.1** *18* **(Ninja Turtle • Raphael), imprinted seal**
This yo-yo was retailed on a blister card, Stock No. 1515.

**1451** *18* **(Ninja Turtle • Leonardo), sculpted**
This yo-yo was retailed on a blister card, Stock No. 1515. Avon also released this same yo-yo on a 4-1/2" x 4-1/2" card.

**1452** *41* **(Nintendo), imprinted seal**
Super Mario Brothers are on one face of this yo-yo; the Legend of Zelda is on the reverse. This yo-yo was retailed on a blister card, Stock No. 1534.

**1453** *08* **(101 Dalmatians), sculpted**
This yo-yo is based on the 1961 Disney animated film. This yo-yo was retailed on a blister card, Stock No. 1574.

**1453.1** *08* **101 DALMATIANS, photo sticker seal**
This is much more common than the sculpted version. This yo-yo was retailed on a blister card, Stock No. 1635.

Continued ...

## SPECTRA STAR *Continued ...*

| | | |
|---|---|---|
| 1454 | *41* | **PEE-WEE BRAND (Herman), sculpted face** |
| | | This is one of the rarest of the Spectra Star yo-yos. This was the original design, but Pee-Wee reportedly didn't like it and the design was changed to #1455. Only a very limited number of this original version were released. |
| 1455 | *41* | **PEE-WEE BRAND (Herman), imprinted face** |
| | | This yo-yo was retailed on a blister card, Stock No. 1511. |
| 1456 | *08* | **(Pocahontas • raccoon), sculpted** |
| | | This yo-yo features Meeko, the raccoon, from Disney's animated film "Pocahontas." This yo-yo was retailed on a blister card, Stock No. 1607. |
| 1457 | *41* | **(Skull and crossbones), sculpted** |
| | | This sculpted skull and crossbones was marketed as the "Bones" yo-yo. It was retailed on a blister card, Stock No. 1533. |
| 1458 | *41* | **(Soccer ball), imprinted seal, ball shape** |
| | | This yo-yo was retailed on a blister card, Stock No. 1527. |
| 1459 | *41* | **SPACE JAM, photo seal of Michael Jordan and Bugs Bunny** |
| | | This yo-yo is from the 1996 Warner Brothers animated film starring Michael Jordan and various Warner Brothers cartoon characters. This yo-yo was retailed on a blister card, Stock No. 1632. |
| 1460 | *18* | **(Spider-man), sculpted** |
| | | This yo-yo is based on the Marvel Comics character. It was retailed on a blister card, Stock No. 1626. |

## STAR TREK "DEEP SPACE NINE"

Each of the following yo-yos were retailed on a blister card, Stock No. 1586.

| | | |
|---|---|---|
| 1461 | *03* | **(Space Station), sculpted** |
| 1462 | *03* | **(Commander Benjamin Sisko), photo sticker seal, sculpted rim** |
| 1463 | *03* | **(Commander Benjamin Sisko), photo sticker seal** |
| 1464 | *03* | **(Major Kira Nerys), photo sticker seal, sculpted** |
| 1465 | *03* | **(Major Kira Nerys), photo sticker seal** |
| 1466 | *03* | **(Quark), photo sticker seal** |

## STAR TREK "THE NEXT GENERATION"

Each of the following yo-yos were retailed on a blister card, Stock No. 1585.

| | | |
|---|---|---|
| 1467 | *03* | **(The Enterprise NCC 1701-D), sculpted** |
| 1468 | *03* | **(Captain Jean-Luc Picard), photo sticker seal** |
| 1468.1 | | **(Captain Jean-Luc Picard), photo sticker seal, sculpted rim** |
| 1469 | *03* | **(Commander William Riker), photo sticker seal, sculpted rim** |
| 1470 | *03* | **(Commander William Riker), photo sticker seal** |
| 1471 | *03* | **(Lt. Cmdr. Data), photo sticker seal, sculpted rim** |
| 1472 | *03* | **(Lt. Cmdr. Data), photo sticker seal** |

| | | |
|---|---|---|
| 1473 | *03* | **(Star Wars Darth Vader), sculpted** |
| | | This yo-yo was based on the character Darth Vadar from the Star Wars Trilogy of films. This yo-yo was retailed on a blister card, Stock No. 1624. |
| 1474 | *03* | **(Star Wars Storm Trooper), sculpted** |
| | | This yo-yo was retailed on a blister card, Stock No. 1623. |
| 1474.1 | | **(Storm), paper sticker seal** |
| | | This yo-yo is from the X-Men series released by Spectra Star. |
| 1475 | | **(Superman), paper sticker seal** |
| | | This yo-yo was based on the DC Comics character. This yo-yo was retailed on a blister card, Stock No. 1637. |
| 1476 | *26* | **(Tasmanian Devil), sculpted** |
| | | This yo-yo features the Warner Brothers cartoon character. The blister card reads "Football" yo-yo, Stock No. 1587. |
| 1477 | | **NASCAR, photo sticker seal** |
| | | This yo-yo was retailed on a blister card, Stock No. 1579. |
| 1478 | *41* | **(The Mask), photo sticker seal** |
| | | This yo-yo is based on the 1994 blockbuster film starring Jim Carrey. The yo-yo was retailed on a blister card, Stock No. 1603. |
| 1479 | *06* | **TOY STORY, paper sticker seal** |
| | | This yo-yo features Buzz Lightyear from Disney's 1995 computer animated film "Toy Story." It was retailed on a blister card, Stock No. 1631. |
| 1480 | | **TREASURE TROLLS, paper sticker seal** |
| | | This yo-yo was retailed on a blister card, Stock No. 1575. |

*Continued ...*

## YO-YO LISTINGS

1481   *18*   **(Wolverine), sculpted**
The Wolverine is a character from Marvel Comics' "X-Men." The yo-yo was retailed on a blister card, Stock No. 1571.

1482   *34*   **(World), sculpted**
"Earth zone" is a sculpted globe shaped yo-yo. It was part of the original Spectra Star series. It was so popular a knock off was made. The Spectra Star version includes the state of Hawaii; the knock off does not. The knock off also does not have "Spectra Star" imprinted inside the string slot lip. This yo-yo was retailed on a blister card, Stock No. 1507.

## SPORT RELATED

Sport related yo-yos are frequently used as promotional items or sold as souvenirs at sporting events. Various companies manufacture them, but Humphrey is the leader. Sports ball shaped yo-yos are given a distinct category in this book. (See Duncan Sportsline, Festival Be-A-Sport series, and Spectra Star for Sports Ball yo-yos.) The following is just a small sampling of sports related souvenir yo-yos.

| | | |
|---|---|---|
| 1483 | | ASTROS, imprint seal, plastic, Humphrey shape, '80s |
| 1484 | | ASTROS, paper sticker seal, plastic, bulge face F-210 shape, '80s |
| • | | Braves - *See Light-Up section #1031* |
| 1485 | | CALIFORNIA ANGELS, paper sticker seal, plastic, tournament shape, metal string slot axle, '90s |
| 1486 | | CALIFORNIA ANGELS, sticker seal, wood, miniature, drilled axle, '60s |
| 1487 | *29* | CINCINNATI REDS, imprint seal, plastic, Humphrey shape, 1989 |
| 1487.1 | | CINCINNATI REDS, paper insert seal, plastic, tournament shape, view lens, '60s |
| | | An example of this yo-yo is shown next to this section's title. This is a Royal TV yo-yo. See #1272. |
| 1488 | | COWBOYS, plastic sticker seal, plastic, butterfly shape, metal axle, '80s |
| 1489 | *29* | COWBOYS, imprint seal, plastic, Humphrey shape, '80s |
| 1490 | *29* | DODGERS, paper sticker seal, plastic, Humphrey shape, '80s |
| 1491 | | (Downhill Skier), paper insert seal, plastic, slimline shape, view lens, '80s |
| | | This yo-yo retailed on a Jackami Ltd. blister card, Stock No. x423K. |
| 1492 | | EXPOS, paper sticker seal, plastic, Humphrey shape, '80s |
| 1493 | *4* | FIRESTONE DRAG 500, embossed seal, plastic, John Hart Co., sculpted, 1975 |
| | | This model, along with Road Runner #1605.1, was the first wheel yo-yo produced. Greg Hart (Englehart) created the wheel design in the early '70s. In 1975, this model was released for a Firestone Tire and Rubber Co. promotion. Two hub styles were used, a honeycomb pattern and a five spoke pattern. These yo-yos were retailed on blister cards in both Canada and the United States. The cards were labeled Item No. 1004 in the U.S. and Item No. 1004F in Canada. The Canadian cards were written in both French and English. Approximately 25,000 units were retailed in the U.S. |
| 1494 | *29* | FLORIDA MARLINS, imprint seal, plastic, Humphrey shape, 1991 |
| 1495 | *30* | GIANTS, paper sticker seal, plastic, Humphrey shape, '80s |
| 1496 | *4* | GOODYEAR RACING, photo sticker seal, plastic, sculpted, '90s |
| | | A series of several wheel shaped yo-yos were made with photo seals of famous drivers and their cars. See also #30. |
| • | | Joe Namath - *See Festival section #548 and #538.* |
| 1497 | | METS, imprint seal, plastic, Humphrey shape, '90s |
| 1498 | | (Michael Jordan), imprint seal, plastic, tournament shape, '90s |
| | | This yo-yo was retailed on a blister card. |
| 1499 | | ORLANDO MAGIC, imprint seal, plastic, sculpted, '90s |
| | | This basketball shaped yo-yo was retailed on a blister card. |
| 1500 | | PHILADELPHIA PHILLIES, paper sticker seal, plastic, Humphrey shape, '80s |
| 1501 | | PHILLIES, imprint seal, plastic, Humphrey shape, '90s |
| 1502 | | PIRATES, paper sticker seal, plastic, bulge face F-210 shape, '80s |
| 1503 | | PIRATES, imprint seal, plastic, Humphrey shape, '90s |
| 1504 | | SAN DIEGO PADRES, paper sticker seal, plastic, Humphrey shape, '80s |
| 1505 | | ST. LOUIS CARDINALS, paper sticker seal, plastic, Humphrey shape, '80s |
| 1506 | | UCLA BRUINS, paper insert seal, plastic, slimline shape, view lens, '80s |
| 1507 | | US OPEN 1993 BALTUSROL, imprint seal, plastic, sculpted, 1993 |

## YO-YO LISTINGS

## SPORTS BALLS

Over the years, many major manufacturers have added new yo-yos to their product lines by creating Sports Ball series. The first companies to produce sports ball shaped yo-yos were Toy Tinkers and HEP in the early '50s. Duncan, Festival, and Spectra Star have all produced Sports Ball yo-yos. Below is a sampling of some other Sports Ball lines that may be found.

- **Be A Sport Series** - *See the Festival section.*
1508 **(Yo Sport! Series), sculpted plastic, '80s and '90s**
This is a series of four sports balls, a basketball, baseball, soccer ball, and football. They were retailed on non-blistered display card, which has a copyright date of 1987.

1509 *34* **(Franklin Sport Yo-Yo Series), imprint seal, plastic, sculpted, '90s**
This is a series of four yo-yos featuring a football, baseball, and two basketball styles. All these yo-yos have "Franklin" imprinted on the seal. The football is a sculpted design, but the others are tournament shapes. The basketball comes in two design styles, one has a thick width, the other a more standard width. The original retail price was $3.99 each and were retailed on a blister card, Stock No. 2402. These yo-yos were produced for Franklin Sports Industries.

- **HEP Magic String Balls** - *See Toy Tinkers section.*
1510 **(Novelty yo-yo Sports Ball series), plastic, sculpted, '90s**
This is the same series as the Franklin series, but without the Franklin logo. These yo-yos were retailed on a Walgreens blister card with price of $1.99.

- **Spalding** - *See Toy Tinkers section #1597 and #1599.*
1511 **(Sports model yo-yo), plastic, sculpted, '70s**
This is a series of three miniature yo-yos: baseball, football, and basketball. All three retailed for $1.98 on one blister card marked No. 6000. Although this three-pack is sometimes advertised as being released in the '60s, most collectors feel it was produced in the '70s or '80s. An example of this pack is shown next to this section's title.

1512 **(Sport yo-yo series), plastic, satellite shape, '90s**
This is a series of three Sports Balls: soccer ball, football, and basketball. Unlike most Sports Ball yo-yos, these are satellite shaped. They were retailed on a blister card, Stock No. 95013.

- **Sportsline (Duncan)** - *See Duncan Sportsline section.*
1513 **(Tin series), lithograph seal, tin, tournament shape, late '90s**
The U.S. Toy Company released this series of four tin yo-yos in 1997. The series features a football, basketball, baseball, and soccer ball. All four yo-yos carry the mark © C-P.

- **Tin Lithograph Sports series** - *See Oriental Trading Company section #1150*
1514 **(Playball yo-yo series), plastic, sculpted, '90s**
This is a series of three Sports Balls: basketball, baseball, and football. These yo-yos retailed on blister card, Stock No. 9117.

*HEP Sports Balls are yo-yos with a re-wind mechanism. They retailed during the '50s, and are believed to be the first manufactured sports ball shaped yo-yos.*

# YO-YO LISTINGS

## STAR RETURN TOP

Star Return Tops were made in the late '50s or early '60s. They were considered a non-promoted line, so there is no documentation of contests or demonstrators. A blister carded play set, featuring a bag of marbles, a wooden top, and the one-piece tournament yo-yo listed below, was made during this time. The play kit had a pre-printed price of 98 cents.

| | | | |
|---|---|---|---|
| 1515 | 15 | (6 jewel), ink stamped seal, wood, tournament shape, three-piece, pegged string, '60s | |
| | | There is still some debate as to the authenticity of this particular model. | |
| 1516 | 39 | STAR TOURNAMENT, paint seal, wood, tournament shape, one-piece, '60s | |
| | | Unlike other models, the star is outlined. An example of this yo-yo is shown next to this section's title. | |
| 1517 | 39 | STAR RETURN TOP (in circle), ink stamped seal, wood, tournament shape, three-piece, pegged string, '60s | |
| 1518 | | (Star play set) | |
| | | This play set came with a top, marbles, and yo-yo. See #1516. | |
| 1519 | 39 | STAR RETURN TOP (without star), ink stamped seal, wood, tournament shape, '60s | |
| 1519.1 | | STAR RETURN TOP, paper sticker seal, wood, tournament shape, three piece, '60s | |

## STYLE 55 CHAMPION

Like the Whirl King yo-yos, the Style 55 Champions were produced by the Fli-Back Company of High Point, North Carolina.

| | | |
|---|---|---|
| 1520 | 22 | Gold leaf stamped seal, wood, tournament shape, pegged string, three-piece, '60s |
| 1521 | 22 | Foil sticker seal, wood, tournament shape, pegged string, three-piece, '60s |
| | | Production of this yo-yo began in the '50s. It was sold loose out of a counter display box with an original price of 10 cents. This foil sticker seal version is harder to find than the gold leaf stamped seal, see #1520. An example of this yo-yo is shown next to this section's title. |

## TIN MISCELLANEOUS

Yo-Yos in this section do not have any manufacturer seals. However, they can be identified by their graphics.

| | | |
|---|---|---|
| 1522 | 35 | (American flag), lithograph seal, tournament shape, '90s |
| | | This model was made in China and was sold as a spinner top in a polybag with cardboard header. |
| 1522.1 | | (American flag with fireworks), lithograph seal, tournament shape, '90s |
| | | This yo-yo was released by U.S. Toy Co. and has © CP on the rim. |
| 1523 | 03 | (Astronaut space walk), lithograph seal, tournament shape, '60s |
| | | This yo-yo was retailed in a polybag with a saddle header card reading "Return Top." |
| 1523.1 | 03 | (Astronaut waving), lithograph seal, tournament shape, '60s |
| 1524 | | (Bear, cub, and chick), lithograph seal, tournament shape, '80s |
| | | The yo-yo has an internal bell and the false Patent Number 533107. It was made in Japan. |
| 1525 | | (Bear playing tennis), lithograph seal, tournament shape, '90s |
| | | The yo-yo is retailed on blister card reading "Spinner Top." |

*Continued ...*

## YO-YO LISTINGS

**TIN MISCELLANEOUS** *Continued ...*

| | | |
|---|---|---|
| 1526 | | **(Button yo-yos)** |

Badge-A-Mint is a company that sells pin backed button kits. Two of their buttons can be fashioned into a yo-yo. These yo-yos are homemade and have little collectible value. These yo-yos can be identified by the words "Badge A Mint" engraved on the inside lip of the rim.

**1526.1** **(Brass yo-yo), no seal, tournament shape**

A variety of different solid brass and brass plated yo-yos have been made. These are frequently sold as gift yo-yos, some made to be engraved. Many have a brass finger ring at the string end. Like other yo-yos without seals, these have limited collectible value unless they are in a dated package with interesting graphics.

**1527** *35* **(Chipmunk), lithograph seal, tournament shape, miniature, '70s**

This is a Japanese produced yo-yo with an internal bell. The seal on the reverse face has a cat-like creature holding a hat.

• **Clown Faces** - *See Clown section.*

**1528** *33* **(Chief • Indian), lithograph seal, tournament shape, midget, '80s**

This is the smallest yo-yo listed in this book with identifiable markings. The diameter of this yo-yo is 3/4 inch. It was made in Japan and carries the fake Patent Number 533107.

**1529** *33* **(Cowboy), lithograph seal, tournament shape, midget, '80s**

An Indian is on the reverse face. The yo-yo has the fake Patent Number 533107 on the rim.

**1530** *19* **DELUXE RETURN TOP, lithograph seal, tournament shape, '80s**

The seal reads "Made in Japan" and has the fake Patent Number 533107. This yo-yo was retailed in the '80s in a polybag with a cardboard saddle header. The header card reads "KoYo" in small letters and shows a boy with a yo-yo.

**1531** *35* **(Dog face), lithograph seal, tournament shape, miniature, '80s**

The seal reads "Made in Japan" and it carries the fake Patent Number 533107. The reverse face shows flags of the world.

**1532** *35* **(Elephant), lithograph seal, tournament shape, miniature, '80s**

**1533** *35* **(Elephant yo-yoing), lithograph seal, tournament shape, jumbo, '70s**

The yo-yo was retailed in polybag with a saddle header card.

**1534** **(Jo Jo), lithograph seal, tournament shape, '70s**

This yo-yo has a checkerboard patterned lithograph rather than a seal. It was retailed in a polybag with a saddle header card reading "Jo Jo."

**1535** *35* **(Juggling bear), lithograph seal, tournament shape, '80s**

A monkey eating a banana is on the reverse face.

**1536** *19* **KITTY CUCUMBERS, lithograph seal, tournament shape, '80s**

This is from a series of cat based yo-yos and miscellaneous items.

**1537** *19* **(Lion), lithograph seal, tournament shape, miniature, view lens, pinball, mid '60s**

This unusual tin yo-yo has a view lens on each face. One face has a mirror, the other features a lion and a pinball game. This retailed for 19 cents in a polybag with a saddle header card from the Toy Merchandising Corporation. There is also a monkey face version of this yo-yo.

**1538** *19* **MAGIC RETURN TOP, lithograph seal, tournament shape, '70s**

**1539** *35* **(Multi-Colored), lithograph seal, riveted disk axle, '60s**

This riveted disk style yo-yo was made in Japan.

**1540** *35* **(Panda riding rainbow), lithograph seal, tournament shape, '80s**

The yo-yo has an internal bell and the fake Patent Number, 533107. It was made in Japan.

**1540.1** *35* **(Panda), lithograph seal, tournament shape, miniature, '80s**

This yo-yo was made in Hong Kong.

**1541** *35* **(Psycho swirl pattern), lithograph seal, tournament shape, '80s**

This yo-yo carries the Stock No. 316.

**1542** *35* **(Skull and crossbones), embossed seal, tournament shape, '70s**

This is an unusual tin yo-yo with a wooden axle. It has more mass than the typical tin yo-yo and plays better. This yo-yo has a peculiar metallic finish. The date and maker of this yo-yo is unknown.

**1543** *37* **(Stainless steel yo-yo), slimline shape**

The word "yo-yo" has been cut into the face. This was made in Japan, however, the date is unknown.

**1544** *35* **(Target), lithograph seal, tournament shape, '70s, '80s and '90s**

The multi-colored concentric circle pattern, or target, is common on yo-yos produced in Japan or Hong Kong. These carry little collectible value unless they are in packaging that can be dated.

**1545** *35* **(Teddy bear playing yo-yo), lithograph seal, tournament shape, '80s**

**1547** *35* **(Tiger), lithograph seal, tournament shape, jumbo, '80s**

This yo-yo was retailed on a blister card.

**1548** *33* **WHOOPEE, lithograph seal, jumbo, riveted disk axle, '30s**

The date and maker are unknown, but this yo-yo is believed to be from the early '30s.

**1549** *34* **(World yo-yo), lithograph seal, tournament shape, '90s**

World map yo-yos are some of the most common of the tin yo-yos. The design, a map of the world, has not changed over the last several years. Even after the breakup of the Soviet Union, these yo-yos are still being reproduced with U.S.S.R. markings. Both, O.T.C. and the U.S. Toy Company release yo-yos of this type. An example of this yo-yo is shown next to this section's title.

*Lucky's Collectors Guide To 20th Century Yo-Yos*

# YO-YO LISTINGS

## TOM KUHN

Tom Kuhn, a San Francisco dentist, is well known among yo-yo collectors as a producer of high quality wooden yo-yos from the '70s, '80s, and '90s. Kuhn's interest in the yo-yo began in Detroit during the '50s. During his teenage years he won several contests, the most memorable being one run by Duncan demonstrator Bob Rule. As a contest winner, Kuhn was awarded a Pearlescence Jeweled Yo-Yo. For twenty years Kuhn didn't think about yo-yos. Then in 1976, he was given a rosewood yo-yo as a gift. Fortunately, the axle broke after some intense play, which resulted in Kuhn's quest for the perfect yo-yo.

Kuhn first came up with a unique design idea, a take-apart yo-yo. The yo-yo not only could be taken apart to loosen tangled strings, but the sides could be reversed to form different yo-yos. The No Jive 3 - 1 could be assembled in three styles: the "Classic" tournament shape, the "Flying Camel" butterfly shape, and the "Pagoda" where one side piggybacks on the other.

Kuhn also experimented with axle diameters. Since he knew that the thinner the axle the longer the sleep time, Kuhn tried to make the thinnest possible axle that wouldn't easily break. The final product was a wooden axle 1/4 inch diameter with a central groove for the string to rest. The central groove was 23/100th of an inch deep. Next he tried to find the best material to use for the axle, a material that would allow the least string friction. He contracted Dr. Walter Goldenrath at NASA who offered him several materials that had a low coefficient of friction. After making over 100 prototypes, Kuhn decided that nothing was better than wood. He improved on his wooden axle by creating a replaceable steel reinforced wooden axle sleeve.

All of the Kuhn line, except the Silver Bullet series are wooden yo-yos.

- 1550  42  **ABERCROMBIE & FITCH**, laser carved seal, tournament shape, '80s
  This advertising yo-yo was sold in a felt pouch. It is no longer being produced and its unlikely that it ever will have another production run. The yo-yo is a take-apart style.
- 1551  42  **CHAMPION**, die stamped seal, tournament shape, '70s
  This early Kuhn Yo-Yo is considered rare. It was designed to represent the popular "demonstrator carved" yo-yos of the '50s and '60s. It is unlikely that it will ever be re-released.
- 1552  42  **CLIFF HOUSE AND SEAL ROCK**, die stamped seal, tournament shape, '80s
  This is a famous tourist site in San Francisco, California. This yo-yo is discontinued.
- 1553  42  **CUSTOM YO-YO**, die stamped seal, butterfly shape, '80s
  This blue and yellow yo-yo is currently out of production, but may be re-released.
- 1554  15  **DIAMOND SPECIAL (maple)**, die stamped seal, tournament shape, '90s
  This four jewel model is still available retail. It is a take apart-style.
- 1555  15  **DIAMOND SPECIAL (walnut)**, die stamped seal, tournament shape, '90s
  The walnut version of this model is discontinued. The maple version is still being manufactured. See #1554. This take-apart yo-yo originally retailed for $25.00.
- 1556  42  **FLYING CAMEL**, laser carved seal, '90s
  This yo-yo is still available retail. It is a take apart-style.
- 1557  42  **GENUINE BEGINNERS SAN FRANCISCO**, die stamped seal, tournament shape, '80s
  This is a discontinued model.
- 1558      **GENUINE UP & DOWN BEGINNERS**, die stamped seal, tournament shape, pegged string, '90s
  This red and black yo-yo is still available retail.
- 1559  42  **GOLD MEDAL LIMITED EDITION (Olympics)**, die stamped seal, tournament shape, brass axle, '80s
  This is a discontinued yo-yo that was produced for the 1984 Olympics. This take-apart yo-yo was produced for a limited time.
- 1560      **GOLD MEDAL LIMITED EDITION (Olympics)**, die stamped seal, tournament shape, silver axle, '80s
  This is a discontinued yo-yo that was produced for the 1984 Olympics. This take-apart yo-yo was produced for a limited time.
- 1561      **KREEGER AND SONS**, die stamped seal, '70s
  This was one of Kuhn's earliest ad yo-yos. It is considered rare.
- 1562      **MANDALA (multi-jewel)**, laser carved, tournament shape
  This is a one of a kind yo-yo produced by jeweler Sydney Mobell. Real jewels are embedded in the face, 75 Diamonds, 25 Sapphires, 25 Rubies and 25 Emeralds. The yo-yo was shown on the Johnny Carson Show. The original asking price was $10,000.00.
- 1563  42  **MANDALA (1 sunburst)**, laser carved seal, tournament shape, '90s
  This yo-yo has a laser carved sunburst design. A butterfly configuration is also made with this pattern. The yo-yo is a take-apart style and still available retail.
- 1564  42  **MANDALA (2 star mix)**, laser carved seal, tournament shape, '90s
  This yo-yo has a laser carved overlapping star design and is still available retail. The yo-yo is a take-apart style.

*Continued ...*

## YO-YO LISTINGS

● **TOM KUHN** *Continued ...*

| | | |
|---|---|---|
| 1565 | 42 | **MANDALA (3 lace)**, laser carved seal, tournament shape, '90s |

This yo-yo has a laser carved lace design and is still available retail. The yo-yo is a take-apart style.

| | | |
|---|---|---|
| 1566 | 42 | **NIEMAN MARCUS**, laser carved seal, tournament shape, '80s |

This yo-yo retailed at Nieman Marcus. It is no longer produced and is unlikely to be re-released. This was one of the first major sales by Kuhn. The yo-yo is a take-apart style.

| | | |
|---|---|---|
| 1567 | 42 | **NO JIVE MAPLE (leaf)**, die stamped seal, tournament shape, '90s |

This yo-yo is still available retail. It is an exact replica of the original No Jive Yo-Yo that Tom Kuhn first produced to give as Christmas gifts. Only 20 of the original No Jive Maple Leaf yo-yos were made.

| | | |
|---|---|---|
| 1568 | 42 | **NO JIVE 3 - 1**, die stamped seal, '90s |

Kuhn applied for the patent on this yo-yo in 1978. He received it, and Patent Number 4,207,701, in 1980. This original take-apart design allowed for easy repair of broken axles and relief from tangled strings. Early models were all natural wood, while recent ones are painted with decorative airbrushed stripes. This yo-yo is still available retail. An example of this yo-yo is shown next to this section's title.

| | | |
|---|---|---|
| 1569 | 42 | **(No Jive • Woody) 15TH ANNIVERSARY CELEBRATION • 1992**, gold leaf stamped seal, tournament shape |

This yo-yo has been discontinued. It was a commemorative yo-yo celebrating Tom Kuhn's 15th anniversary. The reverse face has a No Jive Woody seal #1569B. Unlike other No Jives of this period, it was painted and had an airbrush stripe.

| | | |
|---|---|---|
| 1570 | 42 | **NO JIVE WOODY**, die stamped seal, tournament shape, '90s |

Rumor has it that the No Jive yo-yo got its name this way. Tom Kuhn was playing with a yo-yo he had just made when his paperboy saw him and asked, "What's that?" Kuhn told him it was a yo-yo he had just made. The paperboy replied, "No Jive, you just made that?" Kuhn decided that "No Jive" would be a great name for the yo-yo.

| | | |
|---|---|---|
| 1571 | 42 | **(No Jive Woody) FISHERMAN'S WHARF**, die stamped seal, tournament shape, '80s |

This is a discontinued line. The yo-yo has a No Jive Woody on reverse face. See #1570.

| | | |
|---|---|---|
| 1580 | | **NO JIVE 3 - 1**, jumbo-giant |

This is the world's largest brand name functioning yo-yo. The yo-yo is 50" in diameter and weighs 256 pounds. Playing this yo-yo requires a crane and a 3/4" Dacron rope. The yo-yo was first launched from a 100-foot crane in San Rafael, California in 1979. The yo-yo was mentioned on the Johnny Carson Show and a color photo of it appears in the 1981 issue of the Guinness Book of World Records. This yo-yo can be seen at the National Yo-Yo museum in Chico, California.

| | | |
|---|---|---|
| 1581 | 42 | **ORIGINAL (dove and mountain)**, die stamped seal, tournament shape, '70s |

This is a very rare discontinued line.

| | | |
|---|---|---|
| 1582 | 42 | **POCKET ROCKET**, foil seal, butterfly shape, '90s |

This model is still available retail.

| | | |
|---|---|---|
| 1583 | 42 | **(Reverse) FLYING CAMEL**, laser carved seal, early '90s |

Reverse yo-yos occur when the cutting tool is reversed, producing a mirror image on the face. The Flying camel is the most sought after of the rare reverse yo-yos. See also #1583.1, #1583.2, and #1583.3. It is unlikely that more reversed yo-yos will be produced. This yo-yo is a take-apart style.

| | | |
|---|---|---|
| 1583.1 | 42 | **(Reverse sunburst) MANDALA**, laser carved seal, '90s |

This yo-yo is a factory error and is very rare. The design is a mirror image of the normal Sunburst Mandala. For more explanation, see #1583.

| | | |
|---|---|---|
| 1583.2 | 42 | **(Reverse star mix) MANDALA**, laser carved seal, '90s |

This yo-yo is a factory error and is very rare. The design is a mirror image of the normal Star Mix Mandala. For more explanation, see #1583

| | | |
|---|---|---|
| 1583.3 | 42 | **(Reverse lace) MANDALA**, laser carved seal, '90s |

This yo-yo is a factory error and is very rare. The design is a mirror image of the normal Lace Mandala. For more explanation, see #1583

| | | |
|---|---|---|
| 1584 | 42 | **ROLLER WOODY**, die stamped seal, tournament shape, ball bearing axle, '90s |

This yo-yo is still available retail. The yo-yo is a take-apart style.

| | | |
|---|---|---|
| 1585 | 42 | **SAN FRANCISCO CALIFORNIA (Golden Gate Bridge)**, die stamped seal, tournament shape, '80s |

This is a discontinued line.

| | | |
|---|---|---|
| 1585.1 | 42 | **SAN FRANCISCO CALIFORNIA (trolley cars)**, die stamped seal, tournament shape, '80s |

This is a discontinued line.

| | | |
|---|---|---|
| 1586 | 42 | **SILVER BULLET**, engraved seal, aluminum, flywheel shape, wood axle, '90s |

The original Silver Bullet is no longer manufactured, but may be re-released. Silver Bullet prototypes were made by Don Watson out of plastic with brass rings inserted in the outer rims. These can be seen in the National Yo-Yo Museum in Chico, CA. This yo-yo is a take-apart model.

| | | |
|---|---|---|
| 1587 | 42 | **SILVER BULLET 2**, engraved seal, aluminum, flywheel shape, ball bearing axle, '90s |

This yo-yo is still available retail. It comes in two styles, bronze and black anodized (shown), and standard aluminum. Each Silver Bullet 2 has a serial number and a date. The seal on early models reads, "San Francisco, CA." In the late '90s, the seal reads, "What's Next MFG., Inc."

The SB2 has been clocked with rim speeds of 100 mph. It is reported to have the longest sleep time of any yo-yo. A SB2 was taken on the space shuttle Atlantis on July 1, 1992, and used by astronaut Jeffrey Hoffman in an educational science video. This yo-yo is a take-apart model.

*Continued ...* ▶

## YO-YO LISTINGS

1588         **SLEEP MACHINE**, die stamped seal, tournament shape, ball bearing axle, '90s
                This yo-yo is still available retail. It is one of the few Kuhn models that is painted. The yo-yo comes in pearlescence white paint or in natural wood. This yo-yo is a take-apart model.

1589         **SMO-BRO-YO-YO TOMMY SMOTHERS** (Silver Bullet), engraved seal, aluminum, flywheel shape, late '80s
                This yo-yo is a Silver Bullet with a Smothers Brothers engraving. This yo-yo was retailed at the Smothers' Winery.

1590   *39*   **SMOTHERS BROTHERS SMO • BRO**, laser carved seal, tournament shape, late '80s
                The yo-yo has laser carved autographs of Tom and Dick Smothers. It is no longer produced and is unlikely to be re-released. The yo-yo originally retailed at $22.50 and is the most sought after of the Smothers Brothers yo-yos. This yo-yo is a take-apart model.

1591   *42*   **SMOTHERS FINE WINES CHARDONNAY**, laser carved seal, late '80s
                This yo-yo is no longer produced and is unlikely to be re-released. It was sold at the Smothers' Winery in Sonoma, California. The Smothers' Fine Wines yo-yo came in both tournament and butterfly shapes. There are two different seals, the one shown and the more rare seal with grapes below the word "Smothers." This yo-yo is a take-apart model.

1591.1       **SMOTHERS FINE WINES CABERNET SAUVIGNON**, laser carved seal, '80s
                This yo-yo is no longer produced and is unlikely to be re-released. It was sold at the Smothers Winery in Sonoma, California.

1592   *36*   **(Snoopy • style 1)**, imprint seal, tournament shape, three-piece, '80s
                This is an extremely rare prototype yo-yo that was never released.

1593   *36*   **(Snoopy • style 2)**, imprint seal, tournament shape, three-piece, '80s
                This is an extremely rare prototype yo-yo that was never released.

1594   *42*   **TOM KUHN • DDS • CELEBRATING 25 YEARS**, paint seal, tournament shape, '90s
                This is a white pearlescence painted yo-yo. The seal has a photo of Tom Kuhn. The yo-yo celebrates Kuhn's 25 years as a dentist and master yo-yo maker.

## TOY TINKERS (TOY-O-BALLS)

Toy Tinkers, better known as the "Tinker Toy Company" of Evanston, Illinois, tested the yo-yo market in the mid 1950s with its Toy-O-Ball series. Toy-O-Ball had a rewinding metal reel housed in a plastic shell. These yo-yo balls looked very similar to the HEP Magic String Balls also made in the early '50s. The HEP Balls retailed out of a counter display box of one dozen with an original price of 49 cents. Like the Toy-O-Ball, they came as baseballs, basketballs and footballs. It is unclear whether HEP Magic String Balls were made by Toy Tinkers or the other way around.

Toy-O-Balls and the HEP Magic String Balls were the first automatic rewinding "yo-balls" to be produced. Both have a string that rewinds automatically or manually by twisting a knob. The Toy-O-Balls were individually packaged in a polyethylene bag with a saddle label that had a pre-printed price of 49 cents.

1595   *33*   **CHICAGO FAIR HEP** (baseball), imprint seal, plastic, ball shape
                This was a souvenir from the Chicago Fair in the '50s. The reverse side reads, "I Was There."

1596         **(Baseball) GET HEP**, imprint seal, plastic, ball shape, '50s

1597         **(Baseball) SPALDING TOY-O-BALL**, imprint seal, plastic, ball shape, '50s
                An example of this yo-yo is shown next to this section's title.

1598         **(Basketball) GET HEP**, imprint seal, plastic, ball shape, '50s

1599         **(Basketball) SPALDING TOY-O-BALL**, imprint seal, plastic, ball shape, '50s

1600         **(Football) HEP**, imprint seal, plastic, sculpted football shape, '50s

## YO-YO LISTINGS

### U.S. TOY COMPANY

The U.S. Toy company has been a big supplier of party goods and promotional products since 1953. They market many styles of inexpensive tin and plastic yo-yos. For yo-yos marketed by the U.S. Toy Company, see the Miscellaneous Tin section. Many of the yo-yos distributed by the U.S. Toy Co. carry the mark © C-P.

### WALLACE TOY COMPANY

The Wallace Toy Company is known for distributing two yo-yos; the Bunny Martin Yo-Yo, which is no longer produced, and the Itsy Bitsy Yo-Yo.

| | | |
|---|---|---|
| 1601 | | (Bunny) MARTIN YO-YO, gold leaf stamped seal, wood, tournament shape, '90s |

This yo-yo was available retail long after its production stopped. An example of this yo-yo is shown next to this section's title.

| | | |
|---|---|---|
| 1602 | 34 | ITSY-BITSY YO-YO, sticker seal, wood, midget, '90s |

This yo-yo is retailed loose and on blister cards. Unlike most midget yo-yos, some string tricks can be performed with this model. The Itsy Bitsy is sometimes used as an advertising or souvenir yo-yo.

### WARNER BROTHERS

Warners Brothers is a giant entertainment company best known for movies and cartoon shorts. Looney Toons characters like Bugs Bunny, Sylvester and Tweety, and Daffy Duck were stars of cartoon shorts in the '40s and '50s. Their popularity then and now, in their own TV series, has made them household names and spawned a line of yo-yos.

| | | |
|---|---|---|
| 1603 | 26 | BUGS BUNNY, paper sticker seal, plastic, tournament shape, early '90s |
| 1604 | 26 | BUGS BUNNY, large disk insert seal, plastic, tournament shape, '80s |

This yo-yo retails as the "Six Flags Over Atlantis" theme park.

• Space Jam - *See Spectra Star section #1459.*

| | | |
|---|---|---|
| 1605 | | (Looney Tunes series), paper sticker seals, plastic, tournament shape, '90s |

This is a series of five different yo-yos retailed loose out of a counter display box. The seals feature the Looney Tunes logo, Bugs Bunny, Sylvester, the Tasmanian Devil, and a group picture of Bugs, Daffy, Tweety and the Tasmanian Devil. For an example of Sylvester, see the yo-yo next to this section's title.

| | | |
|---|---|---|
| 1605.1 | 32 | ROAD RUNNER, embossed seal, plastic, John Hart Toys Inc., sculpted, 1974 |

This was the first wheel shaped yo-yo produced. Greg Hart (Englehart) created the wheel yo-yo in the early '70s. He negotiated world rights to used the Warner Brothers characters on yo-yos in 1973. A year later, he released this model based on the Road Runner cartoon.

The yo-yos come in two hub styles, the honeycomb pattern and the five-spoke design. Approximately 50,000 yo-yos were sold in Canada and 50,000 in the United States. The Canadian cards were in French and English without a fixed price. The American version had a pre-printed price of $1.00. Two different blister cards were released. The first shows Wile E. Coyote being run over by a wheel shaped yo-yo. The second is based on the cover of the Gold Key comic book, #90189-310, which shows the Road Runner sitting on a cactus sipping lemonade while yo-yoing with his toe. This Road Runner model is the only yo-yo of the '70s to feature a Warner Brothers cartoon character.

*Continued ...* ▶

## YO-YO LISTINGS

1606   26   **(Tasmanian Devil) MAGIC MOUNTAIN**, plastic, large disk insert seal, tournament shape, '90s
           This was a souvenir from Magic Mountain. See also Spectra Star section #1476.

1606.1      **(Tasmanian Devil)**, sticker seal, plastic, puck shape, light-up, '80s

1607   26   **(Tweety Bird • Sylvester shown)** paint seal, wood, three-piece, tournament shape, '90s
           This boxed yo-yo was retailed in Warner Brothers shops for $5.00.

## WHIRL-KING

Whirl-Kings are three-piece wooden yo-yos originally produced by the Sock-It Company in High Point, North Carolina. Sock-It later became the Fli-Back Company which continued to produce the Whirl-King line. All Whirl-King yo-yos have gold leaf stamped seals. Ordinarily Whirl-Kings are half blue and half red, but solid blues and reds were also produced.

1608   22   **(Small crown)**, die stamped seal, wood, tournament shape, pegged string, 1946-mid '60s
           This is the original Whirl-King made by the Sock-it Co. The production of this yo-yo started in 1946. The lettering on this yo-yo is in a script font. In the early to mid-'60s, it was changed to a non-script font. An example of this yo-yo is shown next to this section's title.

1609   22   **(Large crown)**, die stamped seal, wood, tournament shape, '60s
           This yo-yo replaced the Small Crown style #1608 in the middle '60s. There are two variations of this seal, one with small block lettering, and one with slightly larger block letters.

## WOOD (CARVED)

Demonstrators frequently hand-carved yo-yos in the '40s, '50s, and early '60s to help with sales or to reward contest winners. These carvings are most commonly associated with the Filipino Duncan and Royal demonstrators. An elaborately carved name or scene was the standard. Common scenes include flying birds, palm trees, and sail boats. Because each yo-yo is a unique piece of art, values have to be determined on a piece by piece basis. Ideally carved yo-yos have the carving on the back to keep from damaging the seal. A carving that has resulted in considerable paint loss decreases the value of the yo-yo. Below are a few examples of demonstrator-carved yo-yos.

1610   39   **(5 jewel), 1952 Winner**
           Demonstrators would carve yo-yos, like this, as rewards for contest winners.

1611   39   **(Barbara)**
           It is rare to see a girl's name on demonstrator-carved yo-yo.

1612   39   **(Bobby)**
           To make the carving more visible, the demonstrator rubbed ink into the crevices. Because they don't have paint that chips off, these hand-made natural wood yo-yos are better at standing the test of time. An example of this yo-yo is shown next to this section's title.

1613   39   **(Jack)**
           This yo-yo features a typical scene carved by a demonstrator. The reverse face has a Duncan 77 seal.

1614   39   **(George)**
           Names carved in calligraphy by demonstrators don't decrease a yo-yo's value unless there is severe paint loss or chipping. On the other hand, if a name is just scratched into the yo-yo's face, it is less valuable. This model has the Big "G" Genuine die stamped seal on the reverse face.

## YO-YO LISTINGS

## WOOD MISCELLANEOUS

The following is a list of wooden yo-yos with seals that do not fit in any other category.

1616   39   **105 HYO SILVER LEAF**, die stamped seal, tournament shape, three-piece, pegged string, '50s
This is believed to be a non-promoted brand from the '50s. The origin of this yo-yo is currently unknown.

1617   23   **BANDALORE**, decal seal, tournament shape, one-piece, late '30s
This is a rare yo-yo produced by Pedro Flores in his Rockford, Illinois plant. Flores began producing these yo-yos in the 1930s after he left Duncan. Since he sold the "yo-yo" trademark to Duncan, Flores could not use the name on his toy. He called it a "Bandalore," the generic name used in the U.S. prior to the introduction of the word "yo-yo." An example of this yo-yo is shown next to this section's title.

1618   **(Blank Skipper's yo-yo)**, no seal, midget, '60s
In 1964, this red yo-yo was part of a Skipper outfit set called "Outdoor Casuals." It was later issued in an accessories set for Skipper and her friends. There are no seals or markings on this yo-yo. For Barbie related yo-yos, see #1419 and #1420 in the Spectra Star section.

1619   39   **DOUBLE DOOZER RAINBOW**, die stamped seal, tournament shape, '60s
This was the last yo-yo produced by Cheerio Toys and Games Ltd. in Canada. The Canadian yo-yo trademark trial resulted in Cheerio no longer being licensed to use the word "yo-yo." The former Cheerio Co. produced one last line of yo-yos called the "Doozers." Only one promotion is known to have taken place in Kitchener, Canada. This promotion was unsuccessful and no more Doozers were made. Like the previous Cheerio yo-yos, these yo-yo were made in Sweden.

1620   39   **DOUBLE DOOZER WHISTLER**, die stamped seal, tournament shape, '60s
This yo-yo has a whistle mechanism in the face. See also Double Doozer Rainbow #1619.

1621   17   **EMPRESS SATELLITE**, die stamped seal, satellite shape, three-piece, pegged string, '60s
This is a rare non-Duncan knock off satellite shaped yo-yo made in Japan. Unlike standard Duncan Satellites, this yo-yo is a three-piece model with a pegged string. For satellite collectors, a collection is not complete without this foreign produced imitation. Dell also made a knock off plastic version of the satellite shaped yo-yo called the Astronaut #260.

1622   23   **EXECUTIVE**, foil sticker seal, tournament shape, jumbo.
Various companies have made different jumbo wood yo-yos named "Executive." Some have sticker seals others have brass plates. These can be hard to date unless they have their original packaging.

1622.1   **ELITE YO-YO**, carved seal, tournament shape, '90s
This is a natural maple yo-yo by Woodkrafter Kits that carries a copyright date of 1988. It retailed in a hanging cardboard shadow box, Model 13000.

1622.2   33   **G-YO**, paint seal, multiple geometric shapes, three-piece, late '90s
This series was released by the Holgate Toy Company in 1998. Ten different odd shaped yo-yos are available: heptagon, star, square, flowers, heart, octagon, triangle, shamrock, pentagon, and flower.

1623   43   **GROPER UP-N-DOWN TOP**, decal seal, wood, tournament shape, '50s
This is a non-promoted yo-yo line believed to be from the '50s. This decal seal version is considered more desirable than the die stamped seal version. These yo-yos may have a decorative airbrushed stripe.

1623.1   **GROPER UP-N-DOWN TOP**, die stamped seal, wood, tournament shape, '50s
For more information, see #1623.

1624   **JOCKO • GENUINE YO-YO**, wood, '30s
This is a rare yo-yo retailed in the '30s by Duncan. One of these yo-yos was an exhibit during the Duncan vs. Royal Yo-Yo trademark trial.

1624.1   **JOCKO • BABY**, die stamped seal, wood, tournament shape, miniature, three piece, '30s
Similar to the Genuine Jocko, #1624, except this Jocko does not use the word "Yo-Yo" on the seal.

1625   23   **JEWEL TOURNAMENT FILIPINO SPINNING TOP**, decal seal, tournament shape, one-piece, '50s
The Jewel Philippine Spinning Top Company produced yo-yos out of Brooklyn, New York from 1947 through the early '50s. There is a jewel illustrated on the decal seal, but no rhinestones are mounted on the face.

1626   39   **KOHNER BROTHERS SPINMASTER**, gold leaf stamped seal, tournament shape, pegged string, '50s

1626.1   39   **KOHNER BROTHERS SPINMASTER**, foil sticker seal, tournament shape, '50s
This is much more rare than the die stamped version. See #1626.

1627   39   **KNIGHTS YO-YO**, decal seal, tournament shape, three-piece, pegged string
The manufacturer and date of production are unknown.

1628   **LITTLE WOODY TOP**, paint seal, tournament shape, three-piece
The manufacturer and date of production are unknown.

1628.1   43   **MASTER CRAFT**, gold leaf stamped seal, wood, tournament shape, '50s
Mastercraft was a wood product company based in Canada that produced Cheerio yo-yos from 1952 until around 1962. This yo-yo was made with its own seal, but quantities produced were limited. It is doubtful that this yo-yo was ever promoted.

*Continued ...* ▶

## YO-YO LISTINGS

1629  15  **NADSON TWIRLER TOP (4 jewel)**, ink stamped seal, tournament shape, early '60s
This rare knock off jewel was retailed to capitalize on the popularity of the Duncan Jewel line. The seal shows a marching band major. The jewels may be different colors. The more varied the jewels the more desirable the yo-yo. This model was retailed in a polybag with a saddle header card in the late '50s or early '60s.

1630  **NOVEL SWING TOP**, die stamped seal, tournament shape, '30s
Novel also made a 1937 World's Fair yo-yo. See #1650.

1631  03  **ORBIT**, gold leaf stamped seal, tournament shape, three-piece, pegged string, '60s
This yo-yo has the Gemini space capsule on the seal. The Fli-Back Company made this yo-yo, but it does not say Fli-Back on the seal or the reverse face. These were retailed out of a counter display box that held twenty-four yo-yos.

1632  43  **SPACE WHIRLER**, paper sticker seal, wood, butterfly shape, '60s
This yo-yo was produced by the Sock-It Company and was sold loose out of a counter display box.

1633  **SPIN MASTER • STYLER**, foil sticker seal, tournament shape, three-piece, pegged string, '60s

1634  **TIP-TOP**, wood, '30s
This rare yo-yo was reportedly produced by Little Bear Specialties Company of Kansas City.

1635  43  **(Tom Picou • Constellation)**, tournament shape, '90s
Tom Picou Design in Wood is a custom yo-yo maker that produces exotic wood yo-yos. His yo-yos do not have seals, but some styles are distinctive such as the Constellation (shown) and the Sun and Moon. Sterling silver inserts make up the "stars" in the dipper of the Constellation.

1636  37  **TOP BRASS**, engraved brass seal, tournament shape

1637  37  **TOWLE STERLING 1196**, engraved sterling silver disk seal, tournament shape, three-piece, '70s
Towle produced this model as a gift yo-yo. This yo-yo has a sterling silver disk inlaid in the center of one face. The disk can be engraved. For other Towle yo-yos, see #1078.

1638  22  **TOWN TALK TOURNAMENT**, paint seal, tournament shape, three-piece
The manufacturer and date of production are unknown.

1639  **TUMBLE TOP**, wood, '30s
Illinois Lumber and Tile of St. Louis produced this rare yo-yo.

1640  43  **TWIRLER TOP**, paper sticker seal, wood, tournament shape, '60s
There are two versions of this baseball player seal, one round and one with scalloped edges. This yo-yo is believed to be from the '50s or '60s.

1640.1  43  **WOODCRAFT CO. TWIRLER**, paper sticker seal, tournament shape, '50s
This yo-yo was made in Lawrence, Massachusetts. This yo-yo is believed to have been produced in the '50s.

1641  43  **YOGEE**, paper sticker seal, wood, butterfly shape
The manufacturer and date of production are unknown.

1641.1  31  **(Yoissimo)**, diffraction disk seal, tournament shape, '90s
This is a natural wood maple yo-yo with inlaid diffraction disk inserts. There was no lettering on the disk inserts. At least five different diffraction disk inserts exist. Originally produced by ZBA Inc., this yo-yo is no longer made.

## WOOD (MISCELLANEOUS - UNNAMED)

Yo-Yos without a manufacturer's seal, advertising, or collectible characters on them have little value. A collector only purchases a yo-yo without a seal if it is aesthetically pleasing to that particular collector. Yo-Yos in this category seldom increase in value. Handmade yo-yos, blank yo-yos, and hand-painted yo-yos all fall under this classification. Exceptions would be yo-yos from a known company that are still in the original packaging. If the packaging is opened, the yo-yo looses its value. Below is a listing of yo-yos that do not have seals.

1642  33  **(Bobbin)**, wood and metal, puck shape, standard and jumbo sizes, wood axle, '90s
This is a "homemade yo-yo" created from yarn bobbins used in the textile industry. Although the flanges from the bobbins may be 50-100 years old, many of these yo-yos were recently produced. Old bobbins were still being turned into yo-yos in the '90s and marketed by companies such as Lands' End.

1643  **(Concentric circles • target)**, paint seal, puck shape, '90s
This is a common yo-yo imported from Japan. These yo-yos are three-piece models with fixed strings, sometimes with plastic finger rings. They are often unbalanced and play poorly. These, and similar variations, have little if any collectible value. An example of this yo-yo is shown next to this section's title.

*Continued ...*

## YO-YO LISTINGS

● **WOOD (MISCELLANEOUS - UNNAMED)** *Continued ...*

1644    *34*    **(Exotic wood), tournament shape**
Yo-Yos have been made from more than 40 different types of wood, from African zebra wood to Australian lacewood. Although they are unique and perhaps are rare, the collectable value of these yo-yos is based purely on esthetics. The model shown was custom made by Monarch. It has an inlaid Indian Head Penny, heads on one side, tails on the reverse.

1645    *15*    **(Jewel), tournament shape**
Five jewel multi-color models, like the ones shown, are unusual. Some companies in the '50s and '60s made a variety of wooden jewel model yo-yos without seals. Unfortunately, these all fall in the category of yo-yos without seals. Their value is based solely on the aesthetics of the individual yo-yo.

1646            **(Natural maple), no seal, tournament shape**
Maple has traditionally been the most common wood used in turning quality wooden yo-yos. Many companies have produced natural clear lacquer maple yo-yos without seals. Unless these yo-yo are in original packaging, blank yo-yos have little collectible value.

1647    *33*    **(Stars), tournament shape, jumbo**
Like other yo-yos without seals, jumbo yo-yos have limited collectible value unless they have identifying marks. Because the date and maker are unknown, the collectible value of this yo-yo is based purely on aesthetics. The diameter of this yo-yo is 4-1/2 inches.

 **WORLD'S FAIR**

The 1939 New York World's Fair was arguably the greatest World's Fair. The theme was "The World of Tomorrow." The most remembered exhibit was the Trylon, a 728 foot 3 sided pylon, which sat next to the perisphere, a gigantic white globe. This symbol of the 1939 World's Fair is pictured on most of the souvenirs produced for the exhibition. The fair drew 45 million visitors, and what souvenirs did they take home with them? You guessed it... yo-yos! At least five different souvenir yo-yos were produced for the fair and there may be others.

Yo-Yos have been produced for other World's Fairs. The most collectable are the iron bandalore from the 1904 St. Louis World's Fair and the Space Seattlite Needle yo-yo from the 1962 Seattle World's Fair. Collectors should note there is a yo-yo company that produces a wooden yo-yo called the "Worldsfair." This yo-yo was not associated with a World's Fair. (See Forester section #615 for more details.)

●    **Expo '74 World's Fair** - *See Light-Up section #1033.*
1648    **EXPO '86 VANCOUVER • CANADA, large disk insert seal, plastic, tournament shape**
●    **Space Needle (World's Fair, Seattle 1962)** - *See Duncan Satellite section #452.*
1649    **WORLD'S FAIR 1904 ST. LOUIS, molded seal, metal, riveted disk axle**
Because this yo-yo was produced in the United States prior to the introduction of the word "yo-yo," this model is considered a "Bandalore." It is the first known U.S. "yo-yo" to carry a date.

1650    *40*    **WORLD'S FAIR SWING TOP 1939, ink stamped seal, wood, tournament shape, three-piece**
This is a souvenir yo-yo produced by the Novel Pkg. Corp. for the 1939 New York World's Fair.

1651    *40*    **WORLD'S FAIR 1939, decal seal, wood, tournament shape, three-piece**
This is a souvenir yo-yo from the 1939 New York World's Fair. It has a very similar seal to #1652, but with slightly more detailed graphics. An example of this yo-yo is shown next to this section's title.

1652    *40*    **WORLD'S FAIR 1939, decal seal, wood, tournament shape, three-piece**
This is a souvenir yo-yo from the 1939 New York World's Fair. The seal on this model is less detailed, unlike the one on #1650. There are no buildings outlined on the horizon on this design.

1653    *40*    **WORLD'S FAIR 1939, foil sticker seal, wood, tournament shape, three-piece**
This yo-yo is one of the most difficult to find of the 1939 New York World's Fair Series. The foil sticker seal has a tendency to detach which makes this model rare.

1653.1  *40*    **WORLD'S FAIR 1939, foil sticker seal, wood, tournament shape, one-piece**
Similar to #1653, this is also a very rare 1939 New York World's Fair yo-yo with a foil sticker seal.

1654    *23*    **WORLD'S FAIR 1964, decal seal, wood, tournament shape, jumbo**
This is a rare jumbo-sized wood souvenir yo-yo from the 1964 New York World's Fair. The seal features the 1964 Fair symbol, the Unisphere.

1655    *40*    **WORLD'S FAIR 1982, imprint seal, plastic, tournament shape**
This was a souvenir from the 1982 Knoxville World's Fair. The Duncan Imperial seal is found on the reverse face.

*Continued ...* ▶

*Lucky's Collectors Guide To 20th Century Yo-Yos*

## YO-YO LISTINGS

1656   *40*   **WORLD'S FAIR 1982**, large disk insert seal, plastic, tournament shape
This was a souvenir from the 1982 Knoxville World's Fair.

## YOMEGA

The original Yomega yo-yo was invented and patented by Michael Caffrey in 1982. Production began in 1984 in the basement of Alan Amaral's house. From this modest beginning the company has grown to become one of the largest producers of yo-yos in the world. The innovative key to the original Yomega was a clutch mechanism attached to a free spinning bearing that operated on centrifugal force. As the yo-yo slowed, the spring-loaded clutch clamped down gripping the string, causing the yo-yo to return to the hand. This combination allowed for longer spins with automatic return. Clutch mechanisms are not currently legal in most yo-yo competitions.

Although the original flagship Yomega is referred to on company literature as the "Phantom," it has no seal. Collectors call the Phantoms "Brains" because of the internal clutch mechanism. The colorless translucent yo-yo is made from Dow polycarbonate, a plastic used to make bulletproof glass. The internal clutch mechanism makes the "Phantom" impossible to confuse with other brands of yo-yos. Phantoms were originally retailed in hard clear plastic showcase boxes with felt bases. A small trick book was also enclosed in the showcase box. Early Yomega trick books show the copyright to be by "Caffrey Inc." Later trick books say Yomega Corporation. The Yomega was also the second major company to produce a yo-yo with a take-apart design. (See also Tom Kuhn Yo-Yos.) Take- apart yo-yos allow the player to easily remove string tangles and to adjust the string gap. Yomega has continued to improve their line by adding transaxles (Outrageous and Fireballs), ball bearing transaxles (Raiders), and butterfly designs (Sabre Wings).

Yomegas have been heavily marketed in Japan and some of these yo-yos occasionally turn up in the U.S. These yo-yos have "Bandai 1997" on the seal. The rarest of the Japanese Yomegas are the gold colored award yo-yos that read "Congratulations 1997." These yo-yos were not sold retail. They were only given as awards. Only 500 of these yo-yos were made and it is estimated there are less than five in the United States.

1657        **ALL STAR**, imprint seal, plastic, tournament shape, late '90s
This fixed axle Yomega is distributed by Bandai of America.

1657.1      **AMERICAN EAGLE OUTFITTERS**, imprint seal, plastic, tournament shape, '90s
This is an example of a Yomega yo-yo used for advertising.

1658   *43*   **BANDALORE WITH A BRAIN**, imprint seal, plastic, tournament shape, '90s
Canada still recognizes the yo-yo trademark. Companies and retailers cannot use the word "yo-yo" unless licensed and must identify their products by other names. This is the Canadian version of the Phantom model.

1659        **(California collection)**
These were Phantom "Brains" with colored clutch arms. Standard Phantom "Brain" clutch arms were white. This series was released in 1990.

1660        **(Executive)**, imprint seal, plastic, tournament shape, '90s
This is a Phantom "Brain" that has a clutch mechanism with brass springs and bearings. It retailed in a hard black plastic jewelry box case.

1661        **FIREBALL**, imprint seal, plastic, tournament shape, transaxle, '90s
This is the old Outrageous with a new seal released in 1990.

1661.1      **FIREBALL COLLECTOR'S EDITION**, imprint seal, plastic, tournament shape, transaxle, late '90s
This is a variation of the Fireball released in 1998.

1662   *43*   **FIREBALL JEWEL**, imprint seal, plastic, tournament shape, transaxle, 1994
This is a Fireball variation with one jewel centered on the yo-yo's face. The Yomega jewel line is no longer in production.

1663   *43*   **FIREBALL (glow)**, imprint seal, plastic, tournament shape, transaxle, '90s
This is the same as the Fireball #1661, but made out of glow plastic.

1663.1      **FIRESTORM**, imprint seal, plastic, bulge face shape, transaxle, late '90s
This Yomega is distributed by Bandai of America.

1664   *43*   **GALAXY GLOW**, imprint seal, glow plastic, flywheel shape, fixed axle, '90s

1665   *31*   **(Gem series)**
These were Phantom "Brains" made of colored translucent plastic. The three colors of plastic were Ruby, Sapphire, and Emerald.

*Continued ...* ▶

● **YOMEGA** *Continued ...*

| | | |
|---|---|---|
| 1666 | *34* | **GEMINI GYRO**, imprint seal, plastic, tournament shape, convertible yo-yo top, **1990** |
| | | This is a convertible yo-yo top first released in 1990. It came with colorful removable disk inserts for the face of the top. It was retailed both on a hanging card and in a plastic display box. Yomega no longer produces this yo-yo. |
| 1666.1 | | **METALLIC MISSILE**, paint seal, metal, flywheel shape, ball bearing transaxle, late '90s |
| | | This model was released in 1998. Along with the Wing Force, these are the first metal yo-yos to be made by Yomega. This yo-yo is retailed with rubber rim guards. |
| 1667 | *43* | **(Outrageous)**, imprint seal, plastic, tournament shape, **1989** |
| | | This was Yomega's first transaxle yo-yo. Transaxles allow the yo-yo to spin freely without friction between the string and axle. The Outrageous was replaced by the Fireball. See #1661. For an example of the blister card, see Plate 32, #1667. |
| 1667.1 | | **PANTHER**, paint seal, wood, tournament shape, three-piece, late '90s |
| | | This model was released in 1998. This is the first wood yo-yo released by Yomega and believed to be the first wooden yo-yo to ever be retailed with rubber rim guards. |
| 1668 | | **(Phantom "The Brain")**, imprint seal (if present), plastic, tournament shape, view lens, internal clutch, '90s |
| | | Before transaxles, this yo-yo was considered the flagship of the Yomega line. Phantoms are colorless and translucent with white clutch mechanisms. The yo-yo does not read "Phantom" on the seal, many are made without seals. This yo-yo is now more frequently called "The Brain." Although this yo-yo may be without a seal, its unique appearance easily distinguishes it from other yo-yos. The Executive, California Collection, and Jem series are all spin offs of the Phantom. The Phantoms marketed in Japan are called "Hyper Brains." |
| 1669 | *43* | **POWER SPIN**, imprint seal, plastic, fly wheel shape, '90s |
| | | This is a Teflon transaxle yo-yo made with more mass near the rim for longer sleep time. There are two seal styles, #1669A and #1669B. |
| 1670 | *43* | **RAIDER**, imprint seal, plastic, tournament shape, ball bearing transaxle, '90s |
| | | This yo-yo was first released in 1993 and sold with a black leather yo-yo holster. |
| 1671 | *43* | **RAIDER (glow)**, imprint seal, glow plastic, tournament shape, ball bearing transaxle, '90s |
| 1671.1 | | **RAIDER • JEWEL EDITION**, imprint seal, plastic, tournament shape, ball bearing transaxle, **1994** |
| | | This yo-yo is the rarest of the Yomega Jewel series. It only had a brief production run. Yomega jewels are no longer in production. |
| 1671.2 | | **RBII**, imprint seal, plastic, flywheel shape, ball bearing transaxle, late '90s |
| 1671.3 | | **SABRE RAIDER**, imprint seal, plastic, butterfly shape, ball bearing transaxle, late '90s |
| 1671.4 | | **SABRE WING**, imprint seal, plastic, butterfly shape, late '90s |
| | | This butterfly shaped Phantom was released in 1997. In Japan, this yo-yo is called the "Stealth Brain." |
| 1671.5 | | **SABRE WING FIREBALL**, imprint seal, plastic, butterfly shape, transaxle, late '90s |
| | | This is the Fireball in a butterfly shape, released in 1997. In Japan, this yo-yo is called the "Stealth Fire." |
| 1671.6 | | **STROBE YO**, imprint seal, plastic, late '90s |
| | | This model was released in 1998 and is Yomega's first light-up yo-yo. |
| 1672 | | **TEAM HIGH PERFORMANCE**, imprint seal, plastic, tournament shape, '90s |
| | | This is a Raider model used by the Team High Performance players of Honolulu, Hawaii. These yo-yos were sold retail. An example of this yo-yo is shown next to this section's title. |
| 1673 | *43* | **TOUCH ME YO-YO**, liquid crystal disk seal, plastic, Humphrey shape, **1991** |
| | | On the blister card, this yo-yo is called a "Sunshine Yo-Yo." It was briefly produced by Humphrey for Yomega. It is believed to be the only liquid crystal yo-yo ever produced. |
| 1674 | | **TWIN TRIK**, paper insert seal, plastic, flywheel shape, '90s |
| | | This yo-yo is no longer in production. |
| 1674.1 | *43* | **WING FORCE**, paint seal, metal, butterfly shape, ball bearing transaxle, late '90s |
| | | This model was released in 1998. Along with the Metallic Missile, these are the first metal yo-yos to be made by Yomega. |
| 1674.2 | | **X-BRAIN**, imprint seal, plastic, tournament shape, internal clutch, late '90s |
| | | This Yomega is distributed by Bandai of America, and has a cross shaped internal clutch different from the original "Brain." |
| 1675 | *43* | **YOMEGA • THE YO-YO WITH A BRAIN**, imprint seal, black and yellow plastic, tournament shape, view lens, **1984** |
| | | This is the very first Yomega to be released and is considered the most collectible. The Patent Number 4332102 appears on the face of the yo-yo. This is the first yo-yo to have a clutch mechanism for automatic returns. Unlike later translucent plastic models, this early model was opaque. Only 10,000 of these were made. |

# Plates

166 • P

GRADING YO-YOS EXAMPLES

## GRADING YO-YOS

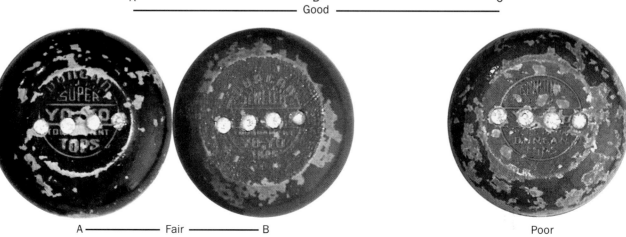

Mint

A ——— Near Mint ——— B

A ——— Excellent ——— B

Fine

A    B    C
———————— Good ————————

A ——— Fair ——— B

Poor

Lucky's Collectors Guide To 20th Century Yo-Yos

PLATE 5

235

236

240

234

241

247

230

252

245

250

246

238

232

227

249

226

229

228

244

251

243

1321

1320

1331

1332

1328

1324

1337

1339

1333

172 • P

DELL • CHEMTOY • IMPERIAL • MISCELLANEOUS PLASTIC YO-YOS

## PLATE 6

| 260.9 | 260.7 | 260.10 | 260.5 | 260.6 |
| 260.4 | 260.11 | 210.2 | 210.1 | 210 |
| 1005 | 1003 | 154 | 1007 | 1227 |
| 1228 | 1219 | 1212 | 1097 | 1217 |
| 1479 | 281 | 280 | 172 | 1213 |
| 1211 | 1210 | 1209 | 173 | 1238 |

Lucky's Collectors Guide To 20th Century Yo-Yos

174 • P

DISNEY • FESTIVAL YO-YOS

## PLATE 8

| 289 |  269 |  297 | 533 |

| 1445 |  1447 |  1418 |  1453 | 1453.1 |

| 1456 |  1432.1 | 517 | 520 | 519 |

| 544 |  541 |  543 | 545 | 547 |

| 549 | 529 | 525 |  554 |  535 |

| 531 | 521 | 523 | 532 | 524 |

Lucky's Collectors Guide To 20th Century Yo-Yos

## PLATE 9

457

462

401

422

515

454

369

353.1

415

423

411

420

314

314

382

The History and Values of Yo-Yos

## PLATE 10

307.38

307.10

307.46

307.16

307.24

307.15

307.30

307.42

307.09

307.52

307.53

307.51

307.36

307.27

307.20

307.59

307.48

307.29

307.54

307.35

307.62

307.37

307.14

307.25

307.50

448.5

307.23

307.61

307.22

448.4

## PLATE 12

331

332

330

327

321A

321B

324A

324B

324C

319

459

460

1309

1292

1305

433

## PLATE 13

283

383

316

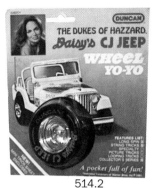
514.2

336

303

430

427

350

450.2

314

495

452

349

451

323

# PLATE 14

DUNCAN MISCELLANEOUS YO-YOS

370

373

344

344.1

421

420

328

419

343

448

448.1

448.3

415

428

446.2

436.2

444

440

443

399

438

445

425

437

Lucky's Collectors Guide To 20th Century Yo-Yos

## PLATE 15

The History and Values of Yo-Yos

182 • P

PLATE 16

Lucky's Collectors Guide To 20th Century Yo-Yos

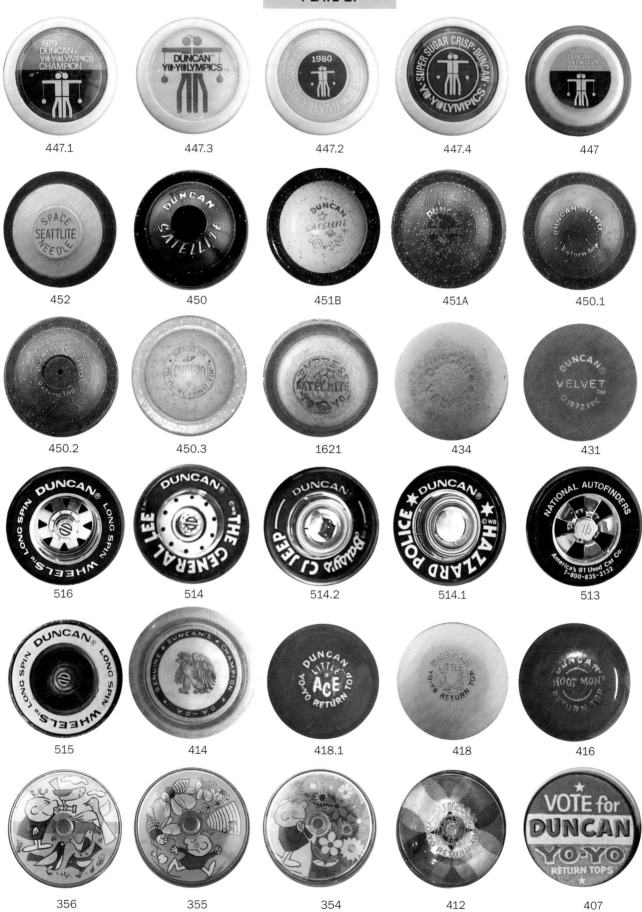

184 • P

PLATE 18

DUNCAN SUPER HEROES • MISCELLANEOUS SUPER HERO YO-YOS

| 470 | 471 | 471 | 473 | 473 |
| 472 | 469 | 469 | 468 | 467 |
| 466 | 467 | 465 | 465 | 464 |
| 155 | 175 | 1443 | 1425 | 1235 |
| 1427 | 1426 | 1424 | 1460 | 177 |
| 1429 | 1481 | 1451 | 1450.1 | 1450 |

Lucky's Collectors Guide To 20th Century Yo-Yos

PLATE 19

The History and Values of Yo-Yos

186 • P

DUNCAN TOURNAMENT YO-YOS

## PLATE 20

| | | | | |
|---|---|---|---|---|
| 491 |  436.1 |  493 | 494 | 501 |
| 449.2 | 449 | 449.1 | 506 | 495 |
|  499 | 503 | 500 | 502 | 511 |
| 507 | 508 | 510 |  507.1 | 509 |
| 489 | 488 | 486 | 487 | 490 |
| 512 | 512.2 | 512.1 | 505 | 504 |

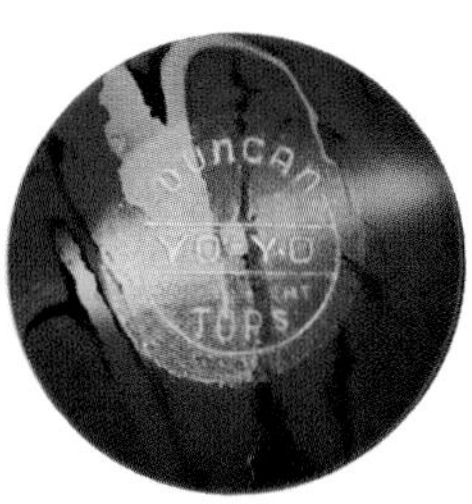

Lucky's Collectors Guide To 20th Century Yo-Yos

## PLATE 21

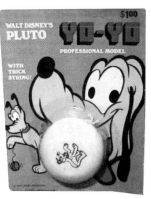 299

 526

 548

 538

 679

 519

 527

 555

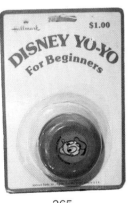 265

 1188

 551

 534

 522

 653

 271

 552

The History and Values of Yo-Yos

188 • P

FLI-BACK • HI-KER • CHICO • MISCELLANEOUS WOOD YO-YOS

## PLATE 22

| 572 | 572 | 573 | 574 | 571 |
| 565 | 564.1 |  564 | 567 | 568 |
| 561 | 570 | 566 |  569 |  575 |
| 1609 | 1608 | 1520 | 1521 | 1064 |
| 671 | 669 | 670 | 673 | 674 |
| 214 | 215 | 213 | 212 | 1638 |

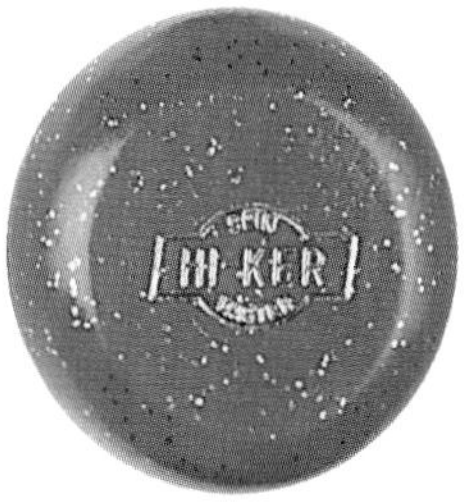

Lucky's Collectors Guide To 20th Century Yo-Yos

## PLATE 23

580

122

1414

1654

1622

7

54

616

617

618

577

583

582

1617

1625

# PLATE 24

Lucky's Collectors Guide To 20th Century Yo-Yos

## PLATE 25

## PLATE 27

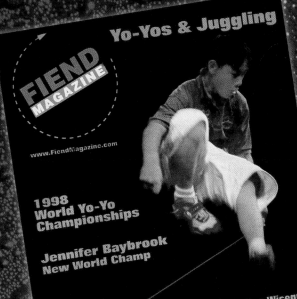

# ARE YOU A FIEND?
### The Magazine for Yo-Yo, Juggling & Tops

- Meet the Pros
- Learn the Latest Techniques
- Up-To-Date Information on:
   Contest & Events Around the World
- The Latest News in Collecting

For More Information on Fiend Magazine:
CALL: **1.888.284.8548**
WEBSITE: **www.fiendmagazine.com**

# LUCKY'S COLLECTORS GUIDE TO 20TH CENTURY YO-YOS
### The definitive guide for collecting yo-yos

- Featuring the History of Yo-Yos
- Over 1000 Photos of Collectible Yo-Yos & Memorabilia
- A Comprehensive Value Guide ($)

For More Information:
CALL: **1.877.969.6728**
WEBSITE: **www.yo-yos.net**

## ORDER NOW!

*The Ultimate Gifts For Any Collector*

PLATE 30

## PLATE 31

471

464

463

1665

680

1032

1112

1033

1641.1

1258

531

536

448.1

182

733

734

Lucky's Collectors Guide To 20th Century Yo-Yos

## PLATE 32

1209

1193

663

210

1027

1256

563

1158

151

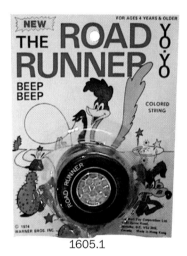
1605.1

260.2

1205

671

749

746

1667

MISCELLANEOUS CARDED YO-YOS

The History and Values of Yo-Yos

NOVELTY YO-YOS

200 • P

## PLATE 33

1104

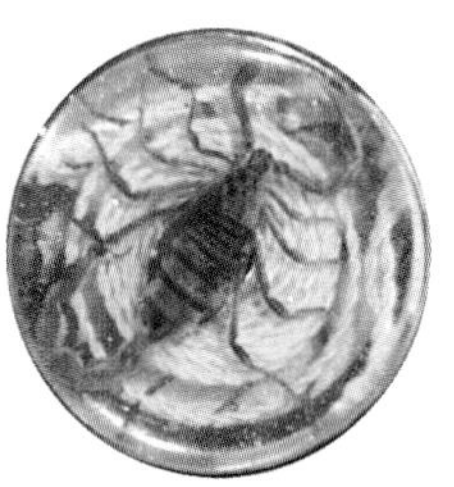

1132

1622.2

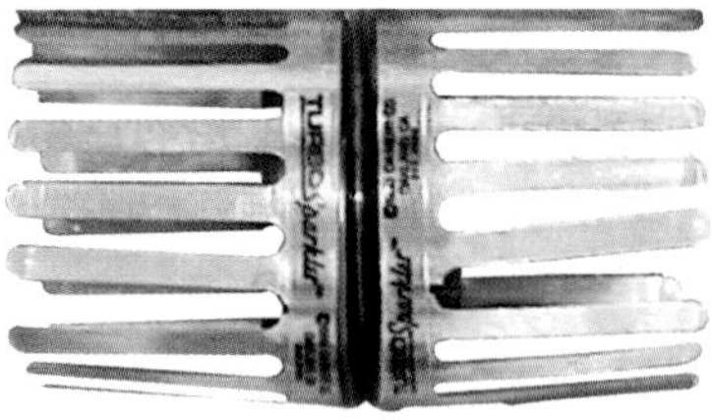

258

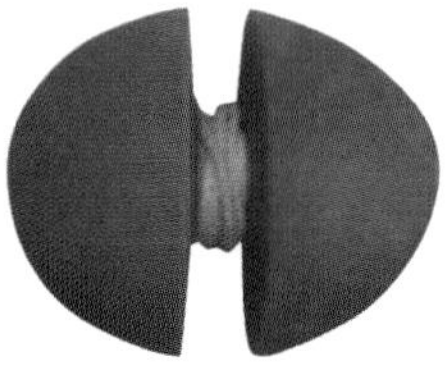

1103

1121

1002

1096

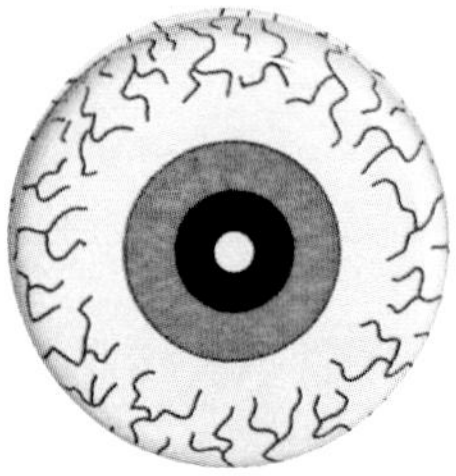

1144.1

1595

1138

1137

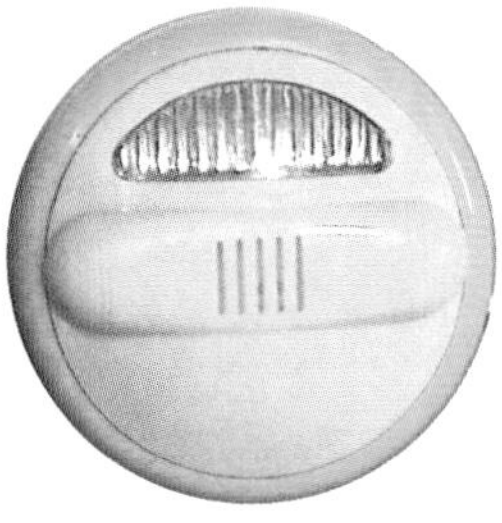

1043

1040

1133

1092

1113

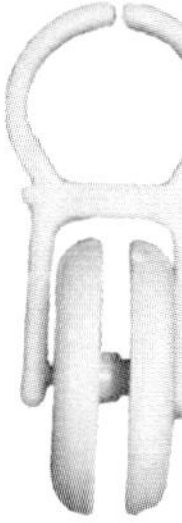

1108

1119

706

1529

1647

1548

1642

1528

Lucky's Collectors Guide To 20th Century Yo-Yos

# PLATE 34

202 • P

## PLATE 35

Lucky's Collectors Guide To 20th Century Yo-Yos

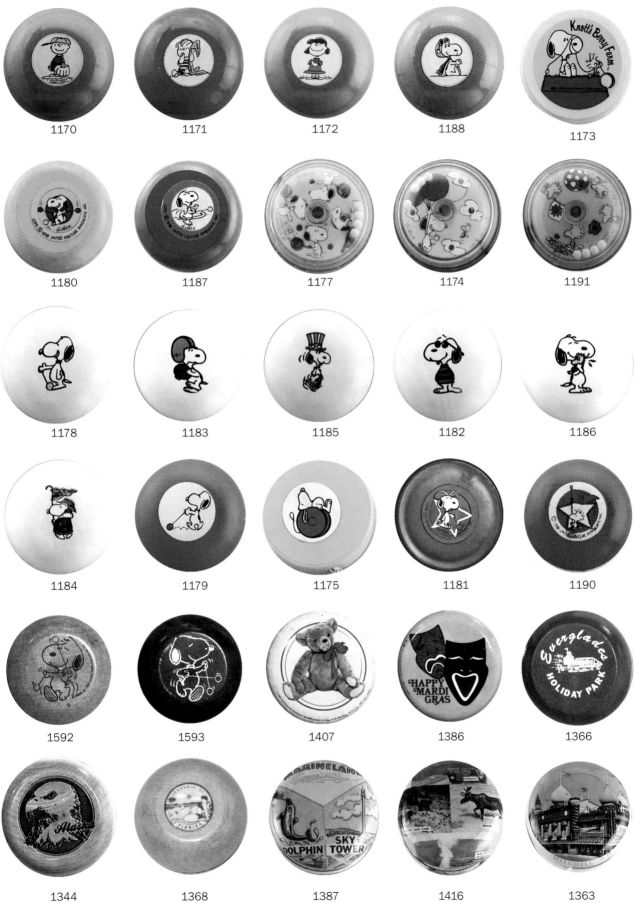

204 • P

PLAYMAXX • NATIONAL • PARKER • CANADA GAMES • MISCELLANEOUS YO-YOS

## PLATE 37

Lucky's Collectors Guide To 20th Century Yo-Yos

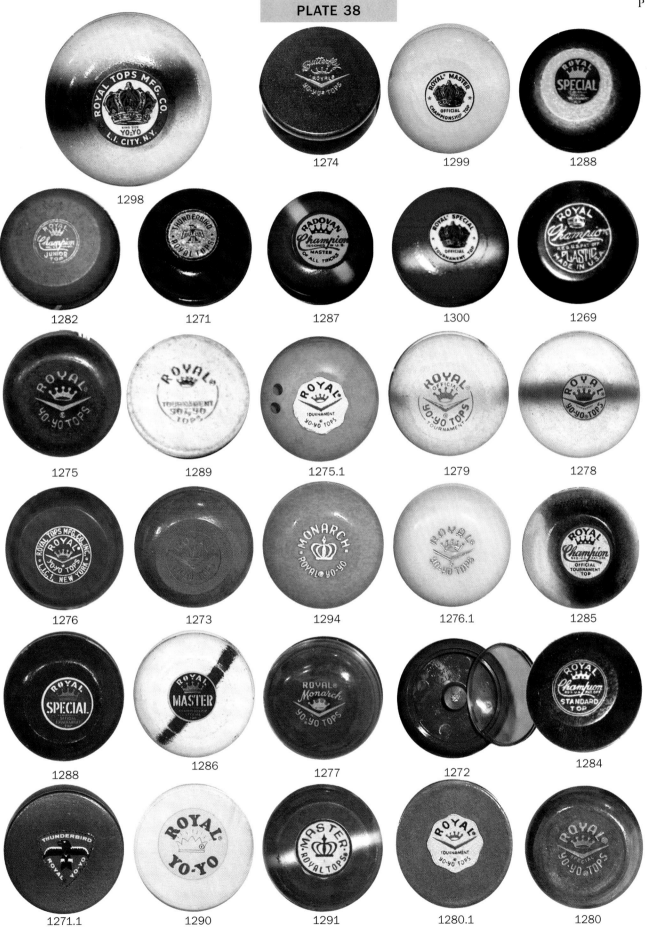

PLATE 39

SMOTHERS BROTHERS • CARVED • BIRD IN HAND • MISCELLANEOUS WOOD • STAR YO-YOS

| 752 | 749 | 751 | 1590 | 1268 |
| 1627 | 1616 | 613 | 614 | 615 |
| 1610 | 1614 | 1611 | 1613 | 1612 |
| 147 | 1084 | 145 | 144 | 143 |
| 1022 | 1023 | 1023.1 | 1626 | 1626.1 |
| 1619 | 1620 | 1516 | 1519 | 1517 |

Lucky's Collectors Guide To 20th Century Yo-Yos

PLATE 40

212 • P

## PLATE 45

1736

662

751

1139

1093

1137.1

1430.1

307.12

406

1728

Lucky's Collectors Guide To 20th Century Yo-Yos

## PLATE 46

1739

1726

1725

1740

1732

1735

1099

664

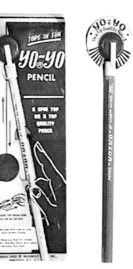
664

66

374

1677

393

MISCELLANEOUS ITEMS

The History and Values of Yo-Yos

## PLATE 47

395

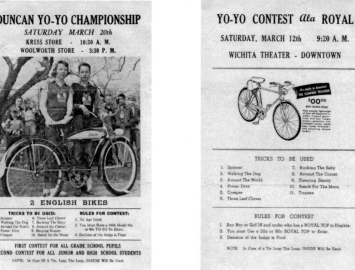

395

395

1886

1887

1884

1722

1717A

397

397  397

## PLATE 48

1717B

1717C

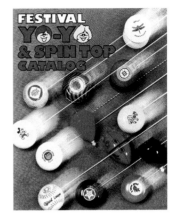
1717D

1717E

1717F

1717G

1717H

1717I

1717J

1733B

1733B

1718

1733B

1733A

The History and Values of Yo-Yos

216 • P

## PLATE 49

AWARD PATCHES

1776

1790

1775

1770

1774

1785

1767

1784

1801

1790.3

1755

1800

1783

1768

1762

1790.1

1771

1787

1789

1795

1798

1799

1769

1750

1757

Lucky's Collectors Guide To 20th Century Yo-Yos

# PLATE 50

The History and Values of Yo-Yos

## PLATE 51

| 1816 | 1809 | 1818 | 1813 | 1811 |
| 1825 | 1823 | 1824 | 1822 | |
| 1826 | 1827 | 1812 | 1819 | 1817 |
| 1829 | 1828 | 1808 | 1815 | 1807 |
| 1683 | 1682 | 1681 | 1723 | |

Lucky's Collectors Guide To 20th Century Yo-Yos

PLATE 52

1837

1839

1838

1836

1737

1830

1841

396

1835

1833

1834            1840

The History and Values of Yo-Yos

## PLATE 53

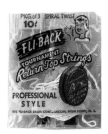    

1883     1861     1909     1894.1     1860

1851     1893     1881     1894     1858

1896     1897     1898     1902

1903     1901     1889     1899

   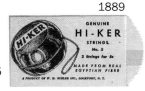

1859     1907     1905     1889

   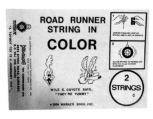

1890     1908     1852     1895

Lucky's Collectors Guide To 20th Century Yo-Yos

## PLATE 54

 1873

 1864

 1866

 1865

 1879

 1869

 1868

 1867

 1870

 1872

 1871

 1874

 1875

 1880

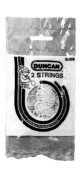 1862

 1876

 1877

 1878

 1853

 1857

 1856

 1854

The History and Values of Yo-Yos

## PLATE 55

1916

1917

1914

1911

1926

1918

1913

1921

1915

1919

1934

1933

1929

1910

1928

1927

1930

Lucky's Collectors Guide To 20th Century Yo-Yos

## PLATE 56

666

204.1

195

209

209.1

208.1

208.2

208

209.2

205.2

205.3

207.1

1846

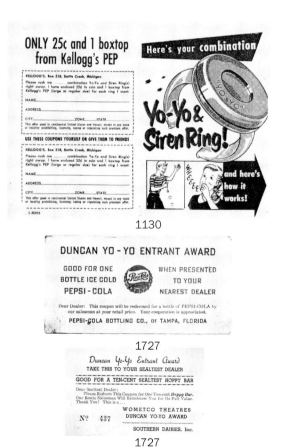
1130

1727

1727

The History and Values of Yo-Yos

# Memorabilia
## Categories

## MEMORABILIA CATEGORIES

### AWARDS

Patches, pins, ribbons, award yo-yos, and trophies were standard fare at many local competitions. Big competitions could yield the victor cameras, bicycles, televisions, stereos, even college scholarships. The awards that collectors find most desirable are trophies with the Mr. Yo-Yo statue attached. The list below is just a sampling of contest awards. For other awards, see the Patches and Pin sections.

| | | |
|---|---|---|
| 1676 | | BOB ALLEN SIDEWINDER (loving cup trophy), '60s |

This award was from the Bob Allen contests held in California in the '60s. See also the Bob Allen section.

| | | |
|---|---|---|
| 1677 | 46 | CHEERIO CHAMPION (beanie hat), '50s |
| 1678 | | (Duncan award sweater with large eagle patch) |

Award sweaters were commonly used by both the Duncan and Cheerio companies. A large eagle patch was given to individuals that could demonstrate proficiency at the eight basic two-handed tricks.

| | | |
|---|---|---|
| 1679 | | DUNCAN CHAMPION (loving cup) |
| 1680 | | (Duncan ribbon) YO-YO CHAMP, '50s |

The ribbon pictures Mr. Yo-Yo "Rocking the Baby."

| | | |
|---|---|---|
| 1681 | 51 | (Duncan Mr. Yo-Yo trophy), '50s |

Early Mr. Yo-Yo statuettes have a free string and yo-yo.

| | | |
|---|---|---|
| 1682 | 51 | (Duncan Mr. Yo-Yo statuette), '70s |

This solid brass statuette was affixed to trophies in the '70s. Unlike the earlier Mr. Yo-Yo trophies, this has the yo-yo string and the yo-yo itself molded to Mr. Yo-Yo's leg. An example of this trophy is shown next to this section's title.

| | | |
|---|---|---|
| 1683 | 51 | DUNCAN YO-YO CHAMPION (plaque), '80s |

This is a Duncan Flambeau award plaque from the '80s.

| | | |
|---|---|---|
| 1684 | | DUNCAN YO-YO CHAMPION (plaque), '60s |

This painted pressed cardboard plaque has the '60s Mr. Yo-Yo engraved on the brass plate. Below Mr. Yo-Yo and the Duncan logos, there is a blank space for engraving the winner's name. The plaque has engraved signatures of both Don and Jack Duncan.

| | | |
|---|---|---|
| 1685 | | (Pen set award • Duncan), '50s |

A pen and brass yo-yo shaped statue as well as a brass plate for engraving are affixed to this wooden plaque.

| | | |
|---|---|---|
| 1686 | | (Mr. Yo-Yo cap • Duncan), '50s |
| 1687 | | (Royal beanie hat), '50s |

This is a felt beanie hat that says, "I'm A Yo-Yo Jr."

| | | |
|---|---|---|
| 1688 | | (Sweater certificate • Cheerio), '30s |

This certificate was awarded to the recipient of an All-Canada Yo-Yo Sweater.

### COUNTER DISPLAY BOXES

From the '20s through the '50s, yo-yos were not retailed on blister cards, but loose out of counter display boxes. These boxes needed to attract buyer's attention so early boxes were covered with eye catching graphics. Display boxes can be difficult to date because some companies used the same box design for decades. New price stickers were frequently placed over pre-printed prices on older boxes in lieu of changing graphics. Collectors should be aware that boxes were not always packed with the style of yo-yo listed on the box. For information on how to calculate the value of a full display box, see the FAQ section.

In the '80s and '90s, the retailing of yo-yos has gone completely to individual packaging. Emphasis on graphics has shifted from the display boxes to the display cards. Although counter boxes are made to hold carded yo-yos, these are typically plain and carry little interest among collectors.

Besides the counter display boxes, some larger companies such as Duncan also market large retail displays. Dump bins, large boxes that hold hundreds of yo-yos, are used and can be folded into stand alone floor displays. Wire display racks and multi-tier variety pack boxes are also common.

*Continued ...* ▶

## MEMORABILIA CATEGORIES

1689    44    **ALOX**
Although the graphics are '30s style, Alox used the same box for decades. They produced the box in house to save money. This box was used into the '80s. The box held 24 yo-yos. See #130.

1690    **COCA-COLA KOOKY KAPS, '60s**
This box holds 48 yo-yos. It is from a '60s promotion by Coke. See #234.

1691    44    **CHAMPION • 10¢ BY SOCK-IT, '50s**
Sock-It and Fli-Back are the same company with different trademarks. See #1520.

1692    44    **CHEERIO CHAMPION, 59¢, '50s**
This box held six standard size models. See #200. The design on this box was used in Canada before and after WWII.

1693    **CHEERIO PRO 99, 39¢, '50s**
This box held twelve Pro 99 yo-yos.

1694    44    **DUNCAN BEGINNER, 19¢, '50s**
This box held 24 Beginner models. The box shown contains Kist soft drink yo-yos from an ad promotion produced by Duncan in the '50s.

1695    44    **DUNCAN BUTTERFLY, '60s**
This box held twelve Butterfly yo-yos. It was used in the late '50s and early '60s.

1696    **DUNCAN FLAMBEAU, '70s, '80s and '90s**
Several styles of this display box exist. These hold carded yo-yos and have limited graphics. Most Duncan Flambeau boxes currently have little collectible value.

1697    **DUNCAN GOLD SEAL • 25¢, '30s**
The graphics on this box show an old man and a little girl playing a yo-yo. The box is marked No. 77 and held 12 Gold Seal yo-yos. This box appears in the well-known "Our Gang" Duncan promotional photo. It is one of the earliest Duncan counter display boxes.

1698    **DUNCAN GOLD SEAL TOURNAMENT • 20-25-35¢, '30s**
The lid of this box features a graphic of a boy doing two-handed tricks in front of a globe. The boy is reportedly Tex Schutz at age 14. Schutz went on to become a professional Duncan demonstrator. The box is marked No. 77 and held 12 Gold Seal yo-yos. It was used for several years. The pre-printed price went up to 25 cents and then to 35 cents. This box has graphics similar to those on #1703. The value listed in the Appendix is for the 20 cent box.

1699    **DUNCAN JEWEL, '50s**
The lid graphics show the '50s style Mr. Yo-Yo with jewels for eyes. The box held 12 Duncan Jeweled yo-yos. See #361. This is one of the more desirable of the Duncan counter display boxes.

1700    44    **DUNCAN JUNIOR NO. 33 • 10¢, '30s, '40s and '50s**
Both big "G" and little "G" Beginner yo-yos were retailed out of this box. The box carries a copyright date of 1935 and holds 36 yo-yos.

1701    44    **DUNCAN METAL, '40s**
This box is marked with the Stock No. 88 and held 12 Duncan tin yo-yos.

1702    **DUNCAN O-BOY • 10¢, '30s**
This box appears in the well-known "Our Gang" Duncan promotional photo which can be seen in Malko's "One And Only Yo-Yo Book." The box is marked No. 44 and held 24 O-Boy yo-yos. This is one of Duncan's earliest counter display boxes.

1703    44    **DUNCAN SUPER TOURNAMENT, '60s**
The box held 12 Duncan Super Return Tops. A price sticker of 49 cents was placed over the original pre-printed price on the box lid. This box has graphics very similar to earlier Duncan Gold Seal Tournament boxes. See box #1698.

1704    44    **DUNCAN YO-YO RETURN TOPS, '60s**
This is rare box that held both yo-yos and string packs.

1705    **FLI-BACK,**
Fli-Back produced a variety of display boxes in the '50s, '60s and '70s. Most Fli-Back counter boxes have little collectible value at this time.

1706    44    **FLORES YO-YO TOPS • 15¢, '50s**
This box is often reported to be from the '20s, but it is in fact from the '50s when Flores restarted his company.

1707    **FRED FLINTSTONE AND YOGI BEAR, '70s**
The box held 12 sculpted head yo-yos, six of each style. See #642 and #659.

1708    **GIANT TEENY YO-YO • 29¢, '70s**
This is a display box from the '70s by Imperial Toy Co.

1709    **HEP MAGIC STRING BALL • 49¢, '50s**

1710    **HI-KER SPINMASTER • 39¢, '50s**
Hi-Ker boxes are uncommon. Spinmaster boxes were used to retail other Hi-Ker styles in addition to Spinmasters.

1711    **LUCK-E JA-DO • 25¢, '50s**
This is a rare display box from a non-promoted yo-yo line by Duncan. See #445.

1712    44    **ROYAL • 49¢**
This box held 12 Royal yo-yos. This box lists the names of champions, some of which were fictitious.

1713    **ROYAL, TOMMY & JOE, 60¢**
An example of this counter display box is shown next to this section's title.

*Continued ...*

*The History and Values of Yo-Yos*

## MEMORABILIA CATEGORIES

● COUNTER DISPLAY BOXES *Continued...*

1714   *44*   **ROY ROGERS DOUBLE R BAR RANCH, '50s**
A large number of these boxes, with yo-yos, were released after being found in a warehouse in the '80s. Each box held 12 yo-yos, individually wrapped in cellophane with an extra string. These boxes are common. See #184.

1715   **SAYCO, '60s - '90s**
The same counter display box has been used for 30 years without a change in graphics. The box holds 12 plastic Sayco yo-yos. Larry Sayegh, the company's founder, designed the box. For more information, see #1218.

1716   **TOM KUHN LASER CARVED, '90s**
This counter display box holds yo-yos that are individually boxed. The graphics are more interesting than on counter display boxes from other companies from the '80s and '90s.

## MISCELLANEOUS ITEMS

### ADVERTISING

Original sales catalogs, brochures and fliers are collectible. Value depends on the company, date and rarity. Values are listed for originals only, not reproductions.

1717A    *47*   (Flores advertising card), late '20s
1717B-E  *48*   (Festival catalogs), '70s and '80s
1717F    *48*   (Ad sheet • Duncan), '80s
A Duncan flasher seal from the Super Hero yo-yo series was attached on to this advertising sheet.
1717G-H  *48*   (Duncan catalogs)
1717I-J  *48*   (Ad sheets • Duncan), '80s

1718   *48*   (Bottle tag) **PEPSI COLA ROYAL YO-YO CONTEST, '50s**
These were hung from the necks of Pepsi bottles during contest promotions. The coupon gave the bearer a discount on a Royal Yo-Yo.

1719   (Cabbage Patch Kids • figurine), plastic, '80s
This 1985 Avon figurine features Cabbage Patch kid Billie Sue playing with a yo-yo.

1720   (Cartoons)
Some collectors collect yo-yo related cartoon clippings from newspapers or magazines. Original artwork for cartoons or animated cells is rare. The value depends on the cartoonist, character and age.

1721   (Cereal boxes)
Cereal boxes sometimes contained yo-yo premiums or offers for mail-in premiums. Cereal boxes that display yo-yo ads are considered collectible. Yo-Yo collectors must compete with cereal box collectors for these items. Two examples of yo-yos that had cereal box ads are Cocoa Puffs #307.14 and Sugar Crisp #307.54. A Sugar Crisp box with a yo-yo promotional offer is shown next to this section's title.

1722   *47*   **CERTIFICATE OF ENTRY, '30s**
For many contests in the '30s, entry required the participants to sell a subscription to the sponsoring newspaper. Participants usually received a free yo-yo and a certificate of entry in the contest. This was a winning situation for the promoters and the newspapers. The newspapers sold a lot of subscriptions and the promoters got free publicity and sold many yo-yos. These contest entry certificates are rare. The certificate shown is from the first major newspaper sponsored contest by a Hearst newspaper around 1930.

1723   *51*   (Clock yo-yo), early '90s
This clock was first released by Hummingbird in 1991. It came in four colors: natural wood, pink, green, and orange.

1724   (Comic) **DAFFY DUCK #143 • THE YO-YO BANDIT, '80s**
In this comic book, issue #143, the lead story is about Daffy Duck fighting the Yo-Yo Bandit.

1725   *46*   (Decal seal strips), Duncan 77 seals, '50s
In the field, demonstrators applied these seals to blank yo-yos.

1726   *46*   (Drinking glass) **YO MAN**
The sides of this plastic yo-yo feature flask shaped drinking glasses and a Yo Man super hero cartoon.

*Continued ...*

## MEMORABILIA CATEGORIES

1726.1        **(Drinking glass • 4-3/4 inch • Little Lulu series), '40s**
Little Lulu was a popular comic book character in the '40s, '50s and '60s. The glass has a picture of Wilbur Van Snobbe playing the yo-yo. It was one in a series of five Little Lulu glasses from the '40s.

1727    *56*    **(Entrant coupons), '50s**
These coupons were given out to contest entrants and were redeemable for free products. The company sponsoring the competitions provided these as awards. The coupons shown are from the '50s.

1728    *45*    **(Holster), leather, Yomega, '90s**

1729       **(Holster), leather, Kuhn, '90s**

1730       **(Holster), plastic, Playmaxx, '90s**
This is a belt clip designed to hold the Pro Yo.

1731       **(Kitty Cucumber series), '80s**
This was a series of novelty items featuring a cat playing a yo-yo. See also #1536. In addition to several yo-yos, there were tins, note pads, matchboxes and other items.

•    **Ornament Yo-Yo** - *See Holiday Christmas section #680.*

1732    *46*    **(Panda playing yo-yo), '60s**
This is a rare wind up mechanical doll that plays the yo-yo when in motion. See also #1735.

1733    *48*    **(Post cards)**
Two types of yo-yo related post cards exist, advertising post cards from manufacturers, see #1733A, and general post cards with yo-yos as the subject matter, see #1733B. The value of the card depends on the age and graphics.

1734       **(Puzzle • Huckleberry Hound with a yo-yo), '60s**

1735    *46*    **(Rabbit playing yo-yo), '60s**
This is a rare wind up mechanical doll that plays the yo-yo when in motion. See also #1732.

1736    *45*    **(Sengpiel supersonic yo-yo pack), '70s**
Linda Sengpiel was the first female professional yo-yo demonstrator. In 1951, she began demonstrating professionally for both Duncan and Pedro Flores. During those early years, she was known as the "Yo-Yo Queen."

In the '70s, Linda was contracted by a large publishing company to package her own trick book with a yo-yo and other items for retail sale. A total of 100,000 units were ordered. However, once the pack had been produced, the contracting company's sales staff was reluctant to promote this new product. Apparently, it was due to the large shelf space necessary to display the set, in combination with a low profit margin. The order was accepted, but before distribution could begin, the company shut down. The packs were held in a warehouse for a year before being returned to Sengpiel.

Shortly thereafter, an agreement was reached with Lawson, a food store chain, to release the Sengpiel Supersonic Yo-Yo Pack in the Northeast. A commercial promoting the pack was produced and aired during this period. The yo-yo kit was on an oversized blister card (16 x 11-1/2 inches) and included a Super Sonic Yo-Yo, three membership cards, four pin back pins, two stickers, a Festival string pack and a trick book, #1928. The kit retailed for $2.98. A less common, smaller version of the kit was also packaged to cut down on shelf space. This small card version had a fixed price of $1.49 and contained a yo-yo, two pin back buttons, and a string pack.

1737    *52*    **(Sign • Minnie Mouse with yo-yo), '80s**
This is a 15-3\4 " diameter wooden sign with the '30s style Minnie Mouse. It is not a licensed Disney product. It is believed to be French in origin and made in the late '80s.

1738       **(Trick cards • Cheerio), late '40s**
During the years 1939 to 1941, Cheerio in Canada made a set of 52 trick trading cards with one trick illustrated on each card. One card was given away in each string pack. A full box of yo-yos was awarded to anyone that brought in a complete set of 52 cards. There were some kids that managed to complete a set and redeem them for a box of Cheerio yo-yos. These cards are considered very rare. At this point, a complete set is not known to exist. For other trick cards, see #405.

1739    *46*    **(Truck • die cast 1955 Diamond-T flatbed ), 1996**
This is a limited edition (2000) 1:34 scale model produced by Eastwood Automobilia. The truck is decorated to resemble a Duncan factory truck. Although the truck is modeled after a 1955 style flatbed, the Duncan logo and phone number on the truck are in a '90s format. A Duncan Flambeau re-released wood Super Tournament Yo-Yo (#509) comes with the truck. The original retail price was $39.99.

1740    *46*    **(Watch • Mickey Mouse playing yo-yo • by Lorus), '90s**
A yo-yo goes up and down as the watch ticks.

1741       **(Wine cork), early '90s**
A miniature Tommy Smothers playing a yo-yo is mounted on this bottle stopper.

1742       **(Yo-Yo playing clown), '60s**
This 9 inch tall, battery operated Yo-Yo Playing Clown was produced by Alps in the '60s.

1743       **(Yo-Yo playing monkey), '60s**
This 12 inch tall, battery operated Yo-Yo Playing Monkey was produced by YM in the '60s.

1744       **(Yo-Yo figurine sculpture) DUNCAN #5077 (from the Tom Clark Collection), 1989**
This is a retired limited edition sculpture from the Tom Clark Gnome Collection. Each piece is numbered and dated. This sculpture is of a gnome named Duncan who is sitting on a yo-yo. The gnome is having trouble eating a doughnut he is holding because he's tied up in the yo-yo string. Twelve coins from different countries are at the base of the sculpture. By following the coins in sequence, you will go "Around the World."

## MEMORABILIA CATEGORIES

### PATCHES

Award patches have been given to contest winners since the 1930s. Patches have remained the most frequent award given to recognize yo-yoing skill. Not only were patches presented to contest victors, but they were also used as awards for increasing skill levels. The Duncan Co. is credited with the first use of award patches, but many other yo-yo companies issued patches.

A variety of styles have been made over the years and collectors find the unusual ones more desirable. Patches with early dates, embroidery vs. silk screening, three or more colors, unusual shapes or unique graphics are considered more desirable. Shield shaped patches are the most common design style. The patches listed are a small sampling of those used over the years. Values listed are for patches in near mint condition.

| # | Value | Description |
|---|---|---|
| 1745 | 50 | 15 GOLD AWARD (chevron shaped), '50s |
| 1746 | 50 | 15 OR 10 SILVER AWARD (chevron shaped), '50s |
| 1747 | | 1936 DUNCAN'S GOLD SEAL WINNER (shield shape), embroidered |
| 1748 | | 1937 DUNCAN WINNER (shield shape) |
| 1749 | | 1938 DUNCAN WINNER (shield shape) |
| 1750 | 49 | 1942 DUNCAN WINNER (shield shape), 5" x 5-3/4" |
| 1751 | | 1949 DUNCAN BRONZE AWARD |
| 1752 | 50 | 1950 DUNCAN EXPERT AWARD (two wings) |
| 1753 | | 1953 DUNCAN SKILL AWARD (one wing) |
| 1754 | | 1954 DUNCAN WINNER (shield shape • one gold star), 4" x 4-1/2" |
| 1755 | 49 | 1954 DUNCAN SKILL AWARD (one wing), 3-1/4" x 4-7/8" |
| 1756 | | 1955 DUNCAN WINNER (shield shape) |
| 1757 | 49 | 1957 DUNCAN CHAMPION (shield shape) |
| 1758 | | 1958 DUNCAN EXPERT (shield shape) |
| 1759 | | 1958 DUNCAN CHAMPION (shield shape) |
| 1760 | | 1994 NATIONAL CHAMPIONSHIP |
| 1761 | 50 | 1996 NATIONAL CHAMPIONSHIP |
| 1762 | 49 | 22 CHAMPION AWARD (star shaped), '50s |
| 1763 | 50 | 22 EXPERT AWARD DUNCAN (crown shaped), '50s |
| 1764 | 50 | 8 OR 5 BRONZE AWARD (chevron shaped), '50s |
| 1765 | 50 | AYYA MEMBER, embroidered, '90s |
| 1766 | 50 | CHEERIO 18 TRICK, '50s |
| 1767 | 49 | CHEERIO 9 TRICK, '50s |
| 1768 | 49 | CHEERIO BRONZE AWARD, '50s |
| 1769 | 49 | CHEERIO DISTRICT CHAMP, embroidered patch, 5-1/2" x 5", '50s |
| 1770 | 49 | CHEERIO SILVER AWARD, '50s |
| 1770.1 | 49 | CHEERIO SCHOOL INSTRUCTOR, '50s |
| 1771 | 49 | DUN CAN CAN YO-YO CAN YOU, '50s |
| 1772 | 50 | DUNCAN AMATEUR (silk screen of boy with a yo-yo), '80s |
| 1773 | 50 | DUNCAN AWARD (miniature pennant) embroidered felt, '50s |
| 1774 | 49 | DUNCAN BRONZE AWARD (star shape), '50s |
| 1775 | 49 | DUNCAN CHAMP AWARD, (non-fluted Olympic torch), embroidered, 4-1/4" x 4-1/4" |
| 1776 | 49 | DUNCAN CHAMPION (shield shape • 5 gold stars at top), 3" x 3-1/4", '60s |
| 1777 | | DUNCAN COCA-COLA CHAMPION (shield shaped), late '50s |
| 1778 | 50 | DUNCAN EXPERT (silk screen of boy with a yo-yo), '80s |
| 1779 | 50 | DUNCAN GOLD AWARD (shield shape), embroidered, 4-1/4" x 4", '60s |
| 1780 | | DUNCAN JUNIOR INSTRUCTOR 20, '50s |
| 1781 | 50 | DUNCAN MASTER (silk screen of boy with a yo-yo), '80s |
| 1782 | 50 | DUNCAN (Olympic Torch Fluted) CHAMP AWARD, '60s |
| 1783 | 49 | DUNCAN YO-YO TOP EXPERT (yo-yo embroidered in center), '50s |
| 1784 | 49 | (Felt pennant • small silk screened yo-yo champion), '60s |
| 1785 | 49 | FIRST PLACE FLAMBEAU (shield shaped), '70s |
| 1786 | | FIRST PLACE RETURN TOP CONTEST (shield shaped), '60s |
| 1787 | 49 | First Royal (shield shaped), silk screened, '70s |
| 1788 | | (Flying eagle champion sweater patch), '50s and '60s |

This is the most desirable award patch. The patch was originally awarded to contestants that could demonstrate proficiency at the eight basic two-handed tricks. It is similar in design to #1789, but much larger.

*Continued ...*

## MEMORABILIA CATEGORIES

1789  49  **DUNCAN, CHAMPION (flying eagle patch), 4" x 4-1/2", '70s - '90s**
The Flambeau version, used in 1971, is shown. This patch has been used by Duncan into the late '90s and has been sold retail with some Duncan Super (re-release) offers. See #509. The pre-Flambeau patch is the same design, but has a blue border instead of black.

1790    49  **HI-KER 5 TRICK, '50s**
1790.1  49  **HI-KER 15 TRICK AWARD, '50s**
1790.2  50  **HI-KER CHAMPION, '50s**
1790.3  49  **HI-KER LOOP AWARD, '50s**
1790.4  50  **HI-KER SENIOR INSTRUCTOR, '50s**
1791    50  **INTERNATIONAL ROYAL YO-YO CLUB (crown in center), '50s**
1792    50  **JUNIOR INSTRUCTOR (half circle), '50s**
1793    50  **I CAN YO-YO CAN YOU?, '50s**
1794    50  **KING OF YO-YOS CHAMPION (flying wings) ROYAL, '50s**
1795    49  **NO JIVE • 3 - 1 YO-YO, embroidered, '90s**
1796    50  **PRO YO CHAMPION, '80s**
1797    50  **ROYAL CHAMP (three colored shield), '50s**
1798    49  **ROYAL CHAMPION AWARD, silk screened, '60s**
1799    49  **ROYAL TOURNAMENT, silk screened, '70s**
1800    49  **ROYAL YO-YO CHAMP (eagle on yo-yo), '50s and '60s**
1801    49  **SECOND PLACE FLAMBEAU (shield shaped), '70s**
1802        **SECOND PLACE RETURN TOP CONTEST (shield shaped), '60s**
1803    50  **THIRD DUNCAN CONTEST, '60s**
1804        **THIRD DUNCAN CHAMPION, '60s**
1805        **THIRD PLACE RETURN TOP CONTEST (shield shaped), '60s**
1806    50  **TOM KUHN CUSTOM YO-YOS, '90s**
These were not award patches. The patches were sold retail in three color styles.

 **PINS**

Yo-Yo contest buttons or pins were given out to participants as early as 1931. Pins were produced as either souvenirs or awards for contests. Pins from earlier Duncan tournaments were produced as awards, but were not retailed.

1807  51  **1989 GREAT NORTHERN YO-YO CONTEST, 2-1/8"**
1808  51  **1990 GREAT NORTHERN YO-YO CONTEST, 2-1/8"**
1809  51  **1990 CALIFORNIA YO-YO CHAMPIONSHIP, 2-1/8"**
1810      **1991 WASHINGTON CHAMPIONSHIP, 2-1/8"**
1811  51  **1992 CALIFORNIA YO-YO CHAMPIONSHIP, 2-1/8"**
1812  51  **1992 TOM KUHN YO-YOS 15TH ANNIVERSARY CELEBRATION, 2-1/8"**
1813  51  **1993 NATIONAL CHAMPIONSHIP, enamel pin**
1814      **1993 NATIONAL CHAMPIONSHIP, 2-1/8"**
1815  51  **1994 NATIONAL CHAMPIONSHIP, enamel pin**
1816  51  **1994 NATIONAL CHAMPIONSHIP, 2-1/2"**
1817  51  **1995 NATIONAL YO-YO CHAMPIONSHIPS**
1818  51  **1996 NATIONAL YO-YO CHAMPIONSHIPS, enamel pin.**
1819  51  **1996 NATIONAL YO-YO CHAMPIONSHIPS, 2-1/8"**
1820      **AMERICAN YO-YO ASSOCIATION YO-YO MEMBER, 2-1/8"**
1821      **DUNCAN PLAYGROUND CHAMP, early '60s**
This Duncan award pin is made of brass.
1822  51  **DUNCAN EXPERT (flying wings), '50s**
This is thought to be the most desirable of the Duncan award pins.
1823  51  **DUNCAN EXPERT LOOP-THE-LOOP, 7/8", '50s**
Duncan produced a series of award pins in the '50s as playground contest prizes. These pins featured Mr. Yo-Yo performing specific tricks. They were awarded to contestants performing those specific tricks. It is unknown how many tricks were part of this pin series. All pins in this series are considered highly collectible.

*Continued ...*

## MEMORABILIA CATEGORIES

**PINS** *Continued ...*

1824  51  **DUNCAN YO-YO CONTESTANT PIN, NO. 77, 7/8", '50s**
This is the most common Duncan award pin. All contestants received this pin. They were not required to perform a perfect trick to be awarded this pin.

1825  51  **DUNCAN YO-YO EXPERT FORWARD PASS PIN, 7/8", '50s**
This is an award pin from the same series as #1823.

1826  51  **GENUINE DUNCAN FAMILY COLLECTION, 1-1/2", '90s**

1827  51  **GENUINE DUNCAN FAMILY COLLECTION, 1-3/4", 1992-1993**
This pin was created to advertise the Duncan Family Collection. It is sold only in the Chico Town Museum.

1828  51  **NO JIVE 3 - 1 YO-YO, 2-1/8", 1977-1997**
This Tom Kuhn pin celebrates the 20th year of the No Jive Yo-Yo.

1829  51  **YOMEGA YO-YO WITH A BRAIN, 2-1/4", '90s**

## POSTERS AND SHEET MUSIC

### POSTERS

Most yo-yo contest posters were of the "On Sale Here!" variety. A poster's value depends on the age, condition, graphics and company. The following list includes just a small sampling of posters produced over the years. For more information, see Contest Flyers under Miscellaneous Items.

1830  52  **1993 NATIONAL CHAMPIONSHIP**
This was the first, modern National Yo-Yo Championship. All modern National Championships have been held in Chico, California. Memorabilia from the first championship in 1993 is considered the most collectible.

1831  **1994 NATIONAL CHAMPIONSHIP**

1832  **CHEERIO, 18" x 7-3/4", red and white**
This poster shows a yo-yo and Cheerio string pack.

1833  52  **COCA-COLA BRAND YO-YOS SOLD HERE**

1834  52  **BUY YOUR DUNCAN YO-YO HERE, 17" x 17-1/2 ", late '50s**

1835  52  **(Duncan Flambeau) HEY KIDS MEET THE DUNCAN PROFESSIONAL HERE!, '70s**
Photos of the Duncan demonstrator hosting the contest were placed in the empty spaces on this poster. The posters with original demonstrator photos are considered more collectible.

1836  52  **DUNCAN • GENUINE DUNCAN YO-YO'S ON SALE HERE, '80s**

1837  52  **DUNCAN, round, 13" diameter, '50s**
This has the Broken Y 77 seal with Mr. Yo-Yo on a pedestal.

1838  52  **DUNCAN, round, 13" diameter, late '60s**
This poster reads "Duncan Yo-Yo on Sale Here."

1839  52  **DUNCAN, round 13" diameter, late '70s**
This poster reads "On Sale Here," and features Flambeau's rosy cheek Mr. Yo-Yo.

1840  52  **DUNCAN • YO-YO CONTEST HERE, early '70s**
This poster was included in the Duncan contest kit #398.1.

1841  52  **DUNCAN 1979 NATIONAL YO-YOLYMPICS**
This poster was included in the Yo-Yolympics contest kit #398.2.

### SHEET MUSIC

The yo-yo has been used as a theme for several songs. Sheet music of these tunes is considered collectible. Interesting cover graphics illustrating yo-yos increases the value of sheet music.

1842  **GO AND GET YOURSELF A YO-YO, Albert Selden, © 1955**

1843  **OH! NO! YO-YO, Winstead and Stafford © 1931**

1844  **THE YO-YO, Madge Williams © 1938**

1845  **YO-YO, Lowery Music Co. © 1971**
An example of this is shown next to this section's title.

*Continued ...*

## MEMORABILIA CATEGORIES

1846  *56*  YO-YO, Milton James no. 3807 © 1947
1847       YO-YO, The Osmonds MGM Records © 1966
1848       YO-YO, Tricks Piano Solo, William Gillock © 1959

### STRING PACKS

String packs made their first appearance in the late 1920s. It is unclear which company first realized the profit potential of replacement strings. Collectors give the credit to either Flores or Duncan for producing the first string packs. Cotton fiber string, having withstood the test of time, is the material of choice for yo-yo strings. Both Flores and Duncan experimented with silk strings so it is possible to find a string pack containing silk strings. Silk tends to burn through the axle quicker making it less suitable than cotton. Nylon strings were tried, but cotton remains the standard.

It is often difficult to date string packs because the artwork on the packaging was used for years without any design changes. The price on the pack can often serve as a guide to the age. The earliest Duncan string packs were three for 5 cents and then two for 5 cents. These prices continued until the mid '50s when prices jumped to three for a dime. By the '90s, string packs routinely sold for over one dollar.

The production of string packs increased profits for both demonstrators and companies. It also supported a small cottage industry of string cutters and packers. Until the advent of automatic string cutters and tiers, these workers made extra income by cutting and packaging string at home. In the factories, string was cut from large spindles, twisted and tied, then packed in the familiar glassine envelopes. These were then stapled, twelve packages per bundle, and sold by the gross to yo-yo companies.

Because string packs are routinely stapled, the small staple holes in the packs do not decrease the pack's value. However, large holes or staple rust marks could decrease the pack's value. The majority of collectors do not care whether or not a string pack contains the original strings. They generally will not pay any more for a full pack than an empty pack. Exceptions to this would be packs that have unusual strings, such as Peanut string packs which contain multi-colored strings.

1851  *53*  ALOX • FREE WHEELING TOP STRING (no price), polybag, saddle header card, '80s
1852  *53*  CANADA GAMES • 5 STRINGS (no price), cardboard folder, '90s
1853  *54*  CHEERIO OFFICIAL • 2 FOR 5¢, glassine envelope, '40s
           This pack carries a copyright date of 1939. This pack was retailed only in Canada.
1854  *54*  CHEERIO • BIG C YO-YO STRINGS • 2 FOR 10¢, paper envelope, '60s
           This string pack was released only in Canada. It is believed to be the last string pack made by the Canadian Cheerio Co.
1855       CHEERIO YO-YO STRINGS • 2 FOR 5¢, glassine envelope
1856  *54*  CHEERIO • OFFICIAL STRINGS (no price), glassine envelope, '40s and '50s
           This string pack was used by the U.S.A. Cheerio Company, Cheerio Toys and Games Inc. These companies produced yo-yos from 1946-54.
1857  *54*  CHEERIO • SUPER TWIST YO-YO STRINGS • 2 FOR 10¢, paper envelope, late '50s
           This is a Canadian Cheerio string pack not retailed in the United States.
1858  *53*  CHEMTOY-YO-YO REPAIR KIT (no price), polybag, '70s
           This pack contained two yo-yo strings.
1859  *53*  CHICO SUPERB • 2 FOR 5¢, glassine envelope, '50s
           This is the only Chico string pack known to have been produced.
1860  *53*  COCA-COLA STRINGS (no price), blister card, '90s
1861  *53*  DELLS BIG D • 2 FOR 10¢, plastic envelope, '60s
           This is the only Dell string pack known to have been produced.
1862  *54*  DUNCAN • 2 STRINGS • #3026 (no price), polybag
1863       DUNCAN • 2 FOR 5¢ (large globe • red), glassine envelope, '30s
           This is believed to be the second Duncan string pack produced. Collectors call this the "large globe pack."
1864  *54*  DUNCAN • 2 FOR 5¢ (large globe • yellow), glassine envelope, '30s
           This yellow color variation is more difficult to find than the red version. See #1863.

*Continued ...*

## MEMORABILIA CATEGORIES

● **STRING PACKS** *Continued ...*

1865  54  **DUNCAN • 2 FOR 5¢ (small globe • red), glassine envelope, late '30s - '50s**
This is the most familiar of the early Duncan string packs. The red pack is the most common color variation, but packs were also made in yellow. This pack was produced from 1937 through the middle '50s. Collectors call this the "small globe pack."

1866  54  **DUNCAN • 2 FOR 5¢ • EGYPTIAN FIBER CORD STRINGS, glassine envelope**

1867  54  **DUNCAN • 2 FOR 10¢, Flambeau, paper envelope, '70s**
This vertical string pack does not have an illustrated logo.

1868  54  **DUNCAN • 2 FOR 10¢ (horizontal pack • Mr. Yo-Yo), paper envelope, early '60s**
This is the most common of the early '60s Duncan string packs. It was used from around 1960 until the Duncan bankruptcy in 1965. This was the last string pack made by the Donald F. Duncan, Inc. All Duncan string packs made following this one were made by Flambeau. These string packs can sometimes be found stapled to a GAY Plastics display card. In the early '60s, GAY Plastics produced some plastic yo-yos for Duncan.

1869  54  **DUNCAN • 2 FOR 10¢ (horizontal pack • Mr. Yo-Yo), glassine envelope, early '60s**
This blue color variation is much more rare than the standard red and orange color. See #1868.

1870  54  **DUNCAN • 2 FOR 10¢ (vertical pack • Mr. Yo-Yo • Flambeau), paper envelope, '70s**
Vertical string packs such as this are less common than the standard horizontal string packs.

1871  54  **DUNCAN • 2 FOR 15¢ (vertical pack), glassine envelope, '70s**
Vertical string packs such as this are less common than the standard horizontal string packs.

1872  54  **DUNCAN • 2 FOR 15¢ (vertical pack), Flambeau, '70s**
There are two versions of this string pack, one in a glassine envelope and the other in a paper envelope.

1873  54  **DUNCAN • 3 FOR 5¢ (large globe • red), glassine envelope, early '30s**
This is believed to be the first string pack produced by Duncan. Collectors call this the "large globe pack"

1874  54  **DUNCAN • 3 STRINGS • WAX, 49¢, blister card, mid '70s**
This is a two-compartment blister card. One compartment held wax and the other held the string. This is the only Duncan pack known to have contained wax. Collectors call this the "Duncan Wax Pack."

1875  54  **DUNCAN • 4 STRINGS FOR 49¢, Flambeau, blister card, late '70s**
1876  54  **DUNCAN • 5 STRINGS FOR 49¢, Flambeau, polybag, #3026, '80s**
1877  54  **DUNCAN • 5 STRINGS (no price), Flambeau, blister card, '90s**
1878  54  **DUNCAN • 5 STRINGS (no price), Flambeau, blister card, '90s**
     ●   **Duncan String Pak Butterfly -** *See Duncan Butterfly section #333.*

1879  54  **DUNCAN SUPER CORD • 3 FOR 10¢, glassine envelope, '60s**
The production of this string pack began in the late '50s and continued until around 1960. This string pack has three different backs.

1880  54  **DUNCAN SWIVEL-EZE FINGER CONTROL ACCESSORY STRING PACK, (no price), Flambeau, '80s**
This was first retailed in 1979 as a blister carded string pack with four strings. A Velcro finger loop attached to a metal swivel is included in this pack. This accessory prevents the string from winding and unwinding. Production continued into the early '80s. L.E. Powell patented a similar device in 1882, but it is unknown if this original swivel ring was ever produced for sale.

1881  53  **FESTIVAL • 2 FOR 10¢, glassine envelope, '70s**
Production of this pack likely started in the late '60s and continued through the '70s. This is the only glassine envelope known to be used by the Festival Company. This string pack can also be found in the Sengpiel Supersonic Yo-Yo pack.

1882      **FLI-BACK • 2 FOR 5¢, cellophane envelope, '60s**
Care should be taken in handling any Fli-Back string packs because the plastic tends to crack easily. The strings should not be removed from this sealed pack.

1883  53  **FLI-BACK • 2 FOR 10¢, cellophane envelope, '70s**
This pack is much more common than the two for 5 cents version.

1884  47  **FLORES • 2 FOR 10¢ (Yo-Yo Wonder Toy logo), 1929**
This is the second string pack made by Flores. It was produced after his move to Los Angeles, California.

1885      **FLORES • 3 FOR 10¢, 1929**
The pack reads "Flores Yo-Yo Corporation, Phone Gladstone 3505, 6301 Sunset Blvd., Hollywood CA."

1886  47  **FLORES • 10¢ (Yo-Yo logo), 1929**
This was the first string pack produced by Flores from his original location in Santa Barbara, California. The price was 10 cents, but the number of strings enclosed was not shown on the pack.

1887  47  **FLORES • 2 FOR 5¢, 1929-1930**
This was the last style of string pack produced by Flores from his Illinois factory.

1888      **GOODY MAGIC TWIRLER STRINGS • 3 FOR 5¢, '30s**
Goody string packs are uncommon. Most three for 5 cents packs from other companies are thought to be from the '30s, but the production dates of this pack are unknown. It is unclear if other Goody string packs were produced. An example of this string pack can be seen in Malko's "One And Only Yo-Yo Book."

1889  53  **HI-KER STRINGS • 2 FOR 5¢, glassine envelope, '50s**
This is the only string pack currently documented to have been released by Hi-Ker.

1890  53  **HUMMINGBIRD • 5 YO-YO STRINGS (no price), paper envelope, '90s**
1891      **JEWEL YO-YO STRINGS • 2 FOR 5¢, glassine envelope, early '50s**

*Continued ...* ▶

## MEMORABILIA CATEGORIES

1892        **KNIBB TOP CORD • 10 YARDS FOR 15¢**, paper envelope, '50s
           Most string pack strings come wound, cut, and tied, but this pack requires the buyers to do it themselves. KNIBB Co. made plastic pinball game yo-yos in the '50s.

1893   *53*   **PARKER REPLACEMENT STRINGS • 2 STRINGS PER PACK** (no price), paper envelope, '80s
           These strings were made by a Canadian yo-yo manufacturer.

1894   *53*   **PEANUTS • 4 FOR 25¢**, glassine envelope, early '80s
           This is one of the few string packs that featured colored strings. Replacement strings were made for the Hallmark Peanuts series of yo-yos which were retailed with colored strings. One string of each color was included, "Fuss Budget Blue," "Good Grief Green," "Red Baron Red" and "Great Pumpkin Orange."

1894.1   *53*   **PRO SPIN STRING, 3 for 49¢**, polybag

1895   *53*   **ROAD RUNNER • 2 STRINGS** (no price), paper envelope, '70s
           These colored string yo-yos were manufactured by the John Hart Toy Co. for the Road Runner Yo-Yo. The pack is written in English on one side and French on the other.

1896   *53*   **ROYAL CHAMPION • 2 FOR 5¢ • JOE RADOVAN** (in crown), glassine envelope, '50s
           This string pack pictures Joe Radovan centered in a crown. Joe Radovan was Royal's founder.

1897   *53*   **ROYAL CHAMPIONS • 2 FOR 5¢, TOMMY AND JOE**, glassine envelope, '50s
           Joe was Royal's founder, Joe Radovan, and Tommy was his son.

1898   *53*   **ROYAL WORLD CHAMPIONS ON PARADE • 2 FOR 5¢**, glassine envelope, '50s
           This is perhaps the rarest of the Royal string packs. This string pack lists the names of champions, some of which were fictitious.

1899   *53*   **ROYAL • 3 FOR 19¢**, glassine envelope, '70s
1900   *53*   **ROYAL • 2 FOR 15¢**, glassine envelope, '70s
1901   *53*   **ROYAL • 3 FOR 10¢**, glassine envelope, '60s
1902   *53*   **ROYAL • 2 FOR 10¢**, glassine envelope, Style 1, '70s
1903   *53*   **ROYAL • 2 FOR 10¢**, glassine envelope, Style 2, '70s
1904        **RUSSELL, COCA-COLA STRINGS • 3 STRINGS** (no price), polybag, '80s
           Russell has made several foreign yo-yo string packs for Coca-Cola.

1905   *53*   **STREAMLINE TOPS • 3 FOR 5¢**, glassine envelope, '30s
           This is the only known string pack for Kaysons yo-yos.

1906        **SAYCO**, glassine envelope, '60s and '70s
           This is a very rare string pack used by Sayco. Although Sayco still produces yo-yos, they only produced string packs in the '60s and '70s.

1907   *53*   **TWIN TWIRLER (Kusan's) 2 FOR 10¢**, glassine envelope, early '60s
           This is the only known string pack for the twin twirlers.

1908   *53*   **GENUINE EGYPTIAN FIBER CORD, YO-YO STRINGS • 2 FOR 5¢ • D. BRAGADO**, glassine envelope, '50s
           This string pack is likely from the '50s. It is not currently known which company used this pack.

1909   *53*   **YOMEGA • REPLACEMENT STRINGS** (no price), paper pack, '90s

## TRICK BOOKS

What can be considered the first trick sheet dates back to the late '20s when Flores illustrated how to throw the yo-yo to simply make it go up and down. As early as the 1930s, yo-yo tricks were given names. "Around the World" and "The Waterfall" are the earliest trick names documented in print. The first true Duncan trick book, published in 1947, illustrated those tricks needed for Duncan competitions. It featured the Mr. Yo-Yo character demonstrating each trick. Since that time, many different companies have released trick books.

1910   *55*   **CHEERIO • THE ART OF YO-YO PLAYING • 15¢**, '50s
           The text and drawings in this book suggest it was produced shortly after Duncan acquired the Cheerio name. The preface and history of the yo-yo, as well as the trick descriptions, are identical to those in the Duncan yo-yo book. The Duncan Yo-Yo Man character is also used in this book. Photographs in this book show kids with Cheerio yo-yos and Cheerio patches. The book is 48 pages long and has a Cheerio Club member card that can be torn off the back of the book, signed, and carried in the wallet. The size is 4-3/4" x 3-3/4".

*Continued ...*

## MEMORABILIA CATEGORIES

● **TRICK BOOKS** *Continued ...*

**1911**  *55*  **DUNCAN • BUTTERFLY YO-YO TOURNAMENT TRICKS, 1963**
This book is included in the personalized Duncan Butterfly kit #401. It is a 16-page book of basic yo-yo tricks. The size is 7-1/8" x 5-1/8".

**1912**  **DUNCAN • CHAMPION YO-YO TRICKS • 10¢, 1961**
This is an illustrated 16 page color booklet released in 1961. The book contains basic yo-yo tricks.

**1913**  *55*  **DUNCAN FLAMBEAU • GIANT BOOK OF YO-YO TRICKS • 50¢, early '70s**
This trick book is similar to the 1961 Duncan Giant Book of Yo-Yo Tricks, #1914. The cover is nearly identical except for the pre-printed price and the change of the name from Duncan to Flambeau. It is believed to be the first trick book released by the Flambeau Corp. This book has fewer yo-yo tricks than the 25 cent version of the Giant Yo-Yo Book, but fills in the extra space with top tricks. There is an ad for Duncan Tops showing the five new tops produced (Imperial, Duncan Twin Spin, Hummer, Sir Duncan, and Beginners). The book has 36 color pages and is 7-1/8" x 5-1/8" in size.

**1914**  *55*  **DUNCAN • GIANT BOOK OF YO-YO TRICKS • 25¢, 1961**
This is a 32 page trick book. It is 5" x 7".

**1915**  *55*  **DUNCAN • SUGAR CRISP YO-YO OLYMPICS, 1980**
This is a 32 page black and white trick book. It contains contest rules for the Sugar Crisp Yo-Yolympics. It also has coupons and an ad for the Super Sugar Crisp Duncan Yo-Yolympics promotion yo-yo.

**1916**  *55*  **DUNCAN • THE ART OF YO-YO PLAYING • 10¢, 1947**
This 52 page book was the first Duncan trick book produced. The cover features a statue of Mr. Yo-Yo and has Duncan's Little "G" Tournament Yo-Yo seal in the upper right hand corner. The cover is in color, but the contents are black and white. The size of the book is 5-1/8" x 4-5/8".

**1917**  *55*  **DUNCAN • THE ART OF YO-YO PLAYING • 10¢, 1950**
This version is similar to the first Art of Yo-Yo Play book, #1916. The cover is the same as the first book with one exception. In the right upper corner, there is a Duncan 77 seal. The tricks and descriptions are the same, but the photographs are different. Like #1916, this book is 52 pages, has black and white contents, and is the same size. It was released in 1950, but continued to be printed for several years.

**1918**  *55*  **DUNCAN • TRICK BOOK, 1962**
This is a 24 page, color, trick book that was given away in Duncan's multi toy variety pack in the early '60s. The pack contains a Little Ace yo-yo, Duncan Tornado top, and a Duncan handball (which is a version of the paddleball). The book features Duncan yo-yo tricks, Spinning top tricks and handball tricks. The size is 5-1/8" x 7-1/8".

**1918.1**  **DUNCAN • TOURNAMENT TRICKS • 10¢, 1961**
This is an illustrated color trick book released in 1961 at the same time as #1912. This book contains basic tricks as well as instructions on more advanced tricks. In addition to trick instructions, the official contest rules and score sheets are included.

**1919**  *55*  **DUNCAN • YO-YO & SPINTOP TRICK BOOK, '80s**
This is a 48 page, color, trick book with a copyright date of 1985. On the upper left hand corner of the cover, there is a pre-printed price of $1.50. This book contains mostly yo-yo tricks with a few top tricks on the last two pages. The size of the book is 5" x 7".

**1920**  **DUNCAN • YO-YO & SPINTOP TRICK BOOK, '90s**
This book is nearly identical to the 1985 version #1919 except there is no pre-printed price on the cover. However, there is a bar code on the back cover. The book carries a copyright date of 1992. The Duncan Company's address was changed to Middlefield, OH. (The old address was Baraboo, WI.)

**1921**  *55*  **DUNCAN • YO-YO TRICK BOOK • YO-YO PLAYERS INTERNATIONAL, early '80s**
This is a 48 page color trick book with a date of 1979.

**1922**  **IT'S YO-YO TIME, 1996**
This 146 page book was written by Stuart Crump, Jr., the editor of "The Yo-Yo Times." It features 300 tricks, a few of them illustrated, culled from the pages of "The Yo-Yo Times." Published by: Creative Communications, Inc., PO Box 1519-IYT, Herndon, Va, 20172.

**1923**  **KLUTZ • YO-YO BOOK, by John Cassidy, '90s**
This is an 83 page instructional book by John Cassidy. It originally came with a wooden tournament shaped yo-yo, with no seal. Later, the yo-yo was changed to have a maple leaf imprint seal followed by a metallic paint maple leaf seal. More recent releases have a plastic, flywheel shaped yo-yo with pocket rocket on the seal. Published by: Klutz Press Inc., 2121 Staunton Ct., Palo Alto, CA 94306.

**1924**  **KUHN • SB2 FLIGHT MANUAL, '90s**
This is a 24 page trick book that came with the original Silver Bullet yo-yo. The book contains black and white photographs. The book size is 4-1/4" x 5-1/2".

**1925**  **MAD JACK YO-YO, by Mike Knowles, 1996**
This publication is one in a series of Mad Jack instructional books. This book uses cartoons and humorous text to describe basic yo-yo tricks. It is in color, 48 pages and retails for $10.95. The book comes with a wood yo-yo attached.

**1926**  *55*  **OH OH TERRIFIC YO-YO • FRANCO-AMERICAN**
This Duncan trick book was distributed during the Franco-American yo-yo promotion. It is a black and white book, size 5" x 7".

*Continued ...*

*Lucky's Collectors Guide To 20th Century Yo-Yos*

## MEMORABILIA CATEGORIES

1927   *55*    **POST • CHAMPIONSHIP TRICKS, '50s**
This rare, black and white, 16 page miniature trick book was produced by Duncan. It came with a wooden Rainbow yo-yo as part of a Post Cereal mail in promotion. It featured the original Mr. Yo-Yo character demonstrating each trick.

1928   *55*    **SENGPIEL • OFFICIAL YO-YO BOOK, 1972**
This is a round "yo-yo shaped" 20 page trick book. It has a color cover with black and white contents. It retailed on a large blister card with a yo-yo, pin back buttons, and string packs. The book is 8 inches in diameter. See also #1736.

1928.1    **THE AMAZING YO-YO, by Ross R. Olney, 1980**
This is one of the few yo-yo trickbooks available in hardback. Yo-Yo expert George Humphreys is featured, in photographs, demonstrating common yo-yo tricks.

1929   *55*    **THE LITTLE BOOK OF YO-YOS, by Stuart Crump, Jr. (Professor Yo-Yo), 1997**
This 126 page miniature book comes with a wooden midget yo-yo. The size of the book is 3-1/4 inches by 3 inches. The yo-yo does not have a seal. Basic and advanced tricks are illustrated in the book.

1930   *55*    **THE ONE AND ONLY YO-YO BOOK, by George Malko, 1978**
There are 20 tricks shown in the appendices of this book, but it was not intended as a trick book. This is the only published book devoted to the history of the yo-yo. It features many black and white photos and prints from the early history of the yo-yo. Of interest to collectors are the many images of yo-yo memorabilia from the Duncan Collection. This book is highly coveted by yo-yo collectors and is difficult to acquire.

The impetus for "The One and Only Yo-Yo" was an article published by Malko in 1970 in the LithOpinon Number 18 titled "Will There Ever Be Another Yo-Yo Champ?" The article was received so well that Avon books agreed to have Malko turn the project into a book. Twelve thousand copies were printed in 1978. As a special promotion for the book, Tom Parks, a Duncan demonstrator, performed in a book store when it was initially released.

1931    **THE ORIGINAL DUNCAN YO-YO AND SPIN TOP TRICK BOOK, 1996**
This is a re-release of the book "The Art of Yo-Yo Playing." See #1916. The cover changed but the contents are the same. This was retailed in conjunction with the re-release of the Duncan Super Tournament #509.

1932    **WORLD ON A STRING, by Helane Zeiger**
This is a 173 page, black and white book with a colored cover. Tricks are demonstrated with black and white photographs. The second printing has an introduction and update by Dr. Tom Kuhn. An example of this book is shown next to this section's title.

1933   *55*    **YANCY'S YO-YO TRICK BOOK, Humphrey, 1975**
This is a four page trick book produced by the Humphrey Corp. in the early '70s.

1934   *55*    **YO-YO SECRETS • BY WORLD FAMOUS MR. YO-YO, 1971**
This is a 20 page black and white trick book produced for the Festival Yo-Yo Company. It was written by Bob Rule, a well-known professional demonstrator. The book contains many photographs and is 5-1/2" x 8-1/2" in size.

# Appendix

# RESOURCES

## ASSOCIATIONS:

### American Yo-Yo Association (AYYA)
This is a National Association promoting yo-yo competitions and collecting. Members include players, collectors, students, adults, manufacturers and yo-yo professionals. The AYYA newsletter and Yo-Yo Times comes with membership. Published since 1994. Annual membership $20.00. American Yo-Yo Association (AYYA), 627 163rd St., Spanaway, Wash., 98387.

## MUSEUMS:

### International Museum of Yo-Yo History & Pro Yo Training Center
Located 2947 E. Grant, Decorators Square Shopping Center, Tucson, Ariz. 85716, this museum exhibits over 1000 yo-yos, and offers yo-yo instructional classes. Open: Mon. - Fri. 10 AM - 6 PM, Sat. 10 AM- 4PM, and Sun. 12 noon - 5 PM. For more information (520) 322-0100.

### National Yo-Yo Museum
This museum is located in the Bird in Hand store, Chico, Calif. It has the largest display of yo-yos including the Duncan Family Collection. The museum also has many one-of-a-kind items. The store retails a wide variety of modern yo-yos and yo-yo related items. All serious collectors eventually make the pilgrimage to Chico. The Museum's curator is Bob Malowney. Open to the public: Mon. - Sat. 10 AM -5 PM, and 11 AM - 4 PM on Sunday. Free admission. Information (916) 893-0545. National Yo-Yo Museum, 320 Broadway, Chico, California.

### Spinning Tops Museum
This museum exhibits over 1000 tops, yo-yos and gyroscopes. The museum's curator is Judith Schulz. Programs are by appointment only. Call (414) 763-3946 for details. Spinning Tops Museum, 533 Milwaukee Avenue, Burlington, Wis., 53105.

## PUBLICATIONS:

### FIEND Magazine
A quarterly publication featuring ultra extreme yo-yo, top and juggling experts. Collector articles are also included in each issue. For subscription information, call (850) 385-6463.

### Yo-Yo Times
This is a quarterly publication edited by Stuart and Jodi Crump. It features all types of yo-yo related information including contests, yo-yo related stories, new tricks, collecting, letters and classified ads. Published since 1988. Annual subscription $18.00. Subscription comes with full membership to the AYYA. Yo-Yo Times, 1519, Herndon, Va., 22070.

### Spin-Offs
This is the publication of The Spinning Top Museum. The magazine has a variety of historic references and stories on tops, yo-yos and gyroscopes. For subscription information, call (414) 763-3946.

### AYYA Newsletter
This is the official newsletter of the American Yo-Yo Association. It is published twice a year and comes with a membership to the AYYA.

### Nobel Disk
This quarterly newsletter, edited by Bill Alton, promotes yo-yo play in the New England area. For subscription information, contact Bill Alton 1, 32 Middle Street # 11, Portsmouth, N.H., 03801.

## WEB SITES:

The following web site is maintained by the author. It provides links to yo-yo related sites of interest to collectors:

www.yo-yos.net

# APPENDIX

 **VALUES**

Values are listed in numerical order by the item's reference number. Values are given for packaged items, those in near mint condition, and those in good condition. For assistance in calculating values of items of different grades, see the Value Formula, explained in the Grading System section of Collecting and History. If the item did not come packaged or if the presence of packaging is unknown, there will be no value listed in the packaged column. If a value is less than one dollar, there will be a "-" in the column. If the yo-yo has been retailed within the last two years, no value will be given, only the word "Retail."

## YO-YO VALUES KEY

| | | |
|---:|---|---|
| **P** | = | Packaged |
| **NM** | = | Near Mint |
| **G** | = | Good |
| **Rare** | = | Limited quantities and unlikely to be sold |
| **Retail** | = | Product has been retailed within the last two years |
| **Varies** | = | Multiple items with varying prices |
| **-** | = | Valued for less than one dollar |

# APPENDIX

**P** = PACKAGED    **NM** = NEAR MINT    **G** = GOOD    **-** = VALUE LESS THAN $1

| ITEM NUMBER | P | NM | G | ITEM NUMBER | P | NM | G | ITEM NUMBER | P | NM | G |
|---|---|---|---|---|---|---|---|---|---|---|---|
| 1 | 16 | 9 | 3 | 56 | | 5 | 1 | 111 | | 10 | 3 |
| 2 | | 38 | 19 | 57 | | 5 | 1 | 112 | | 65 | 30 |
| 3 | | 28 | 8 | 58 | | 5 | 1 | 113 | | 14 | 5 |
| 4 | | 7 | 2 | 59 | | 2 | - | 114 | | 10 | 2 |
| 5 | | 45 | 20 | 60 | | 3 | - | 115 | | 10 | 2 |
| 6 | | 45 | 20 | 61 | 13 | 12 | 2 | 116 | | 12 | 3 |
| 7 | | 300 | 100 | 62 | | 4 | - | 117 | | 14 | 3 |
| 8 | | 300 | 100 | 63 | | 1 | - | 118 | 12 | 10 | 2 |
| 9 | | 45 | 20 | 64 | | 2 | - | 119 | | 10 | 2 |
| 10 | | 45 | 20 | 65 | | 10 | 3 | 120 | | 8 | 2 |
| 11 | | 1 | - | 66 | 38 | 12 | 3 | 121 | | 12 | 4 |
| 12 | | 1 | - | 67 | | 11 | 4 | 122 | | 40 | 20 |
| 13 | | 1 | - | 68 | | 28 | 10 | 123 | | 30 | 15 |
| 14 | | 10 | 2 | 69 | | 33 | 12 | 124 | | 14 | 4 |
| 15 | | 12 | 4 | 70 | | 15 | 5 | 125 | | 30 | 14 |
| 16 | 14 | 12 | 4 | 71 | | 5 | 1 | 126 | | 16 | 5 |
| 17 | | 1 | - | 72 | | 18 | 7 | 127 | | 25 | 10 |
| 18 | | 4 | - | 73 | 12 | 9 | 2 | 128 | | 35 | 14 |
| 19 | | 3 | - | 74 | | 28 | 10 | 129 | | 45 | 18 |
| 20 | | 3 | - | 75 | | 15 | 5 | 130 | | 16 | 5 |
| 21 | | 1 | - | 76 | | 12 | 4 | 131 | | 30 | 6 |
| 22 | | 1 | - | 77 | | 28 | 10 | 132 | Retail | | |
| 23 | | 12 | 4 | 78 | | 12 | 3 | 133 | Retail | | |
| 24 | | 8 | 3 | 79 | 6 | 5 | 1 | 134 | Retail | | |
| 25 | | 6 | 2 | 80 | | 12 | 3 | 135 | Retail | | |
| 26 | | 2 | - | 81 | | 10 | 2 | 136 | Retail | | |
| 27 | | 12 | 4 | 82 | | 10 | 3 | 137 | 12 | 5 | 1 |
| 28 | | 1 | - | 83 | 12 | 10 | 2 | 138 | 7 | 3 | 1 |
| 29 | | 1 | - | 84 | | 10 | 2 | 139 | | 15 | 5 |
| 30 | 8 | 3 | 1 | 85 | | 14 | 5 | 140 | 7 | 3 | 1 |
| 31 | | 1 | - | 86 | | 12 | 3 | 141 | 7 | 3 | 1 |
| 32 | 6 | 2 | - | 87 | | 22 | 9 | 142 | 12 | 5 | 1 |
| 33 | | 5 | 1 | 88 | | 12 | 3 | 143 | | 35 | 15 |
| 34 | | 5 | 1 | 89 | | 12 | 3 | 144 | | 20 | 5 |
| 35 | | 4 | - | 90 | | 12 | 3 | 145 | | 15 | 7 |
| 36 | | 3 | - | 91 | | 18 | 6 | 146 | | 10 | 5 |
| 37 | | 4 | - | 92 | | 12 | 3 | 147 | | 18 | 9 |
| 38 | 5 | 4 | 1 | 93 | | 3 | 1 | 148 | | 50 | 20 |
| 39 | 10 | 6 | 2 | 94 | | 45 | 25 | 149 | 40 | 20 | 6 |
| 40 | | 2 | - | 95 | 12 | 10 | 2 | 150 | 70 | 60 | 30 |
| 41 | | 2 | - | 96 | | 10 | 3 | 151 | 40 | 20 | 6 |
| | | | | 97 | | 6 | 2 | 152 | 50 | 40 | 20 |
| 43 | | 7 | 2 | 98 | | 8 | 2 | 153 | | 8 | 2 |
| 44 | | 5 | 1 | 99 | | 10 | 2 | 154 | 8 | 5 | 1 |
| 45 | | 5 | 1 | 100 | | 12 | 3 | 155 | 25 | 15 | 6 |
| 46 | | 1 | - | 101 | | 10 | 2 | 156 | 12 | 6 | 2 |
| 47 | | 1 | - | 102 | | 3 | 1 | 157 | 12 | 12 | 3 |
| 48 | | 4 | - | 103 | | 10 | 2 | 158 | 16 | 8 | 3 |
| 49 | | 8 | 3 | 104 | | 8 | 2 | 159 | 12 | 6 | 2 |
| 50 | 16 | 12 | 3 | 105 | | 10 | 2 | 160 | 1 | - | - |
| 51 | 14 | 12 | 3 | 106 | | 8 | 2 | 161 | 3 | 2 | - |
| 52 | | 8 | 2 | 107 | | 10 | 2 | 162 | 12 | 8 | 2 |
| 53 | | 8 | 2 | 108 | | 18 | 8 | 163 | | 55 | 20 |
| 54 | | 9 | 2 | 109 | | 22 | 10 | 164 | | 90 | 35 |
| 55 | | 1 | - | 110 | | 12 | 3 | 165 | 5 | 2 | 1 |

Lucky's Collectors Guide To 20th Century Yo-Yos

# APPENDIX

**P** = PACKAGED  **NM** = NEAR MINT  **G** = GOOD  **-** = VALUE LESS THAN $1

| ITEM NUMBER | P | NM | G |
|---|---|---|---|
| 166 | 18 | 18 | 6 |
| 167 | 16 | 10 | 4 |
| 168 | 4 | 4 | 1 |
| 169 |  | 85 | 35 |
| 170 | 22 | 12 | 5 |
| 171 | 90 | 20 | 6 |
| 172 | 4 | 4 | 1 |
| 173 |  | 10 | 2 |
| 174 |  | 10 | 3 |
| 174.1 | 12 | 6 | 2 |
| 175 | 22 | 14 | 4 |
| 176 | 18 | 10 | 3 |
| 177 | 8 | 8 | 2 |
| 178 | 15 | 10 | 3 |
| 179 |  | 12 | 3 |
| 180 |  | 5 | 2 |
| 181 | 12 | 6 | 2 |
| 182 | 35 | 20 | 5 |
| 183 | 18 | 12 | 4 |
| 183.1 |  | 15 | 5 |
| 183.2 | 18 | 12 | 4 |
| 183.3 | 18 | 10 | 3 |
| 184 | 18 | 18 | 5 |
| 185 | 10 | 10 | 3 |
| 186 | 12 | 6 | 2 |
| 187 | 15 | 10 | 3 |
| 188 | 28 | 15 | 5 |
| 188.1 | 14 | 8 | 2 |
| 189 | 10 | 6 | 2 |
| 190 | 25 | 10 | 3 |
| 191 |  | 1 | - |
| 192 | 18 | 8 | 3 |
| 193 | 7 | 6 | 1 |
| 194 |  | 60 | 25 |
| 195 |  | 50 | 20 |
| 195.1 |  | 60 | 20 |
| 196 |  | 55 | 20 |
| 197 |  | 50 | 15 |
| 198 |  | 75 | 25 |
| 199 |  | 80 | 28 |
| 200 |  | 75 | 30 |
| 201 |  | 80 | 32 |
| 202 |  | 185 | 75 |
| 203 |  | 65 | 20 |
| 203.1 |  | 65 | 20 |
| 204 |  | 185 | 70 |
| 204.1 |  | 60 | 20 |
| 205 |  | 175 | 50 |
| 205.1 |  | 185 | 60 |
| 205.2 |  | 175 | 40 |
| 205.3 |  | 175 | 40 |
| 206 |  | 90 | 36 |
| 206.1 |  | 80 | 35 |
| 207 |  | 80 | 32 |
| 207.1 |  | 65 | 22 |

| ITEM NUMBER | P | NM | G |
|---|---|---|---|
| 208 |  | 70 | 25 |
| 208.1 |  | 75 | 25 |
| 208.2 |  | 70 | 25 |
| 209 |  | 80 | 30 |
| 209.1 |  | 70 | 25 |
| 209.2 |  | 70 | 25 |
| 210 | 16 | 8 | 2 |
| 210.1 | 18 | 10 | 3 |
| 210.2 | 14 | 7 | 2 |
| 211 |  | 50 | 20 |
| 212 |  | 35 | 10 |
| 213 |  | 40 | 10 |
| 214 |  | 55 | 22 |
| 215 | 55 | 45 | 18 |
| 216 |  | 50 | 20 |
| 217 | 45 | 40 | 14 |
| 218 |  | 16 | 5 |
| 218.1 | 15 | 10 | 3 |
| 218.2 |  | 28 | 10 |
| 219 | 25 | 16 | 5 |
| 220 |  | 1 | - |
| 221 |  | 2 | - |
| 222 |  | 1 | - |
| 223 |  | 1 | - |
| 224 |  | 8 | 2 |
| 225 |  | 50 | 15 |
| 226 | 14 | 4 | 2 |
| 227 |  | 7 | 3 |
| 228 |  | 5 | 2 |
| 229 |  | 6 | 2 |
| 230 |  | 40 | 18 |
| 231 |  | 30 | 8 |
| 232 | 16 | 6 | 2 |
| 233 | 14 | 4 | 2 |
| 234 |  | 14 | 5 |
| 235 |  | 120 | 50 |
| 236 | Rare |  |  |
| 237 |  | 120 | 50 |
| 238 |  | 6 | 2 |
| 239 |  | 10 | 4 |
| 240 |  | 100 | 40 |
| 241 |  | 100 | 35 |
| 242 |  | 35 | 10 |
| 243 | Retail |  |  |
| 244 |  | 8 | 2 |
| 245 |  | 5 | 2 |
| 246 | 24 | 14 | 5 |
| 247 |  | 15 | 6 |
| 248 | 24 | 14 | 5 |
| 249 |  | 8 | 2 |
| 250 | 22 | 12 | 4 |
| 251 | 17 | 7 | 3 |
| 252 | 28 | 18 | 6 |
| 253 |  | 5 | 2 |
| 254 |  | 7 | 2 |

| ITEM NUMBER | P | NM | G |
|---|---|---|---|
| 255 |  | 8 | 2 |
| 256 |  | 8 | 2 |
| 257 |  | 3 | 1 |
| 258 | Retail |  |  |
| 259 | Retail |  |  |
| 260 | 35 | 8 | 3 |
| 260.1 | 29 | 14 | 5 |
| 260.2 | 29 | 12 | 3 |
| 260.3 | 27 | 12 | 4 |
| 260.4 | 27 | 12 | 4 |
| 260.5 | 27 | 12 | 4 |
| 260.6 | 29 | 12 | 4 |
| 260.7 | 24 | 10 | 3 |
| 260.8 | 27 | 12 | 4 |
| 260.9 | 24 | 12 | 3 |
| 260.10 | 24 | 10 | 3 |
| 260.11 | 35 | 17 | 6 |
| 261 | 12 | 10 | 3 |
| 262 | 8 | 8 | 2 |
| 263 | 20 | 10 | 4 |
| 264 | 20 | 10 | 4 |
| 265 | 30 | 18 | 4 |
| 266 | 10 | 3 | 1 |
| 267 | 7 | 6 | 2 |
| 268 | 25 | 12 | 5 |
| 269 | Retail |  |  |
| 270 | 30 | 20 | 7 |
| 271 | 20 | 10 | 4 |
| 272 | 14 | 7 | 2 |
| 273 |  | 100 | 40 |
| 280 |  | 10 | 3 |
| 281 | 4 | 4 | 1 |
| 282 |  | 10 | 3 |
| 283 | 45 | 25 | 8 |
| 284 | Retail |  |  |
| 285 | 15 | 8 | 2 |
| 286 | 20 | 10 | 4 |
| 287 | 20 | 10 | 4 |
| 288 | 30 | 18 | 8 |
| 289 | Retail |  |  |
| 290 | 45 | 25 | 8 |
| 291 | 27 | 12 | 4 |
| 292 | 25 | 10 | 3 |
| 293 | 285 | 100 | 30 |
| 294 | 55 | 35 | 12 |
| 295 | 15 | 8 | 3 |
| 296 | 30 | 20 | 10 |
| 297 | Retail |  |  |
| 298 | 14 | 7 | 3 |
| 299 | 20 | 10 | 4 |
| 300 | 18 | 10 | 3 |
| 301 | 22 | 12 | 4 |
| 302 | 10 | 5 | 3 |
| 303 | 60 | 40 | 15 |

The History and Values of Yo-Yos

# APPENDIX

**P** = PACKAGED  **NM** = NEAR MINT  **G** = GOOD  **-** = VALUE LESS THAN $1

| ITEM NUMBER | P | NM | G | ITEM NUMBER | P | NM | G | ITEM NUMBER | P | NM | G |
|---|---|---|---|---|---|---|---|---|---|---|---|
| 304 | 65 | 45 | 16 | 307.60 |  | 12 | 4 | 355 | 30 | 20 | 6 |
| 305 | 70 | 50 | 20 | 307.61 |  | 5 | 2 | 356 | 35 | 25 | 7 |
| 306 | 60 | 40 | 15 | 307.62 |  | 5 | 2 | 357 | 45 | 35 | 10 |
| 307 |  | 15 | 6 | 308 |  | 30 | 10 | 358 | 70 | 55 | 20 |
| 307.8 |  | 6 | 1 | 309 | 12 | 6 | 2 | 359 |  | 60 | 20 |
| 307.9 |  | 10 | 4 | 310 |  | 12 | 3 | 360 | 100 | 65 | 25 |
| 307.10 |  | 8 | 2 | 311 |  | 12 | 4 | 360.1 |  | 135 | 50 |
| 307.11 | 4 | 4 | 1 | 312 |  | 14 | 5 | 360.2 | 100 | 70 | 30 |
| 307.12 | 40 | 14 | 4 | 313 | 12 | 6 | 1 | 360.3 | - | 75 | 35 |
| 307.13 |  | 20 | 8 | 314 | 28 | 15 | 6 | 361 |  | 165 | 90 |
| 307.14 | 6 | 6 | 2 | 315 | 65 | 45 | 22 | 362 |  | 60 | 30 |
| 307.15 |  | 12 | 4 | 316 | 17 | 10 | 3 | 363 |  | 85 | 40 |
| 307.16 | 6 | 6 | 2 | 317 | 14 | 8 | 2 | 364 | Rare |  |  |
| 307.17 |  | 3 | 1 | 318 |  | 75 | 25 | 365 |  | 375 | 175 |
| 307.19 |  | 15 | 6 | 319 | 4 | 1 | - | 366 | 18 | 14 | 4 |
| 307.20 |  | 15 | 6 | 320 | 6 | 4 | 1 | 367 | 16 | 12 | 3 |
| 307.21 | 5 | 3 | - | 321 | 10 | 4 | 1 | 368 | 16 | 12 | 3 |
| 307.22 |  | 3 | 1 | 322 |  | 70 | 25 | 369 | 14 | 10 | 2 |
| 307.23 |  | 8 | 2 | 323 | 60 | 55 | 20 | 370 |  | 150 | 85 |
| 307.24 |  | 3 | 1 | 323.1 | 120 | 100 | 40 | 371 | 70 | 55 | 28 |
| 307.25 |  | 5 | 2 | 324 | 4 | 1 | - | 372 |  | 100 | 55 |
| 307.26 |  | 6 | 2 | 325 |  | 95 | 40 | 372.1 |  | 140 | 75 |
| 307.27 |  | 10 | 3 | 326 | 25 | 10 | 3 | 373 |  | 125 | 65 |
| 307.28 |  | 15 | 4 | 327 | 14 | 8 | 2 | 374 | 90 | 75 | 35 |
| 307.29 |  | 16 | 6 | 328 | 65 | 60 |  | 375 |  | 55 | 25 |
| 307.30 |  | 14 | 5 | 329 | 45 | 40 | 10 | 376 |  | 30 | 12 |
| 307.31 | 5 | 5 | 2 | 330 | 16 | 10 | 4 | 377 | 28 | 18 | 8 |
| 307.32 |  | 3 | 1 | 331 | 38 | 25 | 10 | 379 | 20 | 12 | 4 |
| 307.33 |  | 3 | 1 | 332 | 16 | 10 | 3 | 380 | 17 | 10 | 3 |
| 307.34 |  | 10 | 3 | 333 | 16 | 12 | 3 | 381 |  | 60 | 20 |
| 307.35 |  | 14 | 3 | 334 | 17 | 11 | 3 | 382 | 10 | 4 | 1 |
| 307.36 |  | 10 | 3 | 335 | 14 | 8 | 2 | 383 | 30 | 16 | 5 |
| 307.37 | 16 | 10 | 3 | 336 | 20 | 14 | 4 | 384 | 18 | 12 | 4 |
| 307.38 |  | 5 | 2 | 337 | 14 | 8 | 2 | 385 | 14 | 10 | 3 |
| 307.39 |  | 12 | 4 | 338 | 14 | 8 | 2 | 386 | Retail |  |  |
| 307.40 |  | 65 | 25 | 339 | 14 | 8 | 2 | 387 | Retail |  |  |
| 307.41 | 6 | 6 | 2 | 340 | 16 | 10 | 3 | 388 | Retail |  |  |
| 307.42 |  | 12 | 4 | 341 | 16 | 10 | 3 |  |  |  |  |
| 307.43 |  | 30 | 12 | 342 | 14 | 8 | 2 | 392 |  | 8 | 2 |
| 307.44 |  | 10 | 3 | 343 | 8 | 3 | 1 | 393 |  | 15 | 5 |
| 307.45 |  | 15 | 5 | 344 | 24 | 14 | 4 | 394 | 150 | - | - |
| 307.46 | 5 | 5 | 1 | 344.1 | 18 | 8 | 3 | 395 | Varies | 20-30 |  |
| 307.47 |  | 12 | 4 | 345 | 80 | 65 | 24 | 396 | Varies | 30-50 |  |
| 307.48 |  | 45 | 25 | 346 | 40 | 25 | 8 | 397 | Varies | 1-20 |  |
| 307.49 |  | 45 | 25 | 347 | 28 | 14 | 5 | 398 | 350 | - | - |
| 307.50 |  | 14 | 6 | 348 | 40 | 25 | 10 | 398.1 | 250 | - | - |
| 307.51 |  | 12 | 4 | 349 | 14 | 8 | 3 | 398.2 | 95 | - | - |
| 307.52 |  | 12 | 5 | 350 | 25 | 15 | 7 | 399 | 95 | 15 | 6 |
| 307.53 |  | 12 | 4 | 351 | 10 | 4 | 1 | 400 | 250 | - | - |
| 307.54 | 14 | 10 | 3 | 352 | Retail |  |  | 401 | Retail | - | - |
| 307.55 |  | 4 | 1 | 352.1 | Retail |  |  | 402 | 50 |  |  |
| 307.56 |  | 30 | 12 | 352.2 | Retail |  |  | 403 | Varies | 2-20 |  |
| 307.57 |  | 30 | 12 | 352.3 | Retail |  |  | 404 |  | 8 | - |
| 307.58 |  | 4 | 1 | 353 |  | 60 | 40 | 15 | 405 | 35 |  |
| 307.59 |  | 30 | 12 | 354 |  | 30 | 20 | 6 | 406 | 50 | - | - |

# APPENDIX

**P** = PACKAGED  **NM** = NEAR MINT  **G** = GOOD  **-** = VALUE LESS THAN $1

| ITEM NUMBER | P | NM | G |
|---|---|---|---|
| 407 |  | 25 | 15 |
| 408 |  | 8 | 2 |
| 409 |  | 30 | 15 |
| 410 | 5 | 5 | 1 |
| 411 | Retail |  |  |
| 412 | 50 | 40 | 15 |
| 413 | 50 | 40 | 15 |
| 414 |  | 45 | 18 |
| 415 | Retail |  |  |
| 415.1 | 40 | 30 | 12 |
| 416 | 35 | 25 | 12 |
| 417 | 10 | 6 | 2 |
| 418 |  | 17 | 7 |
| 418.1 |  | 17 | 7 |
| 419 | 10 | 5 | 2 |
| 420 | 55 | 45 | 15 |
| 421 | 58 | 48 | 16 |
| 422 | 12 | 4 | 1 |
| 423 | Retail |  |  |
| 424 |  | 50 | 15 |
| 425 | Retail |  |  |
| 426 | 5 | 3 | 1 |
| 427 | 25 | 14 | 5 |
| 428 |  | 28 | 12 |
| 429 | 6 | 3 | 1 |
| 429.1 | 40 | 30 | 12 |
| 430 | 25 | 17 | 5 |
| 431 | 28 | 18 | 6 |
| 432 | 35 | 23 | 7 |
| 433 | 35 | 23 | 7 |
| 434 | 40 | 28 | 8 |
| 435 | 18 | 8 | 3 |
| 436 |  | 65 |  |
| 436.1 |  | 50 | 20 |
| 436.2 |  | 55 | 22 |
| 437 |  | 85 | 40 |
| 438 |  | 125 | 55 |
| 439 |  | 65 | 30 |
| 440 | 65 | 40 | 16 |
| 441 |  | 45 | 18 |
| 441.1 |  | 50 | 20 |
| 442 | Retail |  |  |
| 443 |  | 30 | 12 |
| 444 | 275 | 225 | 100 |
| 445 |  | 60 | 25 |
| 446 |  | 60 | 25 |
| 446.1 |  | 45 | 20 |
| 446.2 |  | 150 | 75 |
| 446.3 |  | 125 | 60 |
| 447 |  |  | .25 |
| 447.1 |  | 18 | 6 |
| 447.2 |  | 16 | 5 |
| 447.3 | 40 | 18 | 6 |
| 447.4 |  | 16 | 5 |
| 448 | 40 | 30 | 10 |
| 448.1 | 20 | 8 | 3 |
| 448.2 | 40 | 30 | 10 |
| 448.3 | Retail |  |  |
| 448.4 |  | 6 | 2 |
| 448.5 |  | 5 | 2 |
| 449 |  | 65 | 30 |
| 449.1 |  | 60 | 25 |
| 449.2 |  | 65 | 30 |
| 450 | 38 | 28 | 10 |
| 450.1 | 38 | 28 | 10 |
| 450.2 | 38 | 28 | 10 |
| 450.3 | 65 | 55 | 25 |
| 451 | 38 | 28 | 10 |
| 452 | 100 | 75 | 35 |
| 453 | 15 | 7 | 2 |
| 454 | 12 | 6 | 2 |
| 455 | 10 | 4 | 1 |
| 456 |  | 38 | 14 |
| 457 | 22 | 8 | 3 |
| 458 | 26 | 12 | 4 |
| 459 | 35 | 18 | 7 |
| 460 | 24 | 10 | 3 |
| 461 | 40 | 12 | 4 |
| 462 | 22 | 8 | 3 |
| 463 | 38 | 23 | 6 |
| 464 | 28 | 18 | 5 |
| 465 | 23 | 13 | 4 |
| 466 | 28 | 18 | 5 |
| 467 | 23 | 13 | 4 |
| 468 | 28 | 18 | 5 |
| 469 | 23 | 13 | 4 |
| 470 | 28 | 18 | 5 |
| 471 | 23 | 13 | 4 |
| 472 | 28 | 18 | 5 |
| 473 | 23 | 13 | 4 |
| 474 |  | 50 | 20 |
| 475 |  | 65 | 28 |
| 476 |  | 155 | 60 |
| 476.1 |  | 155 | 60 |
| 477 |  | 155 | 60 |
| 478 |  | 155 | 60 |
| 478.1 |  | 175 | 60 |
| 479 |  | 125 | 55 |
| 480 |  | 125 | 55 |
| 481 |  | 125 | 50 |
| 482 |  | 110 | 50 |
| 483 |  | 110 | 50 |
| 484 |  | 295 | 95 |
| 485 |  | 65 | 25 |
| 486 |  | 65 | 27 |
| 487 |  | 65 | 27 |
| 488 |  | 65 | 27 |
| 489 |  | 5 | - |
| 490 |  | 80 | 30 |
| 491 |  | 70 | 30 |
| 492 |  | 60 | 27 |
| 493 |  | 55 | 22 |
| 494 |  | 58 | 25 |
| 494.1 |  | 45 | 20 |
| 495 | 38 | 25 | 8 |
| 496 |  | 35 | 15 |
| 497 | 35 | 25 | 6 |
| 498 |  | 85 | 40 |
| 499 |  | 38 | 16 |
| 500 |  | 38 | 12 |
| 501 |  | 50 | 23 |
| 502 |  | 80 | 35 |
| 503 |  | 80 | 35 |
| 504 | 8 | 5 | 2 |
| 505 | 12 | 7 | 3 |
| 506 |  | 75 | 32 |
| 506.1 |  | 110 | 45 |
| 507 |  | 38 | 16 |
| 507.1 |  | 80 | 32 |
| 508 |  | 35 | 14 |
| 509 | 10 | 5 | - |
| 510 |  | 65 | 20 |
| 511 | 15 | 10 | 3 |
| 512 |  | 50 | 22 |
| 512.1 | 16 | 8 | 2 |
| 512.2 | 16 | 8 | 2 |
| 513 |  | 10 | 3 |
| 513.1 | 12 | 5 | 1 |
| 514 | 18 | 10 | 3 |
| 514.1 | 25 | 15 | 5 |
| 514.2 | 22 | 12 | 4 |
| 514.3 | Retail |  |  |
| 515 | Retail |  |  |
| 515.1 | 15 | 10 | 3 |
| 516 | 25 | 10 | 3 |
| 517 | 18 | 8 | 3 |
| 518 | 20 | 10 | 3 |
| 519 | 20 | 10 | 3 |
| 520 | 18 | 8 | 3 |
| 521 | Retail |  |  |
| 522 | 22 | 12 | 5 |
| 523 | 27 | 17 | 6 |
| 524 | 42 | 32 | 10 |
| 525 | 22 | 12 | 5 |
| 525.1 | Retail |  |  |
| 526 | 55 | 12 | 5 |
| 527 | 20 | 10 | 4 |
| 528 | 25 | 15 | 7 |
| 529 | 20 | 10 | 4 |
| 530 |  | 20 | 8 |
| 531 | 22 | 12 | 5 |
| 532 | 32 | 22 | 8 |
| 533 | 32 | 22 | 8 |
| 534 | 18 | 8 | 2 |
| 535 | 22 | 12 | 5 |

The History and Values of Yo-Yos

# APPENDIX

**P** = PACKAGED   **NM** = NEAR MINT   **G** = GOOD   **-** = VALUE LESS THAN $1

| ITEM NUMBER | P | NM | G |
|---|---|---|---|
| 536 | 22 | 12 | 5 |
| 537 | Retail | | |
| 538 | 50 | 12 | 5 |
| 539 | 65 | 40 | 12 |
| 540 | 65 | 40 | 12 |
| 541 | 18 | 8 | 3 |
| 542 | 18 | 8 | 3 |
| 543 | 18 | 8 | 3 |
| 544 | 18 | 8 | 3 |
| 545 | 22 | 12 | 5 |
| 546 | 22 | 12 | 4 |
| 547 | 24 | 14 | 5 |
| 548 | 45 | 30 | 9 |
| 549 | 20 | 10 | 4 |
| 550 | 18 | 8 | 3 |
| 551 | 18 | 8 | 3 |
| 552 | 55 | 45 | 17 |
| 553 | 60 | 50 | 20 |
| 554 | Retail | | |
| 555 | 26 | 16 | 7 |
| 556 | 20 | 15 | 6 |
| 557 | | 45 | 20 |
| 558 | | 45 | 20 |
| 559 | | 30 | 12 |
| 560 | | 45 | 20 |
| 561 | | 45 | 20 |
| 562 | | 55 | 25 |
| 563 | 28 | 10 | 3 |
| 564 | | 10 | 4 |
| 564.1 | | 6 | 2 |
| 565 | | 10 | 4 |
| 566 | | 7 | 2 |
| 567 | | 25 | 10 |
| 568 | | 40 | 18 |
| 569 | | 6 | 2 |
| 570 | | 6 | 2 |
| 571 | | 35 | 14 |
| 572 | | 50 | 22 |
| 573 | | 75 | 25 |
| 574 | | 38 | 15 |
| 575 | | 45 | 20 |
| 576 | | 15 | 4 |
| 577 | | 300 | 150 |
| 578 | | 200 | 100 |
| 579 | | 300 | 150 |
| 580 | | 35 | 15 |
| 581 | | 80 | 30 |
| 582 | | 100 | 40 |
| 583 | | 250 | 100 |
| 591 | | 100 | 40 |
| 592 | | 6 | 3 |
| 593 | | 5 | 1 |
| 594 | | 3 | 1 |
| 595 | 12 | 8 | 2 |

| ITEM NUMBER | P | NM | G |
|---|---|---|---|
| 596 | 18 | 8 | 3 |
| 597 | 7 | 4 | 1 |
| 598 | | 3 | 1 |
| 599 | | 4 | 1 |
| 600 | | 3 | 1 |
| 601 | | 12 | 3 |
| 602 | | 5 | 1 |
| 603 | | 5 | 1 |
| 604 | | 6 | 2 |
| 605 | | 100 | 50 |
| 606 | | 8 | 2 |
| 607 | | 40 | 20 |
| 608 | | 40 | 20 |
| 609 | Retail | | |
| 610 | Retail | | |
| 611 | Retail | | |
| 612 | | 6 | 2 |
| 613 | | 25 | 8 |
| 614 | | 25 | 8 |
| 615 | | 15 | 5 |
| 616 | | 75 | 20 |
| 617 | | 75 | 20 |
| 618 | | 75 | 20 |
| 619 | | 120 | 50 |
| 620 | | 45 | 16 |
| 621 | | 155 | 45 |
| 622 | | 95 | 45 |
| 623 | | 120 | 55 |
| 624 | | 45 | 16 |
| 625 | | 45 | 16 |
| 626 | | 35 | 10 |
| 627 | | 65 | 27 |
| 627.1 | | 65 | 27 |
| 628 | | 80 | 24 |
| 629 | | 80 | 24 |
| 630 | | 200 | 100 |
| 631 | | 155 | 70 |
| 632 | | 155 | 70 |
| 633 | | 85 | 40 |
| 634 | | 75 | 22 |
| 635 | 6 | 3 | 1 |
| 636 | 22 | 12 | 4 |
| 637 | 30 | 20 | 6 |
| 638 | 22 | 12 | 4 |
| 639 | 20 | 10 | 3 |
| 640 | 20 | 10 | 3 |
| 641 | 16 | 6 | 2 |
| 642 | 22 | 12 | 4 |
| 643 | 10 | 6 | 2 |
| 644 | 20 | 10 | 3 |
| 645 | 30 | 20 | 6 |
| 646 | 5 | 3 | 1 |
| 647 | 20 | 10 | 3 |
| 648 | | 3 | 1 |
| 650 | 24 | 14 | 5 |

| ITEM NUMBER | P | NM | G |
|---|---|---|---|
| 651 | 20 | 10 | 3 |
| 652 | 20 | 10 | 3 |
| 653 | 20 | 10 | 3 |
| 654 | 30 | 20 | 6 |
| 655 | 24 | 14 | 5 |
| 656 | 24 | 14 | 5 |
| 657 | 20 | 10 | 3 |
| 658 | 30 | 20 | 6 |
| 659 | 22 | 12 | 4 |
| 660 | 20 | 10 | 3 |
| 661 | 28 | 18 | 6 |
| 662 | 28 | 8 | 2 |
| 663 | 18 | 8 | 2 |
| 664 | 25 | 15 | 6 |
| 666 | | 65 | 30 |
| 667 | 100 | 85 | 26 |
| 668 | 110 | 95 | 28 |
| 669 | 35 | 20 | 6 |
| 670 | 55 | 40 | 16 |
| 671 | 55 | 40 | 16 |
| 672 | 110 | 90 | 28 |
| 673 | 95 | 60 | 22 |
| 674 | 85 | 55 | 22 |
| 675 | | 1 | - |
| 675.1 | | 1 | - |
| 675.2 | | 1 | - |
| 675.3 | | 1 | - |
| 675.4 | Varies | | |
| 676 | | 1 | - |
| 677 | | 1 | - |
| 678 | 5 | 2 | - |
| 679 | 18 | 8 | 2 |
| 680 | 18 | 10 | 2 |
| 681 | | 3 | 1 |
| 682 | | 1 | - |
| 683 | | 1 | - |
| 684 | | 1 | - |
| 685 | | 1 | - |
| 686 | | 1 | - |
| 687 | | 1 | - |
| 688 | | 1 | - |
| 689 | | 2 | - |
| 690 | | 1 | - |
| 691 | | 1 | - |
| 692 | | 1 | - |
| 693 | | 8 | 2 |
| 694 | | 2 | - |
| 695 | | 1 | - |
| 696 | | 1 | - |
| 697 | | 2 | - |
| 698 | | 1 | - |
| 699 | | 1 | - |
| 700 | | 1 | - |
| 701 | | 2 | - |
| 702 | Retail | | |

Lucky's Collectors Guide To 20th Century Yo-Yos

# APPENDIX

**P** = PACKAGED  **NM** = NEAR MINT  **G** = GOOD  **-** = VALUE LESS THAN $1

| ITEM NUMBER | P | NM | G | ITEM NUMBER | P | NM | G | ITEM NUMBER | P | NM | G |
|---|---|---|---|---|---|---|---|---|---|---|---|
| 703 |  | 5 | 2 | 758 |  | 1 | - | 822 |  | 1 | - |
| 704 |  | 6 | - | 759 |  | 1 | - | 823 |  | 1 | - |
| 705 |  | 1 | - | 760 |  | 2 | - | 824 |  | 2 | - |
| 706 |  | 1 | - | 761 |  | 1 | - | 825 |  | 2 | - |
| 707 |  | 1 | - | 762 |  | 6 | 2 | 826 |  | 1 | - |
| 708 |  | 1 | - | 763 |  | 1 | - | 827 |  | 5 | 1 |
| 709 | Retail |  |  | 764 |  | 1 | - | 828 |  | 1 | - |
| 710 |  | 1 | - | 765 |  | 1 | - | 829 |  | 2 | - |
| 711 |  | 3 | - | 766 |  | 1 | - | 830 |  | 2 | - |
| 712 |  | 1 | - | 767 |  | 1 | - | 831 |  | 2 | - |
| 713 |  | 1 | - | 768 |  | 1 | - | 832 |  | 4 | 1 |
| 714 |  | 1 | - | 769 |  | 6 | 2 | 833 |  | 4 | 1 |
| 715 |  | 1 | - | 770 |  | 2 | - | 834 |  | 2 | - |
| 716 | 15 | 12 | 5 |  |  |  |  | 835 |  | 3 | - |
| 717 |  | 3 | - | 781 |  | 1 | - | 836 |  | 3 | - |
| 718 |  | 1 | - | 782 |  | 2 | - | 837 |  | 2 | - |
| 719 | Retail |  |  | 783 |  | 1 | - | 838 |  | 3 | - |
| 720 | Retail |  |  | 784 |  | 2 | - | 840 |  | 10 | 2 |
| 721 |  | 1 | - | 785 |  | 4 | 1 | 841 |  | 1 | - |
| 722 | Retail | - |  | 786 |  | 2 | - | 842 |  | 6 | 2 |
| 723 |  | 1 | - | 787 |  | 1 | - | 843 |  | 3 | - |
| 724 |  | 2 | - | 788 |  | 1 | - | 845 |  | 4 | 1 |
| 725 |  | 3 | - | 789 |  | 5 | 1 | 846 |  | 4 | 1 |
| 726 |  | 1 | - | 790 |  | 3 | - | 847 |  | 1 | - |
| 727 | 13 | 5 | 1 | 791 |  | 3 | - | 848 |  | 2 | - |
| 728 | 13 | 5 | 1 | 792 |  | 2 | - | 849 |  | 4 | 1 |
| 729 |  | 45 | 16 | 793 |  | 3 | - | 850 |  | 1 | - |
| 730 |  | 6 | 2 | 794 |  | 5 | 1 | 851 |  | 1 | - |
| 731 |  | 6 | 2 | 795 |  | 2 | - | 852 |  | 1 | - |
| 732 | 22 | 12 | 5 | 796 |  | 2 | - | 853 |  | 8 | 2 |
| 733 | 13 | 4 | 1 | 797 |  | 10 | 3 | 854 |  | 1 | - |
| 734 | 15 | 5 | 1 | 798 |  | 8 | 2 | 855 |  | 1 | - |
| 735 |  | 20 | 8 | 799 |  | 1 | - | 856 |  | 5 | 1 |
| 736 | 14 | 5 | 1 | 800 |  | 7 | 2 | 857 |  | 3 | - |
| 737 | 14 | 5 | 1 | 801 |  | 4 | 1 | 858 |  | 1 | - |
| 738 | 18 | 8 | 3 | 802 |  | 2 | - | 859 |  | 1 | - |
| 739 | - | - | - | 803 |  | 10 | 3 | 860 |  | 2 | - |
| 740 | 18 | 8 | 3 | 804 |  | 2 | - | 861 |  | 1 | - |
| 741 | 22 | 12 | 3 | 805 |  | 1 | - | 862 |  | 1 | - |
| 742 | 17 | 12 | 3 | 806 |  | 1 | - | 863 |  | 1 | - |
| 743 | 30 | 20 | 8 | 807 |  | 2 | - | 864 |  | 2 | - |
| 744 | 16 | 6 | 2 | 808 |  | 2 | - | 865 |  | 2 | - |
| 745 |  | 45 | 20 | 809 |  | 2 | - | 866 |  | 1 | - |
| 746 | 14 | 5 | 1 | 810 |  | 3 | - | 867 |  | 1 | - |
| 747 | 14 | 5 | 1 | 811 |  | 25 | 8 | 868 |  | 2 | - |
| 748 | 20 | 10 | 3 | 812 |  | 2 | - | 869 |  | 2 | - |
| 749 | 16 | 6 | 2 | 813 |  | 5 | 1 | 870 |  | 3 | - |
| 750 |  | 8 | 3 | 814 |  | 5 | 1 | 871 |  | 6 | 1 |
| 751 | 12 | 5 | 2 | 815 |  | 3 | - | 872 |  | 12 | 3 |
| 752 | 15 | 10 | 4 | 816 |  | 3 | - | 873 |  | 1 | - |
| 753 |  | 65 | 30 | 817 |  | 2 | - | 874 |  | 1 | - |
| 754 |  | 12 | 3 | 818 |  | 2 | - | 875 |  | 2 | - |
| 755 |  | 2 | - | 819 |  | 1 | - | 876 |  | 3 | - |
| 756 |  | 2 | - | 820 |  | 1 | - | 877 |  | 1 | - |
| 757 |  | 1 | - | 821 |  | 2 | - | 878 |  | 1 | - |

# APPENDIX

**P** = PACKAGED  **NM** = NEAR MINT  **G** = GOOD  **-** = VALUE LESS THAN $1

| ITEM NUMBER | P | NM | G |
|---|---|---|---|
| 879 | | 5 | 1 |
| 880 | | 3 | - |
| 881 | | 6 | 1 |
| 882 | | 3 | - |
| 883 | | 3 | - |
| 884 | | 2 | - |
| 885 | | 4 | - |
| 886 | | 3 | - |
| 887 | | 10 | 2 |
| 888 | | 3 | - |
| 889 | | 2 | - |
| 900 | | 3 | - |
| 901 | | 2 | - |
| 902 | | 4 | - |
| 903 | | 2 | - |
| 904 | | 2 | - |
| 905 | | 5 | 1 |
| 906 | | 2 | - |
| 907 | | 2 | - |
| 908 | | 2 | - |
| 909 | | 2 | - |
| 1000 | | 10 | 2 |
| 1002 | Retail | | |
| 1003 | 1 | - | - |
| 1004 | 6 | 2 | - |
| 1005 | | 5 | 1 |
| 1006 | Retail | | |
| 1007 | Retail | | |
| 1008 | Retail | | |
| 1009 | Retail | | |
| 1010 | Retail | | |
| 1011 | 7 | 3 | 1 |
| 1012 | | 3 | 1 |
| 1013 | 6 | 3 | 1 |
| 1014 | Retail | | |
| 1015 | 8 | 3 | - |
| 1016 | 4 | 2 | 1 |
| 1017 | 15 | 5 | 1 |
| 1018 | Retail | | |
| 1019 | Retail | | |
| 1020 | Retail | | |
| 1021 | Retail | | |
| 1021.1 | Retail | | |
| 1022 | | 25 | 10 |
| 1023 | | 45 | 18 |
| 1023.1 | | 55 | 20 |
| 1024 | | 135 | 70 |
| 1025 | | 160 | 75 |
| 1026 | | 130 | 65 |
| 1026.1 | | 130 | 65 |
| 1027 | 35 | 16 | 4 |
| 1028 | 35 | 16 | 4 |
| 1029 | 41 | 20 | 4 |

| ITEM NUMBER | P | NM | G |
|---|---|---|---|
| 1030 | | 8 | 2 |
| 1031 | | 4 | 1 |
| 1032 | 22 | 12 | 4 |
| 1033 | 15 | 8 | 2 |
| 1034 | - | 40 | 15 |
| 1035 | 6 | 5 | 1 |
| 1036 | 6 | 4 | 1 |
| 1037 | Retail | | |
| 1038 | | 12 | 5 |
| 1039 | 6 | 4 | 1 |
| 1040 | Retail | | |
| 1040.1 | Retail | | |
| 1040.2 | Retail | | |
| 1040.3 | Retail | | |
| 1040.4 | Retail | | |
| 1040.5 | Retail | | |
| 1041 | | 12 | 4 |
| 1042 | Retail | | |
| 1043 | 5 | 2 | - |
| 1047 | | 6 | 1 |
| 1048 | 5 | 4 | 1 |
| 1049 | | 25 | 8 |
| 1050 | | 25 | 8 |
| 1051 | | 25 | 8 |
| 1052 | | 25 | 8 |
| 1053 | | 25 | 8 |
| 1054 | | 55 | 20 |
| 1055 | | 6 | 2 |
| 1056 | | 12 | 4 |
| 1057 | | 5 | 1 |
| 1058 | | 3 | - |
| 1059 | | 4 | 1 |
| 1060 | | 25 | 8 |
| 1061 | | 20 | 6 |
| 1062 | | 12 | 4 |
| 1063 | | 40 | 18 |
| 1064 | | 50 | 25 |
| 1064.1 | | 30 | 8 |
| 1065 | 75 | 70 | 35 |
| 1066 | | 35 | 14 |
| 1067 | | Rare | |
| 1068 | | 45 | 20 |
| 1069 | | 60 | 25 |
| 1070 | 65 | 60 | 32 |
| 1071 | | 60 | 25 |
| 1072 | | 60 | 25 |
| 1072.1 | | 50 | 22 |
| 1073 | Retail | | |
| 1073.1 | 100 | 80 | 30 |
| 1074 | | 35 | 15 |
| 1075 | 60 | 40 | 18 |
| 1076 | | 8 | 2 |
| 1076.1 | 40 | 35 | 15 |
| 1077 | 85 | 80 | 30 |
| 1078 | 35 | 30 | 15 |

| ITEM NUMBER | P | NM | G |
|---|---|---|---|
| 1079 | | 20 | 8 |
| 1080 | | 10 | 3 |
| 1081 | | 16 | 6 |
| 1082 | | 10 | 3 |
| 1083 | | 15 | 5 |
| 1084 | | 15 | 5 |
| 1085 | | 8 | 3 |
| 1086 | | 15 | 5 |
| 1087 | | 8 | 3 |
| 1087.1 | | 15 | 5 |
| 1087.2 | | 15 | 5 |
| 1087.3 | | 8 | 3 |
| 1088 | 14 | 12 | 4 |
| 1089 | | 4 | 1 |
| 1090 | | 8 | 3 |
| 1091 | Varies | | |
| 1092 | | 35 | 14 |
| 1093 | 15 | 5 | 2 |
| 1094 | | 2 | - |
| 1095 | Retail | | |
| 1096 | Varies | | |
| 1097 | | 5 | 1 |
| 1098 | | 6 | - |
| 1099 | 40 | 30 | 5 |
| 1100 | 6 | 4 | 1 |
| 1101 | | 75 | 30 |
| 1102 | | 4 | - |
| 1103 | 3 | 2 | - |
| 1104 | 13 | 3 | 1 |
| 1105 | 15 | 5 | 1 |
| 1106 | Varies | | |
| 1107 | 60 | 40 | 12 |
| 1108 | 100 | 50 | 15 |
| 1109 | 50 | 30 | 12 |
| 1110 | 1 | - | - |
| 1111 | 4 | 2 | - |
| 1112 | 60 | 40 | 15 |
| 1113 | 12 | 10 | 3 |
| 1114 | 4 | 1 | |
| 1115 | Varies | | |
| 1116 | | 2 | - |
| 1117 | | 2 | - |
| 1118 | | 5 | 1 |
| 1119 | | 1 | - |
| 1120 | 12 | 4 | - |
| 1121 | | 28 | 12 |
| 1122 | Varies | | |
| 1123 | | 5 | - |
| 1124 | | 10 | 1 |
| 1125 | Retail | | |
| 1126 | | 12 | 4 |
| 1127 | 4 | 2 | - |
| 1128 | 4 | 2 | - |
| 1129 | 10 | 3 | 1 |
| 1130 | 350 | 225 | 100 |

# APPENDIX

**P** = PACKAGED  **NM** = NEAR MINT  **G** = GOOD  **-** = VALUE LESS THAN $1

| ITEM NUMBER | P | NM | G |
|---|---|---|---|
| 1131 | | 85 | 40 |
| 1132 | Retail | | |
| 1133 | 8 | 3 | 1 |
| 1134 | Retail | | |
| 1135 | 16 | 6 | 2 |
| 1136 | 3 | 1 | - |
| 1137 | 17 | 7 | 2 |
| 1137.1 | 45 | 25 | 10 |
| 1138 | Retail | | |
| 1139 | 26 | 16 | 4 |
| 1140 | Varies | | |
| 1141 | 7 | 4 | - |
| 1141.1 | 14 | 8 | 2 |
| 1141.2 | 12 | 6 | 2 |
| 1142 | | 1 | - |
| 1143 | | 1 | - |
| 1144 | | 1 | - |
| 1144.1 | | 1 | - |
| 1145 | | 1 | - |
| 1146 | | 2 | - |
| 1146.1 | | 1 | - |
| 1147 | | 1 | - |
| 1148 | | 2 | - |
| 1149 | | 1 | - |
| 1149.1 | | 1 | - |
| 1149.2 | | 1 | - |
| 1150 | | 1 | - |
| 1150.1 | | 1 | - |
| 1151 | 70 | 60 | 18 |
| 1151.1 | 75 | 65 | 25 |
| 1152 | 13 | 3 | 1 |
| 1153 | 13 | 3 | 1 |
| 1154 | 13 | 3 | 1 |
| 1155 | 16 | 6 | 2 |
| 1156 | 22 | 12 | 5 |
| 1156.1 | 24 | 14 | 5 |
| 1156.2 | 15 | 8 | 3 |
| 1157 | 18 | 8 | 3 |
| 1157.1 | 22 | 12 | 6 |
| 1158 | 18 | 8 | 3 |
| 1159 | 16 | 6 | 2 |
| 1160 | 22 | 12 | 3 |
| 1161 | 22 | 12 | 3 |
| 1161.1 | | 24 | 14 |
| 1162 | 60 | 45 | 20 |
| 1163 | 2 | - | - |
| 1164 | 2 | - | - |
| 1165 | 2 | - | - |
| 1166 | 2 | - | - |
| 1167 | 2 | - | - |
| 1168 | 2 | - | - |
| 1169 | 2 | - | - |
| 1170 | 30 | 15 | 5 |
| 1171 | 30 | 15 | 5 |
| 1172 | 30 | 15 | 5 |
| 1173 | | 6 | 2 |
| 1174 | 20 | 14 | 4 |
| 1175 | | 6 | 2 |
| 1176 | | 6 | 2 |
| 1177 | 24 | 14 | 4 |
| 1178 | 22 | 12 | 4 |
| 1179 | 22 | 12 | 4 |
| 1180 | 35 | 25 | 10 |
| 1181 | 20 | 10 | 2 |
| 1182 | 20 | 10 | 3 |
| 1183 | 20 | 10 | 3 |
| 1184 | 20 | 10 | 3 |
| 1185 | 20 | 10 | 3 |
| 1186 | 20 | 10 | 3 |
| 1187 | 35 | 25 | 10 |
| 1188 | 25 | 15 | 5 |
| 1189 | 22 | 12 | 4 |
| 1190 | 20 | 10 | 3 |
| 1191 | 24 | 14 | 4 |
| 1192 | 6 | 4 | 1 |
| 1193 | 25 | 15 | 5 |
| 1194 | | 10 | 3 |
| 1195 | Retail | | |
| 1196 | 2 | - | - |
| 1197 | | 18 | 6 |
| 1198 | 6 | 3 | - |
| 1199 | 8 | 1 | - |
| 1200 | Retail | | |
| 1201 | 2 | 1 | |
| 1202 | 12 | 7 | 2 |
| 1203 | 7 | 3 | - |
| 1204 | | 45 | 15 |
| 1205 | 10 | 3 | - |
| 1206 | 5 | 2 | - |
| 1207 | 15 | 2 | - |
| 1208 | 5 | 2 | - |
| 1209 | 26 | 14 | 4 |
| 1210 | | 10 | 3 |
| 1211 | 12 | 5 | 1 |
| 1212 | 14 | 8 | 2 |
| 1213 | Varies | | |
| 1214 | | 30 | 10 |
| 1215 | 12 | 7 | 2 |
| 1216 | 12 | 8 | 2 |
| 1216.1 | 10 | 6 | 1 |
| 1217 | 20 | 10 | 3 |
| 1217.1 | | 12 | 4 |
| 1218 | Retail | | |
| 1219 | 5 | 3 | 1 |
| 1220 | | 3 | 1 |
| 1221 | 6 | 3 | 1 |
| 1222 | 11 | 5 | 2 |
| 1223 | 15 | 8 | 3 |
| 1223.1 | 10 | 5 | 2 |
| 1224 | 8 | 5 | 1 |
| 1225 | | 3 | 1 |
| 1226 | | 25 | 6 |
| 1227 | | 35 | 9 |
| 1228 | 5 | 2 | - |
| 1229 | | 3 | 1 |
| 1230 | | 40 | 11 |
| 1231 | 2 | 1 | - |
| 1232 | Retail | | |
| 1233 | 10 | 5 | 1 |
| 1234 | 12 | 6 | 1 |
| 1235 | | 2 | - |
| 1236 | Varies | | |
| 1237 | | 10 | 3 |
| 1238 | | 3 | 1 |
| 1239 | | 3 | 1 |
| 1240 | 3 | 2 | 1 |
| 1241 | | 3 | 1 |
| 1242 | Varies | 8-15 | |
| 1243 | | 8 | 3 |
| 1244 | 12 | 4 | 1 |
| 1245 | 8 | 4 | 1 |
| 1246 | | 6 | 2 |
| 1247 | | 6 | 2 |
| 1248 | 8 | 4 | 1 |
| 1249 | Retail | | |
| 1251 | | 6 | 2 |
| 1252 | | 7 | 2 |
| 1253 | | 7 | 2 |
| 1254 | | 7 | 2 |
| 1255 | Retail | | |
| 1256 | 14 | 4 | 1 |
| 1257 | 16 | 6 | 2 |
| 1258 | 22 | 8 | 3 |
| 1259 | 20 | 7 | 3 |
| 1259.1 | Retail | | |
| 1260 | | 8 | 2 |
| 1261 | | 8 | 2 |
| 1262 | | 8 | 2 |
| 1262.1 | | 8 | 2 |
| 1263 | | 8 | 2 |
| 1264 | | 8 | 2 |
| 1265 | | 8 | 2 |
| 1266 | 12 | 6 | 2 |
| 1267 | Retail | | |
| 1268 | Retail | | |
| 1269 | | 45 | 14 |
| 1270 | 30 | 20 | 5 |
| 1271 | 55 | 45 | 15 |
| 1271.1 | 75 | 65 | 22 |
| 1272 | 15 | 7 | 2 |
| 1273 | | 35 | 15 |
| 1274 | | 85 | 30 |
| 1275 | | 25 | 10 |
| 1275.1 | 15 | 7 | 2 |

The History and Values of Yo-Yos

# APPENDIX

**P** = PACKAGED   **NM** = NEAR MINT   **G** = GOOD   **-** = VALUE LESS THAN $1

| ITEM NUMBER | P | NM | G |
|---|---|---|---|
| 1276 |  | 10 | 2 |
| 1276.1 | 20 | 10 | 2 |
| 1277 | 18 | 8 | 2 |
| 1278 |  | 45 | 18 |
| 1279 |  | 40 | 16 |
| 1280 | 20 | 10 | 2 |
| 1280.1 | 50 | 40 | 16 |
| 1281 |  | 50 | 20 |
| 1282 |  | 28 | 11 |
| 1283 |  | 40 | 14 |
| 1284 |  | 30 | 12 |
| 1285 |  | 65 | 24 |
| 1286 |  | 55 | 20 |
| 1287 | Rare |  |  |
| 1288 |  | 45 | 18 |
| 1289 |  | 55 | 16 |
| 1290 | 15 | 6 | 2 |
| 1291 | 35 | 30 | 8 |
| 1292 | 20 | 10 | 4 |
| 1293 | 16 | 6 | 2 |
| 1294 | 16 | 6 | 2 |
| 1295 | 16 | 6 | 2 |
| 1296 |  | 30 | 12 |
| 1297 |  | 30 | 12 |
| 1298 |  | 100 | 30 |
| 1299 |  | 70 | 24 |
| 1300 |  | 75 | 25 |
| 1301 |  | 50 | 16 |
| 1302 |  | 60 | 20 |
| 1303 |  | 60 | 20 |
| 1304 |  | 60 | 20 |
| 1305 |  | 60 | 20 |
| 1306 |  | 70 | 24 |
| 1307 |  | 60 | 20 |
| 1308 |  | 85 | 30 |
| 1309 | 45 | 5 | 1 |
| 1310 |  | 2 | - |
| 1311 |  | 1 | - |
| 1312 |  | 2 | - |
| 1313 |  | 2 | - |
| 1314 |  | 2 | - |
| 1315 |  | 2 | - |
| 1316 |  | 2 | - |
| 1317 |  | 1 | - |
| 1318 |  | 20 | 8 |
| 1319 |  | 25 | 11 |
| 1320 |  | 20 | 7 |
| 1321 | 30 | 25 | 10 |
| 1322 |  | 3 | - |
| 1323 |  | 5 | 2 |
| 1324 | 12 | 5 | 1 |
| 1325 | 12 | 5 | 1 |
| 1326 | 12 | 5 | 1 |
| 1327 |  | 5 | 1 |
| 1328 | 5 | 3 | 1 |

| ITEM NUMBER | P | NM | G |
|---|---|---|---|
| 1329 |  | 20 | 7 |
| 1330 |  | 1 | - |
| 1331 |  | 10 | 3 |
| 1332 |  | 6 | 2 |
| 1333 | 4 | 1 | - |
| 1334 |  | 3 |  |
| 1335 |  | 4 | 1 |
| 1336 |  | 3 | 1 |
| 1337 | 12 | 5 | 1 |
| 1338 | 12 | 5 | 1 |
| 1339 | 12 | 5 | 1 |
| 1340 | 12 | 5 | 1 |
| 1342 |  | 1 | - |
| 1343 |  | 1 | - |
| 1344 |  | 6 | 2 |
| 1345 | 22 | 18 | 6 |
| 1346 |  | 1 | - |
| 1347 |  | 3 | - |
| 1348 |  | 1 | - |
| 1349 |  | 3 | - |
| 1350 |  | 5 | 1 |
| 1351 |  | 3 | 1 |
| 1352 |  | 1 | - |
| 1353 |  | 1 | - |
| 1354 |  | 1 | - |
| 1355 |  | 15 | 5 |
| 1356 | Retail |  |  |
| 1357 | 3 | 2 | 1 |
| 1358 | 14 | 4 | 1 |
| 1359 |  | 12 | 5 |
| 1360 | 14 | 4 | 1 |
| 1361 |  | 1 | - |
| 1362 |  | 3 | 1 |
| 1363 | 22 | 18 | 6 |
| 1364 |  | 15 | 5 |
| 1365 |  | 6 | 2 |
| 1366 |  | 12 | 3 |
| 1367 |  | 6 | 1 |
| 1367.1 |  | 6 | 1 |
| 1368 |  | 45 | 16 |
| 1369 | 22 | 18 | 6 |
| 1370 |  | 30 | 12 |
| 1371 |  | 1 | - |
| 1372 | Retail |  |  |
| 1373 |  | 1 | - |
| 1374 |  | 30 | 12 |
| 1375 |  | 1 | - |
| 1376 |  | 1 | - |
| 1377 |  | 2 | - |
| 1378 |  | 1 | - |
| 1379 |  | 1 | - |
| 1380 |  | 2 | - |
| 1381 |  | 15 | 5 |
| 1382 |  | 4 | 1 |
| 1383 |  | 6 | 2 |

| ITEM NUMBER | P | NM | G |
|---|---|---|---|
| 1384 |  | 30 | 12 |
| 1385 |  | 3 | 1 |
| 1386 |  | 4 | 1 |
| 1387 |  | 15 | 6 |
| 1388 |  | 2 | 1 |
| 1389 |  | 3 | 1 |
| 1390 | 14 | 12 | 3 |
| 1391 |  | 5 | 1 |
| 1392 | 22 | 18 | 6 |
| 1393 | 5 | 5 | 1 |
| 1394 |  | 12 | 4 |
| 1394.1 |  | 12 | 4 |
| 1394.2 |  | 12 | 4 |
| 1395 |  | 14 | 5 |
| 1396 |  | 8 | 2 |
| 1397 |  | 1 | - |
| 1398 |  | 2 | - |
| 1399 |  | 6 | 2 |
| 1400 |  | 1 | - |
| 1401 |  | 25 | 10 |
| 1402 |  | 1 | - |
| 1403 |  | 1 | - |
| 1404 |  | 1 | - |
| 1404.1 |  | 1 | - |
| 1405 | Retail |  |  |
| 1406 |  | 1 | - |
| 1407 |  | 5 | 2 |
| 1408 |  | 1 | - |
| 1409 |  | 2 |  |
| 1410 | 7 | 5 | 2 |
| 1411 |  | 1 | - |
| 1412 | 3 | 2 | 1 |
| 1413 | 22 | 18 | 6 |
| 1414 |  | 30 | 12 |
| 1415 |  | 15 | 5 |
| 1416 | 22 | 18 | 6 |
| 1417 | 5 | 2 | - |
| 1418 | 6 | 3 |  |
| 1419 | 5 | 2 |  |
| 1420 | 5 | 2 |  |
| 1421 | 5 | 2 |  |
| 1422 | 4 | 1 |  |
| 1423 | 4 | 1 |  |
| 1424 | 10 | 4 | 1 |
| 1425 | 6 | 2 | - |
| 1426 | 7 | 3 | - |
| 1427 | 8 | 3 | - |
| 1428 | 4 | 1 | - |
| 1429 | 7 | 3 | - |
| 1430 | 12 | 6 | 2 |
| 1430.1 | 18 | 8 | 2 |
| 1431 | 4 | 1 | - |
| 1432 | 7 | 2 | - |
| 1432.1 | 5 | 2 | - |
| 1432.2 | 8 | 4 | 1 |

# APPENDIX

**P** = PACKAGED  **NM** = NEAR MINT  **G** = GOOD  - = VALUE LESS THAN $1

| ITEM NUMBER | P | NM | G | ITEM NUMBER | P | NM | G | ITEM NUMBER | P | NM | G |
|---|---|---|---|---|---|---|---|---|---|---|---|
| 1433 | 5 | 2 | - | 1482 | 4 | 1 | - | 1532 | | 3 | - |
| 1434 | 7 | 4 | 1 | 1483 | | 2 | - | 1533 | 15 | 10 | 2 |
| 1435 | 6 | 2 | - | 1484 | | 2 | - | 1534 | 2 | - | - |
| 1436 | 6 | 2 | - | 1485 | | 5 | 1 | 1535 | | 2 | - |
| 1437 | 4 | 1 | - | 1486 | | 5 | 1 | 1536 | | 6 | 2 |
| 1438 | 7 | 3 | - | 1487 | | 2 | - | 1537 | 15 | 12 | 4 |
| 1439 | 4 | 1 | - | 1487.1 | | 20 | 7 | 1538 | 15 | 10 | 3 |
| 1440 | 4 | 1 | - | 1488 | | 2 | - | 1539 | | 5 | 1 |
| 1441 | 7 | 2 | - | 1489 | | 2 | - | 1540 | | 5 | 2 |
| 1442 | 5 | 2 | - | 1490 | | 2 | - | 1540.1 | | 1 | - |
| 1442.1 | 7 | 3 | - | 1491 | 3 | 2 | - | 1541 | | 1 | - |
| 1443 | 7 | 4 | 1 | 1492 | | 2 | - | 1542 | | 8 | 3 |
| 1443.1 | 12 | 7 | 2 | 1493 | 25 | 15 | 5 | 1543 | | 8 | 3 |
| 1444 | 6 | 2 | - | 1494 | | 2 | - | 1544 | Retail | | |
| 1445 | 6 | 3 | - | 1495 | | 2 | - | 1545 | | 3 | - |
| 1446 | 6 | 3 | - | 1496 | 6 | 3 | 1 | | | | |
| 1447 | 5 | 2 | - | 1497 | | 4 | - | 1547 | 8 | 6 | 2 |
| 1448 | 5 | 2 | - | 1498 | | 2 | - | 1548 | | 40 | 12 |
| 1449 | 18 | 7 | 2 | 1499 | 3 | 1 | - | 1549 | Retail | | |
| 1450 | 8 | 4 | 1 | 1500 | | 2 | - | 1550 | 40 | 40 | 16 |
| 1450.1 | 7 | 3 | - | 1501 | | 2 | - | 1551 | | 60 | 24 |
| 1451 | 7 | 3 | - | 1502 | | 2 | - | 1552 | | 40 | 16 |
| 1452 | 5 | 2 | - | 1503 | | 2 | - | 1553 | | 30 | 12 |
| 1453 | 10 | 3 | - | 1504 | | 2 | - | 1554 | Retail | | |
| 1453.1 | 4 | 1 | - | 1505 | | 2 | - | 1555 | | 30 | 12 |
| 1454 | 50 | 30 | 12 | 1506 | | 2 | - | 1556 | Retail | | |
| 1455 | 9 | 5 | 1 | 1507 | | 4 | 1 | 1557 | | 20 | 8 |
| 1456 | 6 | 2 | - | 1508 | 3 | - | - | 1558 | Retail | | |
| 1457 | 6 | 2 | - | 1509 | 7 | 3 | 1 | 1559 | | 75 | 30 |
| 1458 | 4 | 1 | - | 1510 | 2 | - | - | 1560 | | 65 | 26 |
| 1459 | 4 | 1 | - | 1511 | 5 | - | - | 1561 | | 40 | 16 |
| 1460 | 4 | 1 | - | 1512 | 3 | - | - | 1562 | Rare | | |
| 1461 | 7 | 2 | - | 1513 | | 1 | - | 1563 | Retail | | |
| 1462 | 5 | 1 | - | 1514 | | 2 | - | 1564 | Retail | | |
| 1463 | 4 | 1 | - | 1515 | | 50 | 20 | 1565 | Retail | | |
| 1464 | 5 | 1 | - | 1516 | | 30 | 12 | 1566 | | 40 | 16 |
| 1465 | 4 | 1 | - | 1517 | | 20 | 8 | 1567 | Retail | | |
| 1466 | 4 | 1 | - | 1518 | 60 | - | - | 1568 | Retail | | |
| 1467 | 7 | 2 | - | 1519 | | 20 | 8 | 1569 | | 8 | 3 |
| 1468 | 4 | 1 | - | 1519.1 | | 25 | 10 | 1570 | Retail | | |
| 1468.1 | 7 | 2 | - | 1520 | | 12 | 5 | 1571 | | 50 | 20 |
| 1469 | 5 | 1 | - | 1521 | | 18 | 7 | | | | |
| 1470 | 3 | 1 | - | 1522 | | 1 | - | 1580 | Rare | | |
| 1471 | 5 | 1 | - | 1522.1 | | 1 | - | 1581 | | 65 | 26 |
| 1472 | 4 | 1 | - | 1523 | 15 | 12 | 4 | 1582 | Retail | | |
| 1473 | 4 | 1 | - | 1523.1 | 15 | 12 | 4 | 1583 | | 95 | 38 |
| 1474 | 4 | 1 | - | 1524 | | 5 | 2 | 1583.1 | | 75 | 30 |
| 1474.1 | 5 | 2 | - | 1525 | | 1 | - | 1583.2 | | 75 | 30 |
| 1475 | 4 | 1 | - | 1526 | Varies | | | 1583.3 | | 75 | 30 |
| 1476 | 4 | 1 | - | 1526.1 | Varies | | | 1584 | Retail | | |
| 1477 | 8 | 5 | 1 | 1527 | | 3 | - | 1585 | | 50 | 20 |
| 1478 | 12 | 4 | 1 | 1528 | | 5 | 2 | 1585.1 | | 50 | 20 |
| 1479 | 4 | 1 | - | 1529 | | 5 | 2 | 1586 | Retail | | |
| 1480 | 5 | 2 | - | 1530 | 10 | 7 | 2 | 1587 | Retail | | |
| 1481 | 7 | 3 | - | 1531 | | 5 | 2 | 1588 | Retail | | |

## APPENDIX

**P** = PACKAGED  **NM** = NEAR MINT  **G** = GOOD  **-** = VALUE LESS THAN $1

| ITEM NUMBER | P | NM | G |
|---|---|---|---|
| 1589 | | 60 | 25 |
| 1590 | | 45 | 18 |
| 1591 | | 40 | 18 |
| 1591.1 | | 40 | 18 |
| 1592 | Rare | | |
| 1593 | Rare | | |
| 1594 | Retail | | |
| 1595 | | 35 | 10 |
| 1596 | | 30 | 10 |
| 1597 | 38 | 34 | 12 |
| 1598 | | 30 | 10 |
| 1599 | 38 | 34 | 12 |
| 1600 | | 30 | 10 |
| 1601 | 10 | 6 | 1 |
| 1602 | Retail | | |
| 1603 | 6 | 3 | 1 |
| 1604 | | 3 | - |
| 1605 | | 3 | - |
| 1605.1 | 40 | 20 | 8 |
| 1606 | | 3 | - |
| 1606.1 | 10 | 6 | 1 |
| 1607 | 10 | 6 | 1 |
| 1608 | | 18 | 8 |
| 1609 | | 12 | 5 |
| 1610 | - | - | - |
| 1611 | - | - | - |
| 1612 | - | - | - |
| 1613 | - | - | - |
| 1614 | - | - | - |
| 1616 | | 25 | 8 |
| 1617 | | 125 | 55 |
| 1618 | 30 | 1 | - |
| 1619 | | 30 | 12 |
| 1620 | | 30 | 12 |
| 1621 | | 25 | 10 |
| 1622 | Varies | | |
| 1622.1 | 8 | 4 | - |
| 1622.2 | Retail | | |
| 1623 | | 40 | 14 |
| 1623.1 | | 32 | 10 |
| 1624 | | 65 | 20 |
| 1624.1 | | 50 | 20 |
| 1625 | | 50 | 20 |
| 1626 | | 25 | 10 |
| 1626.1 | | 30 | 12 |
| 1627 | | 20 | 8 |
| 1628 | | 8 | 2 |
| 1628.1 | | 55 | 25 |
| 1629 | 65 | 55 | 20 |
| 1630 | | 30 | 14 |
| 1631 | | 25 | 10 |
| 1632 | | 25 | 10 |
| 1633 | | 30 | 12 |
| 1634 | | 40 | 16 |
| 1635 | Retail | | |

| ITEM NUMBER | P | NM | G |
|---|---|---|---|
| 1636 | | 5 | 2 |
| 1637 | 25 | 20 | 8 |
| 1638 | | 20 | 8 |
| 1639 | | 30 | 12 |
| 1640 | | 35 | 12 |
| 1640.1 | | 35 | 12 |
| 1641 | | 35 | 14 |
| 1641.1 | 15 | 5 | 1 |
| 1642 | Retail | | |
| 1643 | Varies | 1-3 | |
| 1644 | Varies | 3-20 | |
| 1645 | Varies | 1-10 | |
| 1646 | Varies | 1-5 | |
| 1647 | | 8 | 3 |
| 1648 | | 4 | 1 |
| 1649 | | 295 | 100 |
| 1650 | | 55 | 22 |
| 1651 | | 60 | 24 |
| 1652 | | 60 | 24 |
| 1653 | | 70 | 32 |
| 1653.1 | | 80 | 40 |
| 1654 | | 60 | 24 |
| 1655 | | 10 | 3 |
| 1656 | | 4 | 1 |
| 1657 | Retail | | |
| 1657.1 | 12 | 10 | 5 |
| 1658 | 12 | 10 | 3 |
| 1659 | 14 | 8 | 3 |
| 1660 | Retail | | |
| 1661 | Retail | | |
| 1061.1 | Retail | | |
| 1662 | 35 | 30 | 10 |
| 1663 | Retail | | |
| 1663.1 | Retail | | |
| 1664 | Retail | | |
| 1665 | 14 | 8 | 3 |
| 1666 | 15 | 8 | 2 |
| 1666.1 | Retail | | |
| 1667 | 20 | 5 | 2 |
| 1667.1 | Retail | | |
| 1668 | Retail | | |
| 1669 | Retail | | |
| 1670 | Retail | | |
| 1671 | Retail | | |
| 1671.1 | 50 | 45 | 14 |
| 1671.2 | Retail | | |
| 1671.3 | Retail | | |
| 1671.4 | Retail | | |
| 1671.5 | Retail | | |
| 1671.6 | Retail | | |
| 1672 | Retail | | |
| 1673 | 10 | 4 | 1 |
| 1674 | 12 | 6 | 2 |
| 1674.1 | Retail | | |
| 1674.2 | Retail | | |

| ITEM NUMBER | P | NM | G |
|---|---|---|---|
| 1675 | 35 | 30 | 8 |
| 1676 | | 75 | 35 |
| 1677 | | 40 | 20 |
| 1678 | | 175 | 75 |
| 1679 | | 35 | 16 |
| 1680 | | 25 | 10 |
| 1681 | | 250 | 100 |
| 1682 | | 75 | 50 |
| 1683 | | 20 | 10 |
| 1684 | | 85 | 40 |
| 1685 | | 150 | 75 |
| 1686 | | 30 | 15 |
| 1687 | | 30 | 15 |
| 1688 | | 18 | 8 |
| 1689 | | 20 | 5 |
| 1690 | | 30 | 12 |
| 1691 | | 30 | 12 |
| 1692 | | 50 | 20 |
| 1693 | | 50 | 20 |
| 1694 | | 40 | 15 |
| 1695 | | 40 | 15 |
| 1696 | Varies | | |
| 1697 | | 75 | 30 |
| 1698 | | 70 | 25 |
| 1699 | | 70 | 30 |
| 1700 | | 35 | 10 |
| 1701 | | 65 | 30 |
| 1702 | | 50 | 20 |
| 1703 | | 45 | 20 |
| 1704 | | 45 | 20 |
| 1705 | Varies | 10-35 | |
| 1706 | | 40 | 18 |
| 1707 | | 25 | 10 |
| 1708 | | 15 | 5 |
| 1709 | | 25 | 10 |
| 1710 | | 40 | 18 |
| 1711 | | 50 | 22 |
| 1712 | | 40 | 18 |
| 1713 | | 40 | 18 |
| 1714 | | 20 | 8 |
| 1715 | Retail | | |
| 1716 | | 12 | 5 |
| 1717 A | | 30 | 8 |
| 1717 B-D | | 25 | 10 |
| 1717 F | | 8 | 2 |
| 1717 G-H | | 25 | 10 |
| 1717 i-J | Varies | 5-15 | |
| 1718 | | 6 | 2 |
| 1719 | 15 | 10 | 3 |
| 1720 | Varies | 1-5 | |
| 1721 | Varies | | |
| 1722 | | 30 | 12 |
| 1723 | 50 | 40 | 15 |
| 1724 | | 5 | 2 |
| 1725 | Rare | | |

# APPENDIX

**P** = PACKAGED  **NM** = NEAR MINT  **G** = GOOD  **-** = VALUE LESS THAN $1

| ITEM NUMBER | P | NM | G |
|---|---|---|---|
| 1726 |  | 8 | 3 |
| 1726.1 |  | 75 | 30 |
| 1727 | Varies | 1-10 |  |
| 1728 | Retail |  |  |
| 1729 | Retail |  |  |
| 1730 | Retail |  |  |
| 1731 | Varies |  |  |
| 1732 | 75 | 55 | 20 |
| 1733 | Varies | 1-10 |  |
| 1734 | 25 | 18 | 8 |
| 1735 | 75 | 55 | 20 |
| 1736 | 30 |  |  |
| 1737 |  | 50 | 25 |
| 1738 |  | 10 | 3 |
| 1739 | Retail |  |  |
| 1740 | 50 | 45 | 20 |
| 1741 |  | 15 | 8 |
| 1742 | 120 | 100 | 50 |
| 1743 | 120 | 100 | 50 |
| 1744 | 110 | 110 | - |
| 1745 |  | 15 | 6 |
| 1746 |  | 15 | 6 |
| 1747 |  | 40 | 18 |
| 1748 |  | 35 | 18 |
| 1749 |  | 35 | 18 |
| 1750 |  | 25 | 12 |
| 1751 |  | 20 | 10 |
| 1752 |  | 40 | 20 |
| 1753 |  | 30 | 15 |
| 1754 |  | 15 | 6 |
| 1755 |  | 30 | 15 |
| 1756 |  | 15 | 6 |
| 1757 |  | 15 | 6 |
| 1758 |  | 15 | 6 |
| 1759 |  | 15 | 6 |
| 1760 |  | 15 | 5 |
| 1761 |  | 15 | 5 |
| 1762 |  | 25 | 12 |
| 1763 |  | 30 | 15 |
| 1764 |  | 15 | 6 |
| 1765 | Retail |  |  |
| 1766 |  | 20 | 10 |
| 1767 |  | 20 | 10 |
| 1768 |  | 20 | 10 |
| 1769 |  | 20 | 12 |
| 1770 |  | 20 | 10 |
| 1770.1 |  | 30 | 15 |
| 1771 |  | 30 | 15 |
| 1772 |  | 10 | 4 |
| 1773 |  | 15 | 6 |
| 1774 |  | 25 | 12 |
| 1775 |  | 35 | 18 |
| 1776 |  | 15 | 6 |
| 1777 |  | 35 | 18 |
| 1778 |  | 10 | 4 |

| ITEM NUMBER | P | NM | G |
|---|---|---|---|
| 1779 |  | 10 | 4 |
| 1780 |  | 25 | 12 |
| 1781 |  | 10 | 3 |
| 1782 |  | 35 | 18 |
| 1783 |  | 30 | 15 |
| 1784 |  | 5 | - |
| 1785 |  | 10 | 3 |
| 1786 |  | 10 | 3 |
| 1787 |  | 10 | 3 |
| 1788 |  | 95 | 50 |
| 1789 |  | 15 | 6 |
| 1790 |  | 15 | 6 |
| 1790.1 |  | 30 | 15 |
| 1790.2 |  | 40 | 20 |
| 1790.3 |  | 20 | 10 |
| 1790.4 |  | 25 | 12 |
| 1791 |  | 30 | 15 |
| 1792 |  | 30 | 15 |
| 1793 |  | 30 | 15 |
| 1794 |  | 40 | 20 |
| 1795 |  | 8 | 3 |
| 1796 |  | 10 | 3 |
| 1797 |  | 35 | 18 |
| 1798 |  | 20 | 10 |
| 1799 |  | 10 | 3 |
| 1800 |  | 25 | 12 |
| 1801 |  | 10 | 3 |
| 1802 |  | 10 | 3 |
| 1803 |  | 10 | 3 |
| 1804 |  | 10 | 3 |
| 1805 |  | 10 | 3 |
| 1806 |  | 7 | 1 |
| 1807 |  | 4 | - |
| 1808 |  | 4 | - |
| 1809 |  | 5 | - |
| 1810 |  | 4 | - |
| 1811 |  | 5 | - |
| 1812 |  | 5 | - |
| 1813 |  | 12 | - |
| 1814 |  | 8 | - |
| 1815 |  | 10 | - |
| 1816 |  | 6 | - |
| 1817 |  | 6 | 2 |
| 1818 |  | 10 | 3 |
| 1819 |  | 6 | 2 |
| 1820 | Retail |  |  |
| 1821 |  | 45 | 20 |
| 1822 |  | 75 | 35 |
| 1823 |  | 35 | 15 |
| 1824 |  | 30 | 12 |
| 1825 |  | 35 | 15 |
| 1826 |  | 5 | - |
| 1827 |  | 5 | - |
| 1828 |  | 4 | - |
| 1829 |  | 3 | - |

| ITEM NUMBER | P | NM | G |
|---|---|---|---|
| 1830 |  | 20 | - |
| 1831 |  | 15 | - |
| 1832 |  | 35 | 12 |
| 1833 |  | 10 | 3 |
| 1834 |  | 15 | 5 |
| 1835 |  | 45 | 12 |
| 1836 |  | 20 | 5 |
| 1837 |  | 45 | 12 |
| 1838 |  | 35 | 10 |
| 1839 |  | 25 | 6 |
| 1840 |  | 15 | 5 |
| 1841 |  | 15 | 5 |
| 1842 |  | 10 | 5 |
| 1843 |  | 20 | 7 |
| 1844 |  | 15 | 6 |
| 1845 |  | 10 | 5 |
| 1846 |  | 15 | 6 |
| 1847 |  | 15 | 6 |
| 1848 |  | 10 | 5 |
| 1851 |  | 7 | 1 |
| 1852 |  | 8 | 2 |
| 1853 |  | 22 | 10 |
| 1854 |  | 14 | 6 |
| 1855 |  | 24 | 10 |
| 1856 |  | 22 | 10 |
| 1857 |  | 14 | 6 |
| 1858 |  | 6 | 1 |
| 1859 |  | 20 | 8 |
| 1860 |  | 5 | 1 |
| 1861 |  | 15 | 6 |
| 1862 |  | 4 | 1 |
| 1863 |  | 22 | 10 |
| 1864 |  | 25 | 12 |
| 1865 |  | 18 | 8 |
| 1866 |  | 14 | 6 |
| 1867 |  | 9 | 4 |
| 1868 |  | 8 | 3 |
| 1869 |  | 12 | 5 |
| 1870 |  | 14 | 6 |
| 1871 |  | 13 | 5 |
| 1872 |  | 7 | 2 |
| 1873 |  | 32 | 15 |
| 1874 |  | 6 | 2 |
| 1875 |  | 5 | 1 |
| 1876 |  | 3 | - |
| 1877 |  | 3 | - |
| 1878 | Retail |  |  |
| 1879 |  | 17 | 7 |
| 1880 |  | 15 | 6 |
| 1881 |  | 15 | 6 |
| 1882 |  | 17 | 5 |
| 1883 |  | 10 | 2 |
| 1884 |  | 45 | 20 |
| 1885 |  | 45 | 20 |

## APPENDIX

**P** = PACKAGED  **NM** = NEAR MINT  **G** = GOOD  **-** = VALUE LESS THAN $1

| ITEM NUMBER | P | NM | G |
|---|---|---|---|
| 1886 | | 55 | 25 |
| 1887 | | 45 | 20 |
| 1888 | | 50 | 25 |
| 1889 | | 20 | 8 |
| 1890 | | 3 | 1 |
| 1891 | | 18 | 7 |
| 1892 | | 4 | 1 |
| 1893 | | 7 | 2 |
| 1894 | | 14 | 6 |
| 1894.1 | | 4 | 1 |
| 1895 | | 22 | 10 |
| 1896 | | 18 | 8 |
| 1897 | | 18 | 8 |
| 1898 | | 22 | 10 |
| 1899 | | 9 | 3 |
| 1900 | | 9 | 3 |
| 1901 | | 9 | 3 |
| 1902 | | 12 | 4 |
| 1903 | | 12 | 4 |
| 1904 | | 12 | 4 |
| 1905 | | 20 | 8 |
| 1906 | | 20 | 8 |
| 1907 | | 22 | 10 |
| 1908 | | 18 | 7 |
| 1909 | | 4 | 1 |
| 1910 | | 40 | 16 |
| 1911 | | 12 | 4 |
| 1912 | | 12 | 4 |
| 1913 | | 15 | 6 |
| 1914 | | 18 | 8 |
| 1915 | | 7 | 2 |
| 1916 | | 28 | 12 |
| 1917 | | 25 | 10 |
| 1918 | | 12 | 4 |
| 1918.1 | | 14 | 5 |
| 1919 | | 4 | - |
| 1920 | Retail | | |
| 1921 | | 8 | 3 |
| 1922 | Retail | | |
| 1923 | Retail | | |
| 1924 | Retail | | |
| 1925 | Retail | | |
| 1926 | | 5 | 1 |
| 1927 | | 15 | 6 |
| 1928 | | 12 | 4 |
| 1928.1 | | 22 | 10 |
| 1929 | Retail | | |
| 1930 | | 50 | 20 |
| 1931 | Retail | | |
| 1932 | | 22 | 10 |
| 1933 | | 2 | - |
| 1934 | | 15 | 7 |

Lucky's Collectors Guide To 20th Century Yo-Yos

## APPENDIX

### ABOUT THE AUTHOR

Lucky Meisenheimer, M.D. is regarded as the foremost authority on collectible yo-yos. He has one of the largest private collections of yo-yos and memorabilia in the world. As a charter member of the American Yo-Yo Association, he has served as a board member and chair of the historical committee. He is the author of numerous articles on the history of the yo-yo and yo-yo collecting and has been a contributing advisor for Schroeder's Collectibles Toy Price Guide for several years.

A world-ranked Masters swimmer, Lucky began collecting yo-yos in the 1980s while attending various swimming competitions around the country. Lucky has competed in the National Yo-Yo Championships and has performed on Nickelodeon's television show, "What Would You Do?" He believes you can never have too many yo-yos!

Lucky is a dermatologist and lives with his wife Jacquie and sons, John and Jake, in Orlando, Florida. Lucky and Jacquie are both active volunteers with the YMCA. They serve as board members of the YMCA Aquatic Center and as coaches for Team Orlando Masters swim team and Orange County Special Olympics swim team.

*Extreme behaviors are part of his nature. In the '70s, the author was in "Ripley's Believe it or Not" for swimming a half-mile with his foot in his mouth.*

## APPENDIX

### ● AUTOGRAPHS

**S**tart your own autograph collection. At the next yo-yo competition, collect the autographs of your favorite yo-yo professionals. Carry this book with you and have them sign right on this page.

The annual National and World Yo-Yo Championships, are great events for meeting other collectors, trading yo-yos, and adding autographs of famous yo-yo personalities to your collection.

For more information on the dates and locations of these events and others, contact the American Yo-Yo Association.